Foundational Theories of Classical and Constructive Mathematics

THE WESTERN ONTARIO SERIES IN PHILOSOPHY OF SCIENCE

A SERIES OF BOOKS IN PHILOSOPHY OF MATHEMATICS AND NATURAL SCIENCE, HISTORY OF SCIENCE, HISTORY OF PHILOSOPHY OF SCIENCE, EPISTEMOLOGY, PHILOSOPHY OF COGNITIVE SCIENCE, GAME AND DECISION THEORY

VOLUME 76

Giovanni Sommaruga
Editor

Foundational Theories of Classical and Constructive Mathematics

Springer

Editor
Giovanni Sommaruga
Dept. of Humanities
Social and Political Sciences
Chair for philosophy
ETH, Zurich
Switzerland
sommarug@ethz.ch

ISSN 1566-659X
ISBN 978-94-007-0430-5 e-ISBN 978-94-007-0431-2
DOI 10.1007/978-94-007-0431-2
Springer Dordrecht Heidelberg London New York

Library of Congress Control Number: 2011922884

Printed on acid-free paper

Springer is part of Springer Science+Business Media (www.springer.com)

Preface

The present book project grew out of the Swiss Society for Logic and Philosophy of Science (SSLPS) annual meeting on "Foundational theories of mathematics" which was held in Freiburg (Switzerland) on October 11/12 2006. John Bell and Gerhard Jäger, both participating in different functions in this meeting, responded to this book project with great enthusiasm and fueled its evolution with recurrent positive feedbacks. I'm happy to have had the opportunity over the years to discuss with them Foundations of Mathematics (FOM) and many other hot topics as well as some less hot ones.

I'm grateful to Oxford University Press for permission to reprint Section I.2 of Penelope Maddy's book *Naturalism in Mathematics* (Oxford: Clarendon Press, 1997) and also to the publishing company Polimetrica for permission to reprint Solomon Feferman's article which first appeared in G. Sica's book *What is Category Theory?* (Monza: Polimetrica, 2006). I'd like to thank Norman Sieroka for helping me to translate several papers from (ancient) Word to (modern) LaTeX. I'd also like to thank Lucy Fleet, Senior Assistant to the Editorial Director—Humanities of Springer, for her patience and her refreshing confidence that this book would eventually see the light of day. I'm very grateful to the IT Service of the ETH Zurich and in particular to its collaborator Dieter Hennig for his very valuable support in my rather dilettante LaTeX-engineering.

Last and most I'd like to thank my wife Andrea Gemma and my little son Beniamino for not having banned me from home before regaining normality, and even more so for their loving care with which they endured and accepted my mixture of mental absence and busy bee existence during the more productive phases of this book's production.

Contents

Part III Between Foundations of Classical and Foundations of Constructive Mathematics

Part IV Foundations of Constructive Mathematics

Introduction

Giovanni Sommaruga

Part I A Retrospective: Some Remarks on the Historiography of FOM[1]

Section 1 : The Meaning of 'FOM' According to Mostowski, Parsons and Wang

Before sketching the development of 120 years of studies in FOM, the following questions ought to be addressed: What makes the studies and results considered, analysed, discussed etc. by Andrzej Mostowski, Charles Parsons and Hao Wang foundational studies or results, that is contributions to FOM? What are the features or characteristics of foundational research according to these specialists and historiographers on FOM?

The most important or at least characteristics highly valued and shared by all these specialists on FOM are the following ones:

1. Some types of conceptual analysis
   For Wang, the business of research in FOM is essentially conceptual analysis. He distinguishes 2 essential ingredients of conceptual analysis in FOM (H.W.: in mathematical logic): (i) reduction, (ii) formalisation. The purpose of this conceptual analysis is to make a concept or a set of concepts or a theory more sharp, more precise.

   concerning (i): Wang characterises reduction as "one way of simplifying a concept [...] by reducing more components to less or by simplifying each separate aspect".[2] He continues to further divide reduction into (i.1) local reduction and (i.2) regional or global reduction (in H.W.'s terms "whole"

[1] This first part of the introduction is entirely based on the historical surveys of studies in Foundations of Mathematics (FOM): A. Mostowski (1965), C. Parsons (2006) and H. Wang (1958).

[2] Wang (1958, p. 468).

reduction). Among the types of local reduction he distinguishes the reduction by definition, the reduction by deduction (i.e. applying the axiomatic method) and something he calls uniform local reduction which is the interpretation of a formalism or formal system by another one. And by a regional or global reduction he means the reduction of (a region or branch or) the whole of mathematics to a (another) region or branch. Wang observes that regional or global reductions are often of greater interest to philosophers than local ones, but at the same time they are often not sound.

concerning (ii): Wang characterises formalisation as putting a concept or a set of concepts into a formal system which makes all the implicit assumptions explicit. He subdivides formalisation into thorough formalisation (a thorough axiomatisation containing predicate logic and the concepts to be formalised which are thereby "implicitly defined") and partial formalisation (no or only partial axiomatisation and other concepts than the ones to be formalised occur).

Hao Wang concludes his reflections on conceptual analysis in FOM with two claims: Claim A: Formalisation rather than reduction is the appropriate method in foundational studies, as the latter are primarily interested in irreducible concepts. Claim B: Formalisation as a method has been mainly practiced in mathematical logic (far more so than in any other branch of mathematics) for the last (and the first) 80 years of studies in FOM.[3]

2. An orientation towards the basic distinction between constructive methods and non-constructive (or classical) methods in mathematics.[4]

   In his discussion of this basic distinction, Wang refers to Bernays' 5 shades of constructive and non-constructive methods in mathematics: in order of decreasing constructivity (i) anthropologism (or finitism in the narrower sense), (ii) finitism (in the broader sense), (iii) intuitionism, (iv) predicative set theory (or predicativism), and (v) classical set theory (or platonism). Wang makes a series of comments on these different domains, a particular and very modern comment being the following one: The domains (i)–(v) should not be treated as rival domains among which one has to choose one (for life), but they should rather be treated "as useful reports about a same grand structure which can help us to construct a whole picture that would be more adequate than each taken alone".[5] He identifies a central irreducible concept of each of these domains: of (i) the concept of feasibility, of (ii) the concept of constructivity, of (iii) the concept of (constructive) proof, of (iv) the concept of (natural) number, and of (v) the concept of set. Then he remarks that a sharp and precise definiteness of these only vaguely characterised domains (i)–(v) may be obtained by a conceptual

---

[3] Reduction and formalisation also play a fairly important role in Mostowski's (1967), however in a more implicit way. Parsons in his (2006) treats the axiomatic method (reduction) and formalisation explicitly in the 1st paragraph.

[4] Mostowski calls the divide: infinitistic or set-theoretical vs. finitistic or arithmetical; Parsons calls the divide: platonism vs. constructivism.

[5] Wang (1958, p. 472).

analysis, and in particular by a formalisation of these 5 central concepts. These formalisations might form the hard core of studies in FOM. But, completely in agreement with Mostowski, he adds that at times various ramifications and cross-overs may be more important than the hard core itself.[6] Hao Wang subsequently presents a short historical characterisation of each of these domains (i)–(v) within the development of research on FOM. And Parsons dedicates a whole paragraph (§3) to these various domains.

3. An intimate relation with mathematical logic in its various subdisciplines (set theory, model theory, proof theory and computability theory)
4. (For Mostowski and Parsons) being a contribution to one of the original 3 movements (schools) in the philosophy of mathematics: formalism (Hilbert's program), intuitionism, and logicism, or later on to their more technical successors: meta-mathematics, constructivism, and set theory resp.[7]
5. (For Mostowski and Parsons) being a solution or a partial solution to a major philosophical problem, such as e.g. the completeness problem, the problem of set-theoretical paradoxes, the decision problem, the problem of impredicative definition etc.
6. (Esp. for Parsons) being a contribution to the solution of the problem of justifying mathematical statements or principles (the so-called epistemological point of view in FOM)

NB. These features or characteristics of studies or results on FOM (according to Mostowski, Parsons and Wang) are obviously not mutually exclusive.

Section 2 : The First 80 Years of Studies in FOM According to Mostowski and Wang

According to Mostowski and Wang,[8] the 1st phase of studies in FOM starts in the 1880s.[9] Its most important elements are: Cantor's so-called naive set theory and the subsequent formalisation of the central concept of set (i.e. the various axiomatisations of Cantorian set theory) as a reaction to the appearance of set-theoretical paradoxes; Frege's classical 1st order logic as a formalisation of all usual methods of mathematical argument of a strictly logical nature; and finally, the 3 well-known movements (schools) in the philosophy of mathematics:

---

[6] Parsons draws the dividing line in a slightly different way from Wang: According to him predicativism belongs to the side of platonism rather than to the one of constructivism in a large sense.

[7] Note that all the three specialists in FOM agree (more or less; cf. the following footnote) that the original 3 movements in the philosophy of mathematics failed and somehow came to an end (by about the 1930s).

[8] Wang presumably exempts intuitionism from the criticism and treatment to which he subjects logicism and formalism; see below. And the same may hold for Parsons as well.

[9] The following historical sketch of the development of studies in FOM concerns what is usually called "the history of ideas" in a broad sense.

1. Logicism as a reduction of mathematics to logic
2. Formalism (called finitism by Wang) as a reduction of mathematics to finitist mathematical methods; it is an endeavor towards a formalisation of the central concept of constructivity.
3. Intuitionism as an endeavor towards a formalisation of the central concept of (constructive) proof

Logicism and formalism are, according to Wang, global reductionist projects; and moreover, they both are failures on several accounts. Hilbert's formalism has, however, in a 2nd phase of studies in FOM given rise to uniform local reductions which are of far greater interest than Hilbert's original global reduction. And Mostowski writes implicitly—and ambiguously—about the end or the decline of the three movements in the philosophy of mathematics in the late 1920s and their great impact on more formal and technical developments in a 2nd phase.

Mostowski identifies a new phase of FOM studies beginning in the 1930s with three particularly influential works, namely: K. Gödel's "Über formal unentscheidbare Sätze der Principia Mathematica und verwandter Systeme I" (1931), A. Heyting's "Die formalen Regeln der intuitionistischen Logik" (1930), and A. Tarski's "Der Wahrheitsbegriff in formalisierten Sprachen" (1936).

In his presentation of Gödel's incompleteness theorems, Mostowski formulates and sketches the proofs of the 1st and the 2nd incompleteness theorems, he discusses some important assumptions of these proofs and mentions some highly influential effects on the subsequent development of foundational studies: first, Gödel's 2nd incompleteness theorem was used as a powerful tool for investigating the relative strength of various axiomatic theories; second, his paper contains various results regarding the decision problem(s); and third, the method he invented of comparing intuitively true properties of mathematical objects with properties expressible in the formal system under consideration turned out to be extremely fruitful in meta-mathematics. His division of reasoning in intuitive meta-mathematics and formal mathematics was a very useful tool for establishing properties of formal systems, e.g. consistency, completeness and decidability.

An important strand in this 2nd phase of foundational studies is the development of intuitionism and *constructivism* in logic and mathematics. Heyting's just mentioned paper is the starting point of this strand: In this paper Heyting presents a formalisation of intuitionistic 1st order logic. Gödel proved a bit later that classical logic can be represented in the intuitionistic logic, and Tarski again a few years later that open subsets of a topological space form a matrix in which all provable formulas of intuitionistic propositional logic are valid. Subsequently, various interpretations were proposed to prove soundness and completeness of intuitionistic logic and intuitionistic arithmetic: (a) Tarski's topological interpretation was extended to yield a soundness and completeness proof for intuitionistic 1st order logic. (b) The Beth or tree models constitute another modification of the classical notion of model providing an adequate interpretation of intuitionistic 1st order logic. Both these interpretations suit pure intuitionistic logic very well, but seem less suitable for the interpretation of intuitionistic arithmetic. Mostowski raises the question: What is the purpose of an interpretation of a formal system? and answers the question as follows: It is supposed to give a precise meaning to concepts which are either incompletely explained or taken as primitive terms in the resp. formal system. The fundamental concept in intuitionism is, as Heyting showed, that of construction which is used by the intuitionists without explication. In (c) Kleene's realizability interpretation of intuitionistic arithmetic, Kleene proposes to explicate this concept by identifying it with partial computable functions. By means of his interpretation, he succeeds in proving the soundness of intuitionistic arithmetic, but Rose later disproved completeness. Hence Kleene's realizability interpretation does not provide an adequate interpretation of intuitionistic arithmetic: There must be intuitionistically acceptable "constructions" which are not reducible to partial computable functions. The principle of (d) Gödel's functional interpretation is similar to Kleene's, but Gödel used a much wider class of admissible constructions, namely the class of primitive recursive functionals. Gödel not only proved the soundness of intuitionistic arithmetic w.r.t. his functional interpretation, but also the relative consistency of intuitionistic arithmetic w.r.t. his axiomatic theory of primitive recursive functionals. Mostowski points out that all the interpretations (a)–(d) try to explain intuitionistic (logical or arithmetical) concepts in classical terms, and that an intuitionist would of course be far more interested in an interpretation explaining classical concepts in intuitionistic terms. Mostowski then puts the constructivist response to the set-theoretical paradoxes in a larger perspective, i.e. he describes other constructive, less "extreme" attempts (than the intuitionistic one) to solve the problem of the paradoxes. One sort of attempt employs quite arbitrary (not even necessarily constructive) means. One attempt of this first sort is computable analysis which restricts all mathematical notions and in particular those occurring in mathematical analysis to computable functions. Other attempts of this first sort concern extensions of computable analysis such as Grzegorczyk's elementarily definable analysis or hyper-arithmetical analysis studied by Kreisel. The second kind of attempt, not merely of mathematical but also of philosophical interest, consists of the theories called strictly finitistic by Mostowski: it is their aim to examine constructive mathematical objects by constructive means. One attempt of this second kind is recursive arithmetic the main idea of which is to develop mathematics as a formal system operating exclusively with equations. Another attempt of this second kind is Markov's algorithmic mathematics which reduces all other mathematical notions to the one of algorithm and which implicitly accepts and uses intuitionistic logic.

Another very important strand in the second phase of research in FOM is the unfolding of *computability theory*. Out of the decision problem for a denumerable class C of objects grew the need to define a class of arithmetical functions whose values can be computed in a finitistic way. Thus, the concept of a computable function from a set of integers to the set of integers served to make precise the concept of definability. In the 1930s several definitions of this concept of computable function were proposed which turned out to all be equivalent (cf. also the Church-Turing Thesis CT). The concept of computable function served likewise to define the concept of a recursively enumerable (r.e.) set. Large parts of computability theory were further developed by Kleene: He introduced the concept of relative computability, he defined the degrees of computability based on this concept as well as the arithmetical hierarchy of sets of integers, and he contributed to the study of recursive well-orderings which are part of a constructivistic program attempting to reconstruct parts of classical set theory in computable terms. The hyper-arithmetical hierarchy of sets of integers was but an extension of the arithmetical hierarchy into the constructive transfinite. And the analytical hierarchy of sets of integers was a further extension of the hyper-arithmetical hierarchy.

Another big subject in computability theory (other than hierarchies) is that of functionals which was first introduced by Gödel: He extended the concepts of primitive recursiveness and computability of functions and sets to objects of higher logical types called functionals. Whereas arbitrary functionals are highly infinitistic objects, Gödel considered the very narrow class of primitive recursive functionals, Kleene the much larger class of partial computable functionals, and Spector an intermediate class of so-called Bar-recursive functionals. The main drives for departing from the ideal simplicity of computable functions and sets and heading towards more and more infinitistic objects (objects of higher hierarchies and functionals) were according to Mostowski on the one hand to round off the theory of computability, and on the other to find objects which would be useful for the realisation of Hilbert's program (of consistency). Mostowski doubts whether these more infinitistic objects still fit into a constructivist philosophical program.

A special area of research in computability theory (or closely related to it) is the one of decision problems. Hilbert's original decision problem was: Is there a method allowing to decide effectively whether any 1st order formula is provable or not? The decision problem: Is there a method allowing to decide effectively whether any given formula of 1st order logic is satisfiable in some domain? could be called (the semantical version of) a Hilbert-type decision problem. Now, several partial Hilbert-type decision problems were found to have positive solutions, and Mostowski mentions several classes of 1st order formulas which are decidable. The decision problem: Is there a method allowing to decide effectively whether a given formula (in the theory's language) is provable in a theory T or not? could be called a Skolem-type decision problem. Skolem and Tarski designed a method, called the elimination method, to tackle this problem, and they used it successfully to solve positively the Skolem-type decision problem for various theories such as e.g. the theory of real closed fields. There are however important negative solutions of Hilbert-type decision problems, the most prominent being Church's negative solution of Hilbert's original decision problem in the 1930s, that is Church's proof of the undecidability of full 1st order logic. The basic method, called the reduction method, used in the proofs of undecidability (negative solutions to a Hilbert-type decision problem) is the reduction of a decision problem for a set of formulas K to the decision problem for another set K_0 for which the solution is known to be negative. Church also solved in the negative the Skolem-type decision problem for 1st order arithmetic. And Rosser even proved the essential undecidability of 1st order arithmetic (that is, the undecidability of every consistent extension of 1st order arithmetic). Mostowski finally distinguishes a third sort of decision problem: Is there a method allowing one to decide effectively whether any given 1st order formula is true in a given model M or not? which could be called a model-type decision problem. The Robinsons proved that in most cases the undecidability of a model M can be obtained if one shows the integers and usual arithmetical operations on integers to be definable in M.

According to Mostowski, the strand of (abstract) *set theory* is of special importance in the history of studies in FOM. In 1940 Gödel made a contribution to the consistency problem of hypotheses in set theory which had a deep influence on meta-mathematical work in the following 20 years. Gödel constructed a model of set theory in which the set-theoretical axioms, the Axiom of Choice AC and the Continuum Hypothesis CH, are valid by extending the arithmetical hierarchy into the (Cantorian) transfinite. A set which can be constructed at one of the finite or transfinite levels of this extended arithmetical hierarchy was called by Gödel "constructible". The constructible sets form this model denoted by **L** and they form a hierarchy. The family of constructible sets represents a realisation of the predicative foundation of mathematics. Gödel's Axiom of Constructibility ACon is the following one: Every set is constructible (where a set in Gödel's sense is a transitive and ground set), for short, $\mathbf{V} = \mathbf{L}$. ACon is considered as a highly dubious statement (even by Gödel). The effect of ACon is to give the not sharply defined concept of an arbitrary subset of a given infinite set a very definite limitation and interpretation. There seems to exist the possibility of 2 equally acceptable set theories: an axiomatic set theory + ACon, and an axiomatic set theory + not-ACon. Now, ACon is not only consistent relative to the other axioms of set theory, it also implies e.g. the Generalised Continuum Hypothesis GCH or the well-ordering theorem. A big philosophical question is: Is ACon true? Since ACon is also provably independent of the other axioms of set theory, there exist indeed 2 mutually contradictory systems of set theory. Mostowski wonders whether the choice is a matter of taste or whether there are compelling reasons for choosing the one set theory rather than the other. The just mentioned question touches on the fundamental problem of truth of set-theoretical hypotheses.

The problem of inconsistencies in Cantor's naïve set theory sparked off a number of axiomatic set theories all of which trying to modify Cantor's original theory in such a way that the inconsistencies disappear. Cantor distinguishes in his theory between "consistent" and "inconsistent sets". While Zermelo-Fraenkel's axiomatic set theory ZF simply ignored the inconsistent sets, the Bernays-Gödel axiomatic set theory BG mimics this distinction by assuming not only sets (as ZF does), but also classes. Whereas there is not much difference in mathematical content between ZF and BG, there are considerable differences between ZF or BG and extensions of these systems: in particular, those extensions adopting Tarski's axiom of the existence of inaccessible cardinals or Levy's axiom schema of the existence of various kinds of inaccessible cardinals, or even a very strong Axiom of Infinity: There are compact regular ordinals $\mu > \omega$. These extensions are essential extensions of ZF and BG and they cannot be proven to be consistent relative to ZF or BG. The big question of course is on which grounds these strong or very strong axioms of infinity can be taken to be consistent (in order not to introduce inconsistencies again through the back door).

Moreover, Cantor (as well as Frege) used in his naive set theory a naive Comprehension Principle CP of set existence: Whenever F is a formula (with one free variable), there exists a set S consisting of all elements a satisfying F. Since CP turned out to be inconsistent, subsequent axiomatic set theories tried to make set theory consistent by modifying CP in 3 different ways: (a) by not accepting CP for all formulas F; (b) by restricting the variability of a; (c) by imposing at the same time restrictions (a) and (b). Chwistek's and Ramsey's Simple Type Theory STT accepts (c). The ZF and BG set theories accept (b). And an axiomatic system due to Quine and referred to by NF accepts (a). Mostowski surmises that the axioms of set theory have not reached their definitive form yet. Another axiom of set theory which at the beginning stirred up a lot of philosophical debate, namely AC, was in the 2nd phase of foundational studies investigated concerning its relative consistency as well as its independence. In the 1960s Cohen introduced a new method allowing him to establish the independence of AC and GCH from practically every axiomatic system of set theory built along the ZF-lines. The success of his method is based on the new meta-mathematical concept of forcing. This concept of forcing is of considerable interest also apart from its applications. Furthermore, Cohen's forcing method suggested the study of an essential ingredient of it, namely of the generic sets. These sets seem to satisfy intuitions underlying Brouwer's intuitionistic conception of sets (of integers), and they can be defined not only for set theory, but also for arithmetic and other theories. Cohen's proof that there are (at least) 2 consistent and mutually incompatible set theories launched some more philosophical questioning: (i) Will mathematics accept the existence of these 2 incompatible set theories? or (ii) Will mathematics try to find new axioms which will eliminate one of them? or (iii) Will mathematics try to limit itself to more finitistic domains? The issue between formalists (option (i)), platonists (option (ii)) and intuitionists (option (iii)) is still open.

Yet another strand in the 2nd phase of development of studies in FOM is *proof theory* which is rooted in Hilbert's program (formalism). The main inputs in proof theory were given by Herbrand and Gentzen. Herbrand's main result contains a certain reduction (although obviously not a complete one) of 1st order logic to propositional logic. It shows that if a formula F is provable in 1st order logic, then there exists a proof of it consisting exclusively of subformulas of F. This result greatly simplifies the study of formal proofs. Herbrand's results were rediscovered and greatly improved by Gentzen who devised a new logical system equivalent to the one of the Hilbert school, but much more flexible. The flexibility of Gentzen's approach is obvious from the fact that it is applicable not only to classical logic, but also to many non-classical logics, esp. to intuitionistic logic. Gentzen's other great result is his conception of a consistency proof of arithmetic based on transfinite induction. Herbrand's and Gentzen's work clearly belong to what Mostowski calls the finitistic or arithmetical trend in the studies of FOM. Subsequent work in proof theory, however, borrowed many ideas from the infinitistic or set-theoretical trend, as witnessed by Bernays' general consistency theorem in which set-theoretical semantical notions are consciously imitated in finitistic terms. Mostowski emphasizes that Herbrand's and Gentzen's methods enable certain particular cases of infinitistic, set-theoretical constructions to be made finitistic. There is thus an intertwining of the 2 trends in the studies on FOM.

A last strand in the 2nd phase of studies in FOM is *model theory* (or "logical semantics") which is the study of relations between expressions of a formal language and mathematical objects (or, more precisely, between sentences of a formal language and a class of objects called models): one of the fundamental relations here is that of satisfaction. NB. The nature of these sentences as well as the nature of these models is fairly arbitrary which makes for the great flexibility of model theory. The systematic development of model theory is due to Tarski, started in the early 1950s, and became henceforth one of the most important parts of research in FOM.

Tarski began by showing that all semantical concepts can be reduced to the fundamental concept of a value of a formula (or sentence). He then used this concept to precisely define other important semantical concepts such as e.g. the concepts of satisfiability, validity, logical consequence and that of definability in a given model **M**. Starting from the observation that under certain conditions it is possible to replace the semantical relation: the value of a sentence F in **M** is $\mathcal{V}$ (= true) (or: model **M** satisfies sentence F), by the arithmetical relation (*) f is the Gödel number of a sentence F satisfied by **M**, Tarski raises two questions: (i) Is the arithmetical relation (*) definable in **M**? (ii) If (*) is not definable in **M**, what new relations should be added to **M** to ensure the definability of (*) in the extended model? The answer to question (i) is Tarski's well-known undefinability theorem: The set of sentences true in **M** is not definable in **M**. Tarski's undefinability theorem applied to PA yields Gödel's 1st incompleteness theorem. Question (ii) cannot be answered in a uniquely determined way: Various relations can be added to the model **M** in such a way that the arithmetical relation (*) becomes definable in **M**. A remarkable result here is the following theorem connecting model theory with the theory of inductive definitions: There exist types of inductive definitions which are not reducible to ordinary inductive definitions in a purely arithmetical way.

A particular, typically model-theoretical and highly important problem is the so-called completeness problem. Mostowski presents two formulations of it: (a) Consider an uninterpreted formal system described in a purely syntactic way. Try to find a semantical interpretation, i.e. a model for it satisfying all and only the sentences of that system. (b) Assume given an interpreted language. Try to find a formal system with purely syntactic proof rules allowing one to prove exactly the true sentences of the language. Two different methods were devised to solve the completeness problem of type (a) for 1st order logic. According to the first method which has an algebraic character and is due to Sikorski, Tarski and Rieger, this completeness problem is solved if one shows the existence of a maximal filter A with the property (**) for every formula F with one free variable, if the sentence $\exists x F(x)$ belongs to A then so does at least one sentence of the form $F(t)$. (As a matter of fact, this solution is a consequence of the fundamental theorem of Boolean filter theory.) According to the second method clearly influenced by Gentzen-style formalisations of logic and due to Beth, Hintikka and Schütte, the completeness problem for 1st order logic is solved by looking systematically for a possible counter-example to a given sentence F. Mostowski observes that underlying the completeness problem is a philosophical question concerning the relations between formal systems and their interpretations or models; but despite this philosophical origin the completeness problem has found many purely mathematical applications especially in algebra.

Mostowski continues to sketch two main results of the model theory for the (elementary) language L of 1st order logic. Before doing so, he introduces three important model-theoretical concepts: a model $\mathbf{M_1}$ is a submodel of a model $\mathbf{M_2}$; a model $\mathbf{M_1}$ is an elementary submodel of a model $\mathbf{M_2}$; two models $\mathbf{M_1}$ and $\mathbf{M_2}$ are elementarily equivalent. (1) the different Skolem-Löwenheim theorems, that is, the downward Skolem-Löwenheim theorem and the upward Skolem-Löwenheim theorem. The upward Skolem-Löwenheim theorem is a logical consequence of the compactness theorem (which in turn is a consequence of Gödel's completeness theorem for 1st order logic). Two particularly interesting applications of the Skolem-Löwenheim theorems are on the one hand the characterisation of the so-called spectra of sets X of sentences of L, and on the other hand the discovery of new methods for the solution of the completeness problem: One method due to Vaught is based on the concept of elementary equivalence of two models, and the other one due to A. Robinson on the new concepts of diagram $\mathbf{D(M)}$ of a model $\mathbf{M}$ and model-completeness (not to be confused with the concept of completeness). (2) the (semantical version) of the Craig interpolation lemma. On the one hand this lemma was used by Addison to explore the field of logical and set-theoretical separation principles, on the other hand it was applied by Beth in the theory of definitions. Mostowski then turns to the model theory for non-elementary (higher-order and infinitistic) languages; he briefly discusses the language $\mathsf{Q_a}$, the language $\mathrm{L}^{\mathrm{II}}_{\mathrm{a}}$ of weak 2nd order logic, the language L^{II} of strong 2nd order logic, the sequential 2nd order language $\mathrm{L}^{\mathrm{s}}_{\emptyset}$, the weak and strong higher order languages $\mathrm{L}^{(\mathrm{n})}_{\mathrm{a}}$ and $\mathrm{L}^{(\mathrm{n})}$, the infinitistic languages $\mathrm{L}_{\omega_1,\omega_\emptyset}$ and $\mathrm{L}_{\omega_\mu,\omega_\nu}$ as well as their resp. models. The main difference between the model theory for the language L and the model theory for the languages $\mathsf{Q_a} - \mathrm{L}_{\omega_\mu,\omega_\nu}$ is the failure of the compactness theorem in most of the latter (namely in $\mathrm{L}_\emptyset$, $\mathrm{L}^{\mathrm{II}}_{\mathrm{a}}$, L^{II}, $\mathrm{L}^{\mathrm{s}}_{\emptyset}$ and most of the languages $\mathrm{L}_{\omega_\mu,\omega_\nu}$). The downward Skolem-Löwenheim theorem is valid (with some modifications) w.r.t. all the languages $\mathsf{Q_a} - \mathrm{L}_{\omega_\mu,\omega_\nu}$, whereas the upward Skolem-Löwenheim theorem fails for almost all these languages (due to the fact that it follows from the compactness theorem). Because of the failure of the upward Skolem-Löwenheim theorem, the structure of the spectra in these languages is much more complex than in the case of language L. The study of analogues of the completeness theorem of 1st order logic for non-elementary languages has produced many interesting problems but only few solutions to these problems.

To draw attention to the great flexibility of model theory mentioned above, Mostowski points to a special algebraic construction in model theory, i.e. to the construction of a model as a direct product of a certain family of models. A special case of this very fruitful model construction is the model called the reduced direct product of a certain family of models. Feferman and Vaught applied the direct product model construction to several decision problems. Major applications of the new concept (or construction) of reduced direct product are: (1) a new and simple proof of the compactness theorem for 1st order logic; (2) the following theorem in abstract set theory: If A is a σ-multiplicative filter, then the reduced direct product of well-ordered models is itself a well-ordered model. This theorem forms the basis for a number of results on denumerably additive filters in Boolean algebras of all subsets of a set; (3) in arithmetic, it provides a simple method for constructing non-standard models of arithmetic; (4) in the theory of real numbers, Robinson used it to construct a non-standard model of analysis containing infinitesimals, and he moreover showed how to get in this way a completely rigorous theory—called non-standard analysis—which is equivalent to classical or standard analysis.

Section 3 : The Subsequent 40 Years of Studies in FOM According to Parsons

In his account of the subsequent 40 years of studies in FOM, Charles Parsons sketches first the more technical mathematical-logical development and then the more philosophical development.

Since he considers computability theory and model theory to have become almost purely mathematical, with hardly any foundational-philosophical impact, they drop out of his account completely.

As for proof theory, Parsons distinguishes two new proof-theoretical programs: (i) The analysis of strong subsystems of classical analysis (2nd order arithmetic) by means which could still be thought of as constructive, but are much more powerful and abstract than the means applied in the 2nd phase of development of FOM. (ii) The attempt to reconstruct classical analysis predicatively: it was shown by Harvey Friedman, Stephen Simpson and others that suitable reformulations of standard theorems of analysis can be proved in weak systems of analysis (cf. the method of reverse mathematics).

In Parsons' opinion the most striking foundational results were obtained in set theory: By means of Cohen's forcing method many more independence results were found in set theory and its applications. This discovery of new important independent statements sparked off a search for new set-theoretical axioms along the lines suggested by Gödel in the 1940s. And some progress has been made in this search by developing the consequences of two sorts of new axioms: (a) strong axioms of infinity, (b) special cases of the game-theoretical Axiom of Determinacy AD. It was discovered in particular that strong axioms of infinity implying PD (i.e. the assumption asserting that the axiom AD holds for projective sets of real numbers) have the convenient feature that their consequences in 2nd order arithmetic cannot be altered by forcing. And W. Hugh Woodin's approach to CH aims at extending this result to a higher level. This can be conceived of as an important step in the solution of the problem of whether CH has a determinate truth-value.

On the philosophical side, Parsons draws attention to a number of new philosophical conceptions of foundational interest:

(1) A kind of neo-logicism: This conception was inaugurated in the early 1980s by Crispin Wright who defined what he called Frege Arithmetic FA by 2nd order logic plus Hume's Principle HP plus a Fregean number operator (N_xFx) and

then proved the axioms of 2nd order PA from FA using Frege's definitions (this is nowadays called Frege's theorem since it was essentially already proved by Frege). Wright's and Bob Hale's proposal is to take FA as basic arithmetic and to argue that the notion of cardinal number is a logical notion and that HP is a logical principle. This proposal has led to a lot of discussion and debate about the status of abstraction principles like HP. Parsons notes that "[t]he program of axiomatizing parts of mathematics by abstraction principles is of independent logical interest, and work has been done on analysis, and preliminary work on set theory".[10]

(2) A kind of default platonism: Parsons presumes that "[t]aking the language of classical mathematics at face value, as implying the existence of abstract mathematical objects, even forming uncountable and still larger totalities, and allowing reasoning using both the law of excluded middle and impredicative definitions, is probably a default position among philosophers and logicians".[11] He doubts that any strong (decisive) philosophical arguments can be given for a stronger kind of platonism than the default one, a conception he dubs "robust platonism". Wang and Penelope Maddy accept default platonism as "the limit of what one should claim about the determinateness of the reality described by mathematical theories".[12] This somehow corresponds to the application to mathematics of Quine's naturalistic position, however, without Quine's privileging of empirical science.

(3) One way of rejecting default platonism is by adopting a constructivist stance (constructivism): Parsons observes that in the 40 years of studies in FOM under consideration constructivism has declined significantly as a general approach to FOM competing with classical mathematics. The most remarkable constructivist appearances in this phase of the development of studies in FOM are on the one hand Per Martin-Löf's constructive type theory CTT, and on the other hand Errett Bishop's and Douglas Bridges' constructive approach to mathematics.

(4) Another way of rejecting default platonism is by adopting a nominalist stance (nominalism): The traditional way refuses to take the language of mathematics at face value and tries to reformulate it in such a way that commitment to abstract mathematical objects is avoided. A more radical way was worked out by Hartry Field who rejected "the view that statements of classical mathematics, taken at face value with regard to meaning, are true and even that mathematics aims at truth. He sought to account for the apparent objectivity of mathematics by viewing it instrumentally, as a device for making inferences within scientific theories".[13]

[10] Parsons (2006, p. 49).

[11] Ibid. p. 49.

[12] Ibid. p. 51.

[13] Ibid. p. 50.

(5) Structuralism: Two related intuitions about modern mathematics are fairly common: (a) modern mathematics is a study of (abstract) structures; (b) "mathematical objects have no more of a nature than is expressed by the basic relations of a structure to which they belong".[14]

The structuralist conception of mathematical objects is an elaboration of intuition (b). Its relation to default platonism is ambiguous and admits of at least 2 different positions: (i) eliminative structuralism: it refuses default platonism' s taking the language of mathematics at face value and proposes paraphrases eliminating reference to mathematical objects or at least to the most typical mathematical objects. (ii) non-eliminative structuralism: it rather constitutes an ontological gloss on default platonism and uses the structuralist conception as an explication of what the reference to mathematical objects in mathematical language amounts to.

"One version of structuralism would allow sets as basic objects. This would be a natural way of developing the first intuition [viz. (a)] , understanding structures as set-theoretic constructs. But a general structuralist view of mathematical objects would naturally aim not to exempt sets from structuralist treatment. At this point modality has been introduced."[15] With his system of Modal-Structural mathematics MS, Geoffrey Hellman worked out a version of eliminative structuralism based on this idea.

Parsons ends off his sketch of structuralism with the following critical remark: What the (eliminative) structuralist constructions accomplish depends on the status of 2nd order logic. And this question arises equally for neologicism and for nominalism. There has been much debate concerning this question.

(6) Naturalism and a Gödelian epistemological view : Whereas the philosophical conceptions (2)–(5) often have a strong ontological character, the conceptions (1) and (6) are of a strong epistemological type.

In the early 1970s Paul Benacerraf raised the following problem: If default platonism is true, how is it possible to have mathematical knowledge? Parsons generalises this problem in the following way: Is it possible to provide an epistemology for mathematics which is naturalistic? After stating that not much of philosophical-foundational interest has resulted from these questions, he turns to a Gödelian epistemological view which he takes to be possibly more interesting: "Gödel's view apparently was that much of mathematics (including some higher set theory) could be seen to be evident in an *a priori* way, not contaminated by evidence derived from application in empirical science. However, particularly in higher set theory axioms could obtain additional justification through the theories constructed on their basis, and such justification would be possible for stronger axioms, such as the stronger large cardinal axioms that have been proposed, where a convincing intrinsic justification is not available". [16] Parsons considers this Gödelian epistemological view to be of

[14] Ibid. p. 50.

[15] Ibid. p. 50/51.

[16] Ibid. p. 52.

great interest for the justification of assumptions applied in the accepted solution of the classical problems of descriptive set theory (e.g. in the justification of the axiom AD) as well as for the justification of any possible solution to the problem of CH to be expected in the future.

Part II The Present Perspective : Analytical Summaries of the Present Contributions[17]

Section 1

In his contribution **Foundational Frameworks** Geoffrey Hellman starts off by characterising the sort of questions asked in FOM: they are questions of justification (as opposed to questions of discovery), and moreover questions of an ideal, an on principle possible justification (as opposed to questions of actual justification). Thus, FOM according to Hellman is neither hermeneutics of mathematics, i.e. claiming to tell what working mathematics really is (reminiscent of Shapiro's "philosophy-last principle"), nor a cultural revolution of mathematics, i.e. advocating the replacement of working mathematics by a certain favored mathematical system or scheme (reminiscent of Shapiro's "philosophy-first principle").

Hellman then continues to enumerate important desiderata for any foundations (foundational frameworks) of mathematics (FOM): there are on the one hand the following traditional desiderata: (1) standards of proof, (2) means of expressing mathematical structures and their interrelations, (3) identification and explanation of the logical and mathematical primitives; on the other hand the modern (or post-modern) desiderata: (4) preservation of past gains, (5) accomodation of multiple approaches, i.e. providing for pluralism, (6) extendability of universes of discourse for mathematics. NB. Only some of these desiderata are actually requirements. A minimal requirement for a foundational framework is the following: providing a resolution of the conflicting tendencies of "creative progress" (desideratum 6 and perhaps desideratum 5) and striving for "all-embracing completeness" (desiderata 1–3). (Zermelo)

After laying down the desiderata of a foundational framework, Hellman goes about assessing set theory and category theory in terms of these desiderata.

First, he considers the prevailing set theory ZFC and many of its variants:

desiderata 1-3: ZFC is a major success story, may be the least w.r.t. desideratum 1(c), which is the most philosophical one. 1(c) concerns the epistemological sense of the foundations of mathematics.

desideratum 5: this desideratum is not met very well by ZFC: if a fixed-universe ontology is assumed for ZFC, then respecting other foundations of mathematics becomes problematic.

---

[17] These analytical summaries of the various contributions have all been approved by their resp. authors.

desideratum 6: On a fixed hierarchy view of ZFC, the problem of providing for extendability without exceptions is especially intractable: In a certain sense, ZFC does comply with extendability, namely by treating models of (consistent) theories as sets and recognizing sets and models of arbitrarily higher cardinality. But the unresolved problems are:

- A possible distortion of intended meanings
- The problem raised by the presumed fixed universal, hence maximal background of sets and ordinals (which thereby precludes the possibility of still further objects, other than sets and ordinals)

Second, Hellman deals with category and topos theory (CT & TT):

desiderata 5–6: CT & TT are through Bell's "many toposes" perspective a great success story (unlike set theory). Characteristic of this perspective is a plurality of universes.

desiderata 1–3: Here is where the most serious problems with CT & TT arise. 2 main category-theoretic axiom systems have been proposed with a foundational role in mind, namely: ETCS (the Elementary Theory of the Category of Sets) and CCAF (the Category of Categories As a Foundation).

as for CCAF: desideratum 2: considered as a system of Fregean-style axioms. These axioms are satisfied by "structures" (or at least "interrelated things"). Hence there is some dependence on a background explicating satisfaction of sentences by structures, and this background is not CT itself (it may be a dependence on a background set theory or on a higher-order logic or on a mereology + plural quantification). The conclusion is that w.r.t. the desiderata 1–6, CCAF is not yet adequate.

as for ETCS (or ECTS + R (Replacement)): desiderata 5–6: ETCS(+R) inherits some of the same problems affecting ZFC (cf. ibid.). desiderata 1–3: First, it seems that ETCS(+R) can only be really understood given a prior understanding of general notions such as "collection", "operation" or "relation" (Feferman). Second, a fundamental dilemma concerning categorical foundations of mathematics is the following: if desiderata 2–3 are met by categorical set theory, then desiderata 5–6 will not be met (due to limitations analogous to those of membership-based set theory); if however desiderata 5–6 are met by categorical set theory, then desiderata 2–3 won't be met. The conclusion in this case is that w.r.t. desiderata 1–6, ETCS(+R) is not adequate (yet) either.

Hellman goes on to assess modal-structural mathematics (MS) in terms of the desiderata of a foundational framework.

The modal-structural mathematics MS interpretation of mathematics has 2 components:

1. A hypothetical component: a translation pattern sending any ordinary mathematical sentence to a sentence asserting what would hold in any structure of appropriate type that there might be.
2. A corresponding categorical component: structures of the relevant type are (logico-mathematically) possible.

The Core System of MS is the following:

- The (monadic) logic of plurals
- 2 comprehension principles of mereology
- As an improvement over his original presentation *Mathematics without Numbers*, this presentation adopts both an Extendability Principle for structures for set theories (or precursors) and certain instances of a Modal Reflection Principle suitable to the MS framework, based on the idea that the logico-mathematical possibilities of structures should be "indescribable" by 1st or 2nd order sentences in a specific sense explained by him. From these 2 principles applied within the realm of finite structures, Hellman then can derive the compossibility of infinitely many objects and a "set" of them as needed for reconstructing classical analysis. Thus, it is not necessary to postulate an axiom of infinity separately as it is in set theory and category theory.

desiderata 1–2: MS' coverage of mathematics as practiced is almost as "complete" as that of set theory and category and topos theory; that is, these desiderata are about as well satisfied by MS as by ZFC and CT & TT.

desideratum 1(c): MS provides an interesting epistemological alternative to the acceptance of actual infinities without being as restricted or restrictive as strict finitism.

desiderata 5–6: These are MS' fortes. None of the alternative foundations of mathematics seems to be capable of satisfying them as well as MS.

desideratum 3: Hellman has done some work to make MS meet it; but more remains to be done.

Hellman rejects a strong foundationalism, i.e. which seeks to found all of mathematics on certain or self-evident assumptions, but he adopts a "modest, well-tempered foundationalism", i.e. the search for foundations providing a measure of epistemic order and a balance of unity and diversity (and perhaps, as Hellman puts it, even some insight into the nature of mathematics).

Bob Hale begins his contribution **The Problem of Mathematical Objects** by distinguishing two senses of 'foundations of mathematics' FOM:

1. The logical sense of FOM: foundations consist of a single, unified set of principles from which all or at least a large part of mathematics can be derived.
2. The epistemological sense of FOM: foundations consist of an account explaining how standard mathematical theories can be known to be true or at least be justifiably believed. Foundations in this sense cannot be mathematical theories, but have to be philosophical accounts of how working mathematics is getting known.

Hale's interest is clearly in the epistemological sense of FOM—independently of whether the search for such a foundation is a legitimate ("right") or possible endeavor.

Hale's point of departure is what he calls the problem of mathematical objects, and more specifically the problem whether one can be justified in believing in an infinity of objects of any kind. He subsequently distinguishes two approaches towards a solution of this problem:

(a) The so-called object-based approach: This is an approach arguing directly that it is possible to have access to or knowledge of an at least potentially infinite sequence of objects.
(b) The so-called property-based approach: This sort of approach argues indirectly for an infinity of objects by making the latter depend on an underlying infinity of properties.

According to Hale, Charles Parsons' study of mathematical intuition presents the most clear and convincing example of the object-based approach:

After distinguishing between intuition of objects and intuition that *p* where *p* is a proposition, Parsons makes it clear that the objects of intuition have to be restricted to the concrete and the quasi-concrete (and do not extend to the abstract). He claims that intuition of quasi-concrete stroke-string types can ground propositional knowledge concerning the system of stroke-string types, and that it can provide knowledge of analogues of the elementary Dedekind-Peano axioms. Hence it is possible to have intuitive knowledge of the existence of potentially infinitely many objects.

Hale starts his criticism of Parsons' object-based approach by pointing out that some of the (Parsons') Dedekind-Peano axioms are general and that it is hard to see how intuition of objects can yield knowledge of general truths. He carries on by emphasizing another, more fundamental problem in Parsons' approach: The difficulty concerns what is taken to be required for the existence of a stroke-string type: Is it (i) that there exists at least one token of that type? (ii) that it exists totally independently of any actual, possible or imaginative instantiation? or (iii) that there could exist at least one token of that type?

(i) already requires knowing that there are infinitely many concrete objects. This option is no good. (ii) implies rampant Platonism. This option runs counter to Parsons' attempt to exhibit intuitive knowledge of quasi-concrete stroke-string types. (iii) seems to imply that given any perceived or imagined stroke-string token, a single stroke-string extension of it is imaginable. But spelling out what 'imaginable' means just seems to get Parsons into further troubles.

Since Hale cannot see how any object-based approach could get around some appeal or other to some such grasp of possibilities, he infers that such an approach cannot work.

Question: Can a property-based approach to the problem of mathematical objects do better? Hale's intention is to show that indeed it can. G. Frege sought to prove the existence of successors for all finite numbers given Hume's principle (HP). His purported proof exemplifies according to Hale the so-called property-based approach. For Frege, numbers are essentially numbers belonging to concepts (i.e. a number exists only if there is a concept which it essentially belongs to), and Hale interprets this as saying: numbers essentially belong to properties. Given HP, Frege not only succeeds in proving that the sequence of finite numbers is infinite, but also in proving the Dedekind-Peano axioms in 2nd order logic (the latter is nowadays called Frege's theorem). Whereas Hale considers these proofs to be of considerable philosophical importance, other philosophers of mathematics have raised doubts concerning that importance. The first doubts or objections to be tackled by Hale are Michael Dummett's. According to Hale's analysis, Dummett objects:

i. That Frege can procede in his *Grundlagen der Arithmetik* only on the assumption that there exist inifinitely many objects, but that this assumption cannot be grounded in logic only.
ii. That there is some sort of a vicious circle in Frege's procedure (in Frege's definition of number or in HP) due to the fact that the numbers themselves are taken to belong to the domain over which the individual quantifiers range. Hale claims that his objection can be interpreted in two ways: the alleged circularity can be definitional or it can be epistemological.

Hale dismisses the definitional vicious circularity objection as it seems to imply that for any specific kind of object lying in the range of some (individual) quantifiers one must possess the concept of that kind of object (e.g. here the concept of number). But this, so he argues, is not the case. Hale's rejoinder to Dummett's epistemological vicious circularity objection as well as to Dummett's objection i. is basically the same: According to Dummett, Frege makes the assumption that numerical terms have reference. For Hale there is no such assumption, at least not when HP is put forward as an implicit definition. HP is an instance of a general abstraction principle whose instances are biconditionals which are to be so understood that their truth is consistent with their ingredient abstract terms (e.g. in HP the number terms) lacking reference. In the particular case of HP, there are instances the truth of whose right-hand sides is indeed a matter of logic.

Dummett's objection could be strengthened and would be quite closely related to an objection of Charles Parsons':

iii. Even if there is no explicit assumption to the effect that there exist infinitely many objects (or which boils down to the same thing, that number terms have reference), there is such an assumption implicit in the use of HP: HP quantifies over properties and makes the assumption of the existence of infinitely many objective properties.

Hale admits that if the assumption of existence of infinitely many properties were as problematic as the assumption of existence of infinitely many objects, Dummett's strengthened objection would be devastating. He, moreover, concedes that the first assumption would indeed be as problematic, if properties were conceived of in an extensional way (that's essentially Parsons' objection: there is no philosophical advantage of higher-order logic over set theory). Hale concludes first that the strengthened objection is justified if directed against Frege, and second that any property-based approach appears to be a waste of time if properties are indeed treated extensionally.

But as Hale puts it: "[i]f one thinks instead of properties as individuated non-extensionally, there is at least some chance of philosophical advantage" of higher-order logic over set theory and there is hope for a property-based approach. The assumption of the existence of infinitely many properties may be significantly weaker and so epistemologically less problematic than the assumption that there are infinitely many objects; it may, Hale suggests, even be weak enough to form part of a foundation of mathematics. A big task of the property-based approach will be to clarify what a non-extensional conception of properties amounts to. He provides a few hints at what such a conception might look like.

In her contribution **Set Theory as a Foundation**, Penelope Maddy recalls that set theory as a FOM goes back to the founders of set theory and has become part of contemporary orthodoxy in the philosophy of mathematics and even in mathematics itself. She briefly shows how natural numbers, integers etc. up to the reals and complex numbers can be represented as set theoretic entities and points out that all its standard theorems can be proved from ZFC. This is a remarkable mathematical fact. The philosophical question is what this fact means. What does it show? Maddy presents and discusses 6 interpretations of this fact. They will be called the 6 senses of (set theory as a) 'foundations of mathematics' FOM.

1. The metaphysical sense (attributed to a so-called strong reading of Frege's project): The current set theoretic versions of numbers, functions, spaces etc. show what numbers, functions, spaces, etc. really are; they exhibit the true nature of the various mathematical entities.

   Benacerraf objected to this metaphysical interpretation of FOM that there is not a unique or clearly priviledged identification of natural numbers with certain (pure) sets, but that many different identifications seem equally good. And the same holds in an analogous way for identifications of integers, reals, functions etc.

   According to Maddy, there is a weaker reading of Frege's according to which Frege was merely concerned about the just mentioned mathematical fact without any associated metaphysical ambition.
2. The ontological sense (Quine's ontological reduction): The current set theoretic versions of numbers, functions, spaces etc. admit of an ontological economy: it suffices for a mathematical ontology to merely accept the existence of these current set theoretic versions of the various mathematical entities. Quine advocated such an "ontological reduction", i.e. a replacement of a world view countenancing both numbers, functions, spaces and sets, with a world view countenancing only the sets.
3. The methodological sense (e.g. Moschovakis): the current set theoretic versions of numbers, functions, spaces, etc. are set theoretic surrogates of mathematical entities sharing with them the same mathematically relevant features.

Maddy appears to view a certain order in these senses. These senses are all mutually exclusive and there's an order of decreasing strength in them: the metaphysical sense is the philosophically strongest sense, the ontological sense is somewhat weaker, and the methodological sense is the philosophically weakest one (making the least philosophical presuppositions).

To each of these senses is attached a certain benefit (and it appears that such a benefit ought to be attached): in the case of the metaphysical sense, it is a metaphysical insight; in the case of the ontological sense, it is an ontological economy; in the case of the methodological sense, the benefit is mathematical rather than philosophical: the set theoretic foundations play a unifying role, bringing all mathematical objects and all mathematical structures into the arena of the universe of sets; the set theoretic axioms do not only reproduce the existing mathematics, but have strong consequences for existing fields and produce a new and very fruitful theory in its own right; the arena of the universe of sets "provides a court of appeal for questions of mathematical existence and proof"; etc.

Maddy adopts this methodological sense for herself as well as the logical sense (cf. below) intertwined with the methodological one. These senses 1–3 touch upon ontological issues and can be considered as possible answers to the question how set theory as FOM relates to ontology in the philosophy of mathematics.

Maddy then wants to tackle some corresponding epistemological issues.

4. The epistemic sense: The set theoretic foundations is a way of installing mathematics on a safe consistent (axiomatic) basis (to avoid the paradoxes and contradictions in set theory and mathematics). NB. the "negative" epistemic role of foundations as opposed to Descartes' positive one: foundations in Descartes' sense is an indubitable, certain, self-evident base; here foundations are, what Mac Lane calls, a "security blanket", they should prevent the mathematician from falling down, from falling into contradiction and into a crisis.

   Gödel's 2nd incompleteness theorem constitutes a formidable objection to this epistemic sense of FOM. Maddy argues that the methodological sense of FOM can be upheld even in the face of the failure of the epistemic sense.
5. The logical sense (in Russell's sense): The set theoretic foundations provide a logical basis of mathematics, that is the logically simple principles from which all of existing mathematics (and more) can be deduced.

   Mary Tiles objects to this sense that it is viciously circular. Maddy's reply is that Tiles confuses the logical with the epistemic sense: Tiles' objection only applies to the epistemic sense.

As for the benefits: in the case of the epistemic sense, the benfit is a secure given starting point; in the case of the logical sense, the benefits are an organisation of our knowledge, the great potential of mathematical insight and discovery contained in the principles, etc.

6. The "professional sense": The "professional sense" of set theoretic foundations means that "all mathematicians might as well be set theorists, restricting themselves to the methods characteristic of set theory".

   Maddy rejects the professional sense of set theory as FOM and insists that the methodological sense of FOM by no means implies the professional one.

Maddy's conclusion is then: Set theory as FOM should not be understood in either a metaphysical or ontological or epistemic or professional sense. It is however especially from a mathematical point of view, which is philosophically "harmless", a most prolific and fruitful endeavor to take set theory as FOM in a methodological and in a logical sense.

Stewart Shapiro's contribution **Foundations: structures, sets, and categories** analyzes and discusses several senses of 'foundations of mathematics' FOM:

1. Ontological senses

   a. The ontological (or weak ontological) sense: FOM describes or otherwise provides for the objects of mathematics. Examples are: set theory, category theory, Scottish neo-logicism, ante rem structuralism, etc. The ontological sense of FOM presumably implies an ontological realism of some sort in the philosophy of mathematics.
   b. The metaphysical (or strong ontological) sense: FOM provides the (ultimate) subject matter of mathematics, the true objects of mathematics. A descriptive thesis attached to this metaphysical sense of FOM is: FOM describes, states the true nature of mathematical objects, delimits the underlying fabric of mathematical reality. And a stipulative (conventional and non-descriptive) thesis attached to this metaphysical sense of FOM is: FOM is proposed, stipulated to provide or to present the true nature of mathematical objects (for various practical or pragmatic reasons).
   c. The methodological (or minimal ontological or, in Shapiro's terms, mathematical) sense: FOM provides surrogates for (perhaps) all mathematical objects, by isolating the mathematically relevant features of corresponding mathematical objects. The ontological implication or aspect of this methodological sense is that it can contribute to settle existence problems of certain types of mathematical objects "internally" (e.g. and prominently by a set-theoretic proof of satisfiability). An interest in this methodological sense of FOM is fuelled by structuralism.

2. Epistemic senses

   a. The epistemic (or weak epistemic) sense: FOM constitutes the epistemic basis or the justification of mathematical propositions: The grounds for true or at least knowable mathematical propositions can be found in the proposed foundation. Proving some mathematical proposition from some FOM is a sufficient condition for knowing it.
   b. The strong epistemic sense: Proving a mathematical proposition from some FOM is a necessary condition for knowing it. Shapiro considers this strong epistemic sense of FOM to be rather absurd.

   Stewart Shapiro then distinguishes between 3 possible epistemological goals of Frege's logicism:

   i. The epistemological-mathematical goal: The goal is to prove what can be proved; and the propositions (and axioms) of e.g. arithmetic can be proved from some FOM (e.g. set theory), hence they ought to be proved. Shapiro, however, questions this goal: How can proving something about some surrogates of mathematical objects count as proving something about these objects? And he remarks: From this epistemological-mathematical point of view, different FOMs are not mutually exclusive; it is perfectly compatible with a pluralism of FOMs.

ii. The logico-Cartesian goal: The goal is to achieve perfect mathematical knowledge, i.e. knowledge possessing absolute self-evidence, certainty and clarity. According to Frege, the logical source alone is capable of producing such knowledge. Shapiro comments that set theory can't be a FOM according to this logico-Cartesian point of view, nor can Scottish neo-logicism.

iii. The knowledge-of-sources goal: The goal is to "determine the epistemic pedigree of [mathematical] propositions"—that is to determine whether they are analytic or synthetic, apriori or aposteriori. Shapiro points out that despite Frege's use of epistemic terms such as 'proof' or 'justification', his relation of dependence of propositions is as much metaphysical as it is epistemic. NB. This third point is very similar to Maddy's (Russell's) logical sense of FOM. Shapiro introduces the term 'proper foundational knowledge' for a sort of knowledge which is based on objective grounding relations among the known propositions. A FOM in this perspective is an ultimate or objective ground or justification of mathematical propositions. Shapiro then evaluates set theory, category theory and Scottish neo-logicism as to whether they claim to lead to proper foundational knowledge or not.

3. The linguistic or semantical sense
Presupposition: Every mathematical theorem is of the form

(∗) if such-and-such is the case, then so-and-so holds.

S. Awodey interprets this presupposition in a way which distinguishes it clearly from the old-fashioned "if-thenism". The linguistic or semantic sense: FOM is a good theory or language in which to formulate the various "such-and-such" antecedents and "so-and-so" components of (*) and perhaps also substantial logical theses about how to go from the one to the other. For short: FOM is a good language in which to formulate mathematics.

Shapiro observes first that the linguistic or semantic sense is totally independent from and hence does not imply the ontological sense of FOM: Since from a linguistic perspective a mathematician only specifies (or presupposes) some conditions on properties or relations and investigates what holds in systems meeting these conditions, he or she by no means presupposes a once-and-for-all realm of sets, structures, toposes or whatever, from which all mathematical objects are to be or can be constructed. Second the linguistic or semantic sense is also totally independent from the epistemic sense of FOM: According to Awodey the methods of reasoning in various parts of mathematics are not "global" or uniform across fields or even between theorems, but are themselves "local" or relative. That is, the logic for a given piece of mathematics is conventional or obvious or will be explicitly stated as part of the theorem.

Awodey has 2 theses:

Thesis 1: The theorems of mathematics have the schematic character articulated in (*).
Thesis 2: Category theory is a FOM in the linguistic or semantic sense and it is the best one known.

Shapiro takes Awodey's thesis 1 in a descriptive sense (as it is by the advocates of Awodey's approach). He then introduces the term 'assertory' to refer to those sentences which express propositions about a fixed subject matter, with a fixed truth value. Assertory sentences are to contrast with schematic ones, a schematic sentence having the form (*). Shapiro then raises the question: Can everything recognizable as mathematics be interpreted in the schematic way, i.e. as expressible by schematic sentences? He ends off his contribution by indicating two reasons for a possibly negative answer to his question:

Reason 1 for a possibly negative answer: In metamathematics, on occasion questions concerning the existence of certain structures or the coherence of various mathematical theories arise. Suppose one is given a schematic sentence of the form (*). One may ask whether there are or could be any such-and-such's? Or consider a statement Π that a given specification S is coherent. Then Π looks like a mathematical sentence. Can it be put into the schematic form (*)? What would the "such-and-such"-antecedent be? If a mathematician succeeds in finding a model of S in the iterative hierarchy, one could have a statement of the form (*), namely:

$$(i) \qquad \text{If } \mathsf{ZFC}, \text{ then } \Pi$$

To apply modus ponens to (i), one would not think of (i) as schematic, but rather as assertory: If one asserts that ZFC is true or coherent, one can rightly infer that Π. Or other stock meta-mathematical results such as the completeness theorem of classical 1st order logic or the incompleteness theorem of arithmetic are taken by Shapiro to be assertory rather than schematic. NB. The advocate of the schematic perspective is free to insist that even these theorems or these sentences of existence or coherence have only the schematic form indicated.

Reason 2 for a possibly negative answer: Consider the statements that isomorphic structures are equivalent, and that if Φ is a sentence in the language of 2nd order arithmetic, then Φ is a model-theoretical consequence of (classical) 2nd order PA or $\neg\Phi$ is such a consequence. Shapiro takes these sentences to be assertory. That's why they can be invoked in cases like Wiles' proof of Fermat's last theorem, to settle arithmetic questions via richer contexts. The question then is according to Shapiro: How can the schematic account of mathematics account for this feature of contemporary mathematics, or failing to do so, how can it account for why the common wisdom concerning it is wrong?

Section 2

In his contribution **From Sets to Types to Categories to Sets** Steve Awodey takes there to be three different common styles of FOM, namely set theory, type theory and category theory. He sets out to answer two questions:

1. How do these 3 styles of FOM relate?
2. What are their resp. advantages and disadvantages especially w.r.t. the program of structuralism?

He proceeds in 3 steps to answer the first of these two questions:

Step 1 from sets to types: that is the familiar idea of set-theoretical semantics for a syntactic system
Step 2 from types to categories: that is what categorical logicians call the construction of a "syntactic category"
Step 3 from categories to sets: that is a very recent logical mathematical construction

Step 1: The particular type theory considered by Awodey is an intuitionistic system of higher-order logic with "power types", called IHOL by him.The theorems of IHOL always include the general laws of intuitionistic higher-order logic. Given a type theory T, the familiar way of interpreting it in set theory is by providing a model of it, i.e. an interpretation of its types (basic and constructed) and terms (basic and constructed) satisfying its theorems. A type theory T can be modeled in set theory in different ways. The formulas of T that always come out true under every interpretation will contain the general laws of intuitionistic higher-order logic (usually specified by a deductive system). Given a system of set theory S, there is a distinguished type theory T(S) with a distinguished interpretation in S. The construction of T(S) is quite straightforward and T(S) captures all the type-theoretic information of S.

Step 2: From a type theory T, a category $\mathcal{E}$(T) will be constructed by identifying certain terms as the objects and arrows. This category is of a special kind known as "topos" and $\mathcal{E}$(T) is in particular called a "syntactic topos": As the construction will show, the latter is comprised of syntactical material of T. The general construction of a syntactic topos out of a type theory also demonstrates a completeness theorem for general deductive higher-order logic w.r.t. topos models.

Step 3: It is possible to extract a set theory from a topos. The resulting set theory will have sets and functions which are essentially the objects and arrows of the original topos, and its theorems all hold in the topos.

The idea behind this extraction is motivated by a somewhat similar "construction" of a set theory from a type theory: One can "sum the types" of a type theory T to obtain a universal type which, if the "summing" is carried out the right way, will also contain the powertype of the universal type. This universal type admits of an untyped membership relation modeling an elementary set theory.

Given a topos $\mathcal{E}$, one can really construct the so-called ideal completion of $\mathcal{E}$, referred to by $\mathsf{Idl}(\mathcal{E})$, which is the category of all ideals in the topos $\mathcal{E}$. $\mathsf{Idl}(\mathcal{E})$ contains a particular object, the so-called total ideal, which plays a role analogous to the one of the universal type mentioned above. Thus, $\mathsf{Idl}(\mathcal{E})$ contains a model $(U, \in_U)$ of an elementary set theory, the sets and functions of which form a category equivalent to $\mathcal{E}$. The elementary set theory modeled in this way by every topos is a variant of ZF set theory, called BIST by Awodey, for Basic Intuitionistic Set Theory. Remarkable about BIST is that it is not only sound, but also deductively complete w.r.t. topoi, modeled in their ideal completions. According to the particular topos $\mathcal{E}$, the set theory $\mathsf{S}(\mathcal{E}) = (U, \in_U)$ in $\mathsf{Idl}(\mathcal{E})$ may also model more set-theoretical formulas than the mere deductive consequences of BIST.

Awodey then briefly discusses 3 possible compositions of 2 steps each:

Step 4: from sets to categories
Step 5: from types to sets
Step 6: from categories to types

and he subsequently considers the possibility of what it means to go in a full circle, "once around" to the categories corner, and he observes that this way of going around in a circle results in an equivalence of categories.

Awodey finally addresses the second of his two questions asked at the beginning and thereby draws essentially four conclusions from his previous analyses and observations:

(a) The 3 systems of foundations, set theory, type theory and category theory, are mathematically equivalent, since they allow for the presentation and proofs of the same "mathematical content", just represented in different ways. They are therefore said to be different styles of FOM. He reckons that elementary set theory at least as strong as BIST, type theory in the form of higher-order logic, and category theory as represented by topos theory all represent the same mathematical content.
(b) Awodey claims that "the objects of type theory and set theory are structured by the operations of their respective systems in certain ways that are not mathematically salient". This additional information is essentially what drops out of the just mentioned "mathematical content" of any system of FOM. In contrast, categorical structure is closer to this "mathematical content" and is not lost in translation. But, some other specifically "foundational content" is involved in the type- and set-theoretical presentations, and it is lost in translation.
(c) The advantage of category theory as a style of FOM is thus that it is more stable, more robust, more invariant than type- or set-theoretical constructions: The fact that the translations preserve essentially all the categorical structure (because of the equivalence mentioned above) proves that the presentation in category theory is more"pure": it presents the mathematical content in a direct, invariant, stable way without any additional, irrelevant, "foundational" structure.

    An advantage of type theory might lie in its more concrete, "nominalistic", constructive character (which holds to a much lesser degree for impredicative systems).

And an advantage of set theory consists of being some sort of a "third way" between type theory and category theory: it is much less concrete and constructive than type theory, while unlike category theory having some degree of control over its objects due to its conception of a systematic, iterative generation of its objects "from below".

(d) Awodey subsumes set theory and type theory under heading of "logical foundations". To a certain extent, he ascribes a complementary role to categorical and logical foundations: On the one hand categorical foundations (category theory as an approach to FOM) present the invariant "content of logical foundations". Logical structures in foundational work on the other hand often provide the additional data for specifications and calculations needed to facilitate mathematical constructions and proofs of results which are invariant.

The upshot of Awodey's comparison of the three approaches to FOM is: This comparison proves that category theory is better for describing mathematics in the modern, structural way, i.e. modern mathematics.

In his contribution **Enriched Stratified Systems for the Foundations of Category Theory** Solomon Feferman continues his intermittent explorations (dating back to 1969) of the set-theoretical foundations of category theory. His aim is to account for certain constructions in"naïve" category theory which, though mathematically natural, do not have a direct account within the usual systems of axiomatic set theory such as ZFC. Category theory is initially illustrated by concrete examples of the category of all structures of a given kind such as the category of all groups, the category of all topological spaces, etc., for which the morphisms are the structure preserving mappings. One ought then to be able to consider the category of all categories—whose morphisms are simply the functors between categories—as well as the category of all functors between two given categories—whose morphisms are the natural transformations between functors. These lead one to consider objects which are somehow "too large" to deal with directly in such systems as ZFC. One familiar strategy for a set-theoretical reduction, due to Mac Lane and Grothendieck, is to relativize the notions involved to one or more universes, i.e. sets of sets satisfying the axioms of the underlying set theory; a refinement due to Feferman [1969] [18] employs the reflection principle so as to deal with universes that are surrogates of the class of all sets.

As an alternative to such reductions it has been argued by a number of workers in the field that category theory doesn't need such foundations, since it provides the proper foundations for all of mathematics, including itself. Feferman, however, firmly rejects this position for reasons given in his [1977] (cf. also Hellman [2003]) and not repeated here. But he adds, among other things, that it is not clear how it proposes to handle the above principles of naïve category theory. Note that Feferman does not pursue the set-theoretical foundations of category theory because he is a proponent of set theory—on the contrary he is an opponent on philosophical grounds—but rather "because it is currently widely accepted, its ins and outs are

[18] For the full references of this and all the other publications referred to below, cf. Feferman's contribution.

well understood, and it has dealt successfully with the problems surrounding objects that are 'too large'."

Already in 1977, Feferman suggested the first three of the following four requirements on a system S for the foundations of category theory:

(R1) S should allow for the construction of all mathematical structures of a given kind, including also the category $\mathcal{CAT}$ of all categories.
(R2) S should allow for the construction of the category $\mathcal{B}^{\mathcal{A}}$ of all functors from $\mathcal{A}$ to $\mathcal{B}$ where $\mathcal{A}$ and $\mathcal{B}$ are any categories.
(R3) S should allow for the establishment of the usual basic mathematical structures and the carrying out of the usual set-theoretical operations.

Finally, in order to achieve a set-theoretical reduction (as was implicit in [1977]),

(R4) S should be shown consistent relative to currently accepted systems of set theory.

Feferman then evaluates the usual proposals mentioned above w.r.t. the requirements (R1)–(R4):

(a) Mac Lane's initial proposal was to work with the Bernays-Gödel system of sets and classes BG: (R1) and (R2) are satisfied only in a modified or restricted form (e.g. for small categories). (R3) and (R4) are met in full.
(b) Grothendieck's proposal was to work with ZFC + a strong axiom of "universes". Again, (R1) and (R2) are satisfied only in a restricted form (e.g. for categories lying in a universe U), unlike (R3) and (R4). (Mac Lane later modified this to one universe.)
(c) Feferman's proposal in [1969] and [2004] was to work with ZFC with one or more additional constant symbols for universes U satisfying the reflection principle asserting that every property relativized to U holds if and only if it holds in unrelativzed form; (R1)–(R4) are met in the same way as in Grothendieck's proposal.

As to (c), Feferman takes the main advantage of his over Grothendieck's proposal to be a conceptual one: His U looks from the point of view of the set-theoretical language exactly like the universe V of all sets and thus serves as a stand-in for it (which is not at all assured for the Grothendieck universes).

The conclusion about the proposals (a)–(c) is that on the one hand, they are appropriate for usual applications of category theory, but on the other hand none of them satisfies (R1) and (R2) unrestrictedly. By contrast, Feferman's aim in this piece is to show that the requirements (R1) and (R2) can be fully met by working in certain systems of set theory extending Quine's idea of stratification. It can be shown that (R4) is likewise met by these systems, and so is (R3) to a considerable extent. But there are two ubiquitous set-theoretical constructions which can't be carried out in these systems without ad hoc modification; moreover, certain basic results of category theory can't be formulated unrestrictedly. This is the price to be paid for having these stratified systems of set theory fully satisfying (R1) and (R2). It is an open question whether there is a solution to the question of a system S satisfying

(R1)–(R4) without any such trade-offs. What the systems of set theory to be presented illustrate is that there exist systems which at least meet almost all of the basic requirements without restriction.

The axioms of Quine's system NF of "New foundations for mathematical logic" are Extensionality Ext and Stratified Comprehension SCA, i.e. NF = Ext + SCA. It is still an open question whether NF is consistent. If Ext is replaced by a weakened form of extensionality Ext' allowing urelements, one obtains the system NFU = Ext' + SCA which was shown by Jensen [1969] to be consistent. In order to get off the ground with a stratified theory of relations and thence of structures, one needs a stratified pairing operation. That is, in assigning type levels to variables in a formula for the comprehension principle, the type level of a pair (x, y) should be the same as that of both x and y. Quine had shown how to define a stratified pairing operation within NF by means of Ext plus the assumption of an axiom of infinity; but the latter is not provable in NF (if consistent). A simple alternative, which is the one taken by Feferman, is to extend NFU by a Pairing Axiom P and extend the notion of stratification so that a formula is stratified if pairing is treated in a stratified way. The extended system is $\mathsf{NFU_P}$ = Ext' + SCA + P. Like NFU, $\mathsf{NFU_P}$ can be shown to be consistent by Jensen's methods.

Working within $\mathsf{NFU_P}$ Feferman defines single-sorted and many-sorted 1st order structures and explains how to treat categories as two-sorted 1st order structures. He goes on to show how the category of all structures satisfying a 1st order condition, and in particular the category $\mathcal{CAT}$, exists as an object in this system. Moreover, it is verified that the structure $\mathcal{CAT}$ of all categories with the functors as morphisms belongs to the class of all categories CAT, i.e. $\mathcal{CAT} \in CAT$, whence we can truly speak of $\mathcal{CAT}$ as the category of all categories, and (R1) is satisfied. And then he demonstrates that $\mathcal{B}^{\mathcal{A}} \in CAT$, for any categories $\mathcal{A}$ and $\mathcal{B}$ whence (R2) is equally satisfied by $\mathsf{NFU_P}$.

As for requirement (R3), Feferman partly shows and partly indicates how a great many standard set-theoretical constructions can be carried out quite smoothly in $\mathsf{NFU_P}$. And so he concentrates on operations which can't be carried out without ad hoc adjustments, namely the passage to equivalence classes under an equivalence relation and the formation of the Cartesian product of an indexed family of classes. Moreover, the fact that some basic results of category theory such as the cartesian closedness of the category of all sets and Yoneda's Lemma can't be formulated unrestrictedly in $\mathsf{NFU_P}$ is a serious handicap for using $\mathsf{NFU_P}$ as a foundation of category theory. These problems all involve pairs (x, y) where x is assigned a type n and y is assigned a type $n + 1$. Feferman emphasizes that there is no obvious modification of the notion of stratification for systems with pairing that allows pairs (x, y) of mixed type and is consistent; and there is no easy way out through some sort of "unification" of these mixed types either.

In order to get a system of set theory meeting (R3) in a fuller way, the system $\mathsf{NFU_P}$ is boosted to incorporate ZFC in a certain way. The resulting system S^* is an extension of $\mathsf{NFU_P}$ as well as of ZFC, and S^* can be shown to be consistent relative to ZFC plus the assumption of two strongly inaccessible cardinals. S^* does fix some of the defects of $\mathsf{NFU_P}$. Feferman's work on the systems $\mathsf{NFU_P}$ and S^* dates back to

1974 and was never published, but was circulated informally at the time in a lengthy MS that contained a full proof of the consistency of S*. The methods used for that proof are here only outlined in an appendix.

Work on and with S* still leads back to the kinds of distinctions that the proposal for a direct foundation of "naïve" category theory was meant to avoid. Thus, though this approach using stratified systems does succeed in satisfying (R1), (R2) and (R4), as well as (R3) to a considerable extent, Feferman suggests that it may point to a dead-end in this line of research.

By comparison, more recent research (than that of Feferman's from the mid 1970s) on stratified systems has shown that the kinds of type-shifting problems related to the development of category theory in NFU$_P$ may be avoided by restriction to strongly Cantorian classes. But such a restriction implies giving up requirement (R1), as the collection of strongly Cantorian classes does not form a class. Feferman ends his contribution by discussing advantages and drawbacks and other properties of various extensions of NFU such as NFUA, S* and a modified system S**.

In his contribution **Recent Debate Over Categorical Foundations** Colin McLarty suggests that in order to discuss or rediscuss category theory as a FOM, one ought to go back to the origins, i.e. to the positions of W. Lawvere and S. Mac Lane. For example, Lawvere and Mac Lane's position w.r.t. G. Hellman's question: Are category-theoretic axioms "asserted as true or merely posed as abstract axioms"? is the following: Both (as is the case with ZFC as a FOM); sometimes the one, sometimes the other. The or a difference between categorical foundations and ZFC as foundations is: ZFC is essentially (McLarty: only) used in foundations whereas category theory is used all over mathematics. According to Lawvere's and Mac Lane's position there can be no opposition between set theory and category theory as FOM. The first categorical foundation was a set theory, namely the "Elementary Theory of the Category of Sets" (ETCS). The opposition rather is between membership-based set theories including ZFC and categorical foundations including categorical set theory. In ETCS, Lawvere presents 1st order axioms for the category of sets and he considers them under 2 perspectives:

Perspective (a): Lawvere's perspective (a) assumes that sets, groups, topological spaces etc. all exist prior to any FOM. On this perspective, the 1st order ETCS axioms are abstract and admit of multiple interpretations (already by the mere fact of being 1st order). They are then taken to be non-foundational axioms.

Perspective (b): On Lawvere's perspective (b), these 1st order axioms are asserted as an explicit, self-contained account of sets. This account makes it possible to state and prove the standard results of mathematics as 1st order theorems. They are then taken to be foundational axioms.

Lawvere also presents a stronger, more comprehensive 1st order foundation of mathematics FOM, namely the "Category of Categories As Foundation" (CCAF). A FOM for Lawvere simply is "a single system of 1st order axioms in which all usual mathematical objects can be defined and all their usual properties proved". CCAF too can be considered under the 2 perspectives just mentioned.

Colin McLarty goes on to criticize Hellman's criticism of category theory as FOM on a number of points:

i. Hellman objects Lawvere's and Mac Lane's positions as contradicting the common categorists' reading of their own systems. McLarty admonishes that Hellman fails to see the distinction between foundational and non-foundational axioms in their position which he advocates for himself.
ii. Hellman, endorsing Feferman on this point, argues that a FOM must provide some systematic account of notions such as collection or operation, something which set theory does, but category theory does not. McLarty agrees with Hellman, but points out that neither ETCS nor CCAF is general category theory. ETCS and CCAF are two powerful explicit axiom systems not presupposing sets or any other foundation, but each providing a foundation for category theory and for the rest of mathematics. Thus, Hellman and Feferman fail to distinguish foundations (e.g. ETCS or CCAF) from applications (category theory) in need of foundations.
iii. Hellman suggests that McLarty's reading of ETCS might involve a commitment to fixed foundations. Mc Larty emphasizes that his own as well as Lawvere's and Mac Lane's original positions are as open-ended as Hellman's own position. They merely differ in that their open-endedness is motivated in a different way to Hellman's own motivation.
   Colin Mc Larty briefly discusses the different conceptions of open-endedness held by Lawvere, Mac Lane, Awodey and himself. A FOM for Mac Lane simply is a proposal for the organisation of mathematics. Mc Larty seems to agree with him that foundations are necessarily provisional and will change and improve if mathematics endures into the future.
iv. Hellman believes that the problem with CCAF is that its axioms do mean something else than what they say: They speak of objects and arrows of a metacategory of categories, but what they really mean are structures and the satisfaction of sentences by structures. Mc Larty can't see what Hellman wants to say and claims that the 1st order axioms of CCAF just mean what they say and nothing else.

Mc Larty continues to criticize Feferman's criticism of category theory as FOM:

(a) Feferman criticizes that it is not clear what is meant by categorical foundations for category theory and how category theory handles the problem of the category of all categories and the category of arbitrary functor categories. Mc Larty only comments that Feferman applies the same sort of critique to set-theoretical foundations.
(b) Feferman raises a mathematical objection to category theory as FOM which had previously been raised by Rao: there are some cutting edge methods of topology which Rao does not see how to formulate in categorical terms. Mc Larty replies that, while the basic ETCS axioms obviously do not have the same proof-theoretical strength as ZFC, Lawvere's first paper on the subject gave a reflection principle R s.t. extending ETCS by R gives a set theory provably equivalent to ZFC. So Rao's concern cannot be correct. There is a brute force

translation of Rao's mathematics into ETCS+ R which "would be identical verbatim to the ZFC version at most points, and clumsy at some points". Of course, a practical category theory approach to this problem wouldn't use a brute force translation, but that's a different point.

Colin Mc Larty finally raises the following question: If categorical set theory (e.g. ETCS) and membership-based set theory (e.g. ZFC) do not differ (1) in their technical relations to metatheories, (2) in that the one formally requires the other, but not vice versa, (3) in that the one succeeds in describing structures which the other does not, what then do they differ in? And he answers this question as follows:

Difference (1): ETCS and ZFC differ in their choice of the primitives: ETCS takes composition of functions as primitive, ZFC takes membership as primitive. Mc Larty interprets this difference in Lawvere's way: By this choice, Mc Larty argues, "ETCS is closer to the working methods of mainstream mathematics than ZFC is".

Difference (2): Ontologically, ETCS rests on form and isomorphism invariant structure whereas ZFC rests on substance and membership. Lawvere and Mac Lane differ however in their ontological interpretation of this claim.

Difference (3) (and the greatest difference according to Mc Larty:)The membership-based set theory in all its variants constitutes some sort of monism: All of them deal with essentially the same idea of a set as a collection of elements. Category theory on the other hand gives way to pluralism: There is ETCS taking sets as fundamental, there is CCAF taking categories as fundamental, there is a foundation taking smooth spaces as fundamental, and there is e.g. Lambek's and Scott's free topos. According to Mc Larty, it is on these alternative foundations for mathematics that the future research on FOM should concentrate.

Section 3

In his contribution **The Axiom of Choice in the Foundations of Mathematics** John Bell starts off with the observation that the axiom of choice (AC) is probably the most interesting and the most discussed axiom of mathematics (after the axiom of parallels). Before entering into a discussion of AC's role in FOM, he sketches the history of AC:

- The 1st formulation of AC by Ernst Zermelo in terms of what Zermelo called coverings dates from 1904. John Bell reformulates this 1st formulation in terms of a choice fct. (he calls this reformulation AC1) and in terms of relations (called by him AC2).
- The 2nd formulation of AC by Zermelo in terms of a choice set dates from 1908. Bell calls Zermelo's second formulation the combinatorial axiom of choice and refers to it by CAC.

- Bell briefly skips over the partially very critical readings and the partially very sympathetic reactions to Zermelo's AC.
- The main point of departure for his own discussion of AC in FOM is provided by Paul Bernays' and Martin-Löf's highly interesting analyses of AC.

  1. Bernays' analysis of AC: Bernays considered AC as the result of a natural extrapolation of what he called "extensional logic", valid in the realm of the finite, to infinite totalities. According to Bernays, AC is entitled to a special position only to the extent that the concept of fct. is required for its formulation. Bell draws special attention to another of Bernays' assertions, namely that the concept of fct. in turn receives an adequate implicit characterisation only through AC. Whereas the first of Bernays' assertions might be interpreted as being remarkably similar to the constructivist's justification of AC (due to a constructive interpretation of the quantifier), the second of Bernays' assertions says that the existence of a fct. may be asserted without the ability to provide it with an explicit definition. This latter assertion in turn is incompatible with stronger versions of constructivism. In Bell's terms: "Bernays and the constructivists both affirm **AC2** through the claim that its antecedent and its consequent *have the same meaning*. The difference is that, while Bernays in essence agrees with the constructive interpretation in treating the quantifier block $\forall x \exists y$ as meaning $\exists f \forall x$, he interprets the existential quantifier in the latter *classically*, so that in affirming "there is a function" it is not necessary, as under the constructive interpretation, actually to be *given* such a function."
  2. Martin-Löf's analysis of AC: Martin-Löf compared the constructive affirmability of Zermelo's AC (taken by Bell in the version AC2) with CAC. This comparison takes place within Constructive Type Theory CTT. In CTT the primitive relation of identity of objects of same type is intensional, whereas in set theory it is extensional. In CTT sets are conceived of as intensional, and sets in the usual set-theoretical sense are called extensional sets. It is possible to formulate within CTT a version of AC for extensional sets, called EAC by Martin-Löf. Martin-Löf then proves in CTT that EAC and CAC are equivalent. To formulate his own problem, Bell observes that Martin-Löf likewise proves the equivalence of EQ plus AC2 with CAC, where EQ refers to the assertion that unique representatives can be picked from equivalence classes of any given equivalence relation.

Bell's problem is the following: Can Martin-Löf's proof of equivalence of EAC and CAC non-trivially be presented in set theory? Bell's answer is: Yes, it can. He reformulates his problem by making use of his observation above. Bell's problem reformulated: Can the proof that EQ plus AC2 imply CAC non-trivially be presented in set theory?

Bell's solution: He first casts AC2 into a constructively valid set-theoretical formulation by means of the so-called "propositions-as-types" interpretation underlying CTT. He then reformulates CAC in standard intuitionistic set theory to obtain a certain assertion on a doubly-indexed family of sets. And last, he derives this assertion from the constructively valid set-theoretical reformulation of AC2 plus EQ.

Question: What do Martin-Löf and Bell show? Note that in 1975 Diaconescu proved that in extensional frameworks such as topos theory or set theory, the usual formulations of AC imply LEM (Law of Exluded Middle), thus making logic classical. Moreover, it is wellknown that in intensional constructive frameworks (such as CTT), AC is compatible with intuitionistic logic. Martin-Löf's analysis shows that the imposition of some form of extensionality on AC in CTT will make Diaconescu's theorem applicable. And Bell shows that Martin-Löf's argument not only holds inside an intensional constructive framework, but also in an extensional classical framework such as set theory. And he generalises his argument in the following ways:

(a) In 2nd order logic (with intuitionistic background logic) LEM can be derived from AC by means of the 2 principles of Predicative Comprehension and Extensionality of Functions. Hence, in systems of constructive mathematics affirming AC but not LEM, one of these 2 principles must be given up.
(b) In Hilbert's Epsilon Calculus, LEM can be derived from the logical ε-axiom (i.e. the analogue of AC in Hilbert's calculus) by means of the Principle of Extensionality for ε-terms (i.e. the analogue of the principle of Extensionality of Functions of 2nd order logic in Hilbert's calculus).
(c) If a 1st order weak set theory WST is bolstered with extensionality principles, it is possible to derive LEM from AC.

John Bell's final question is: What modification on an intensional constructive framework is required in order to pass on to a set- (or topos-) theoretical, i.e. extensional, classical interpretation of AC?

An answer due to M.E. Maietti can be given within the general framework of dependent type theories by means of so-called monotypes. That monotypes correspond to monic maps can be illustrated through the use of the category $\mathcal{INDSET}$ of indexed sets and the category $\mathcal{SET}^{\rightarrow}$ of bivariant sets. These two categories can actually be shown to be equivalent. Now, Bell considers $\mathcal{SET}^{\rightarrow}$ as a topos. Under the topos-theoretical interpretation in $\mathcal{SET}^{\rightarrow}$, propositions (formulas) correspond to monic arrows which in turn correspond to monotypes (or mono-objects) in $\mathcal{INDSET}$. If these correspondences are entirely considered in $\mathcal{INDSET}$, one gets the sought modification of an intensional constructive framework: The "propositions-as-types" framework is turned into a "propositions-as-monotypes" framework in which propositions correspond to monotypes (mono-objects) rather than arbitrary types (objects). Bell finally observes: Reconsidering AC under the "propositions-as-monotypes" interpretation within the category $\mathcal{SET}$ of ordinary sets boils down to asserting AC in its usual form which leads to classical logic—contrary to the "propositions-as-types" interpretation of AC.

Jim Lambek and Phil Scott begin their contribution **Reflections on the Categorical Foundations of Mathematics** with a series of observations: Most mathematicians see no need for a foundations of their subject, i.e. for a FOM. Among those who do see a need,

- Most pick set theory, i.e. an axiomatic treatment of the membership relation in 1st order logic
- Some others pick classical impredicative type theory, or rather, classical higher-order arithmetic
- Still others pick category theory or topos theory (the categorical theory of "variable sets", i.e. sheaves).

They then turn to the questions: What are the connections between some of the foundational choices, in particular the choice between higher-order arithmetic and topos theory?

Lambek and Scott present a type theory, that is an extension of intuitionistic higher-order arithmetic, by an inductive definition of types and terms. They call a type theory analytic if it contains no types or terms other than those it must contain. Pure type theory L_0 is the analytic type theory containing no theorems other than those following from the inductive definition. Lambek and Scott then present a topos as a cartesian closed category (ccc) with pullbacks, a subobject classifier Ω and a Natural Numbers Object (NNO) N. They continue by comparing the category of intuitionistic type theories and the category of toposes. To this purpose they introduce functors between the two categories as follows: One functor L assigns to any topos $\mathcal{T}$ its "internal language" $L(\mathcal{T})$ which is an intuitionistic type theory; the other functor T assigns to any intuitionistic type theory L the topos $T(\mathsf{L})$ "generated" by it. "Intuitively, $T(\mathsf{L})$ is the category of 'sets' and 'functions' formally definable within the type theory L". An intuitionistic type theory L may be interpreted in a topos $\mathcal{T}$ by means of a translation of languages $\mathsf{L} \to L(\mathcal{T})$ where such an interpretation may be conceived of as a "model" of L in $\mathcal{T}$. In the case of pure type theory L_0, there is even a unique translation $\mathsf{L}_0 \to L(\mathcal{T})$, for any topos $\mathcal{T}$, and thus a unique logical morphism $T(\mathsf{L}_0) \to \mathcal{T}$, where $\mathcal{F} = T(\mathsf{L}_0)$ is known as the free topos.

Jim Lambek and Phil Scott decide to follow Henkin and use a more narrow concept of model: A model of an intuitionistic type theory L is a so-called local topos, i.e. a topos $\mathcal{T}$ with the 3 properties: consistency, disjunction and existence property. But Henkin and others dealt with classical rather than intuitionistic type theory—and classical type theory is intuitionistic type theory plus LEM (Law of Excluded Middle). And accordingly, a model of a classical type theory is a Boolean local topos which is shown to have the 2 properties: consistency and universal property. Henkin's completeness theorem for a classical type theory can then be expressed as follows: A proposition of a classical type theory L holds in the topos $T(\mathsf{L})$ generated by L iff it is true in all models of L, i.e. in all Boolean local toposes. Lambek and Scott then raise the question: What about Gödel's more famous incompleteness theorem? They reformulate it for the classical as well as intuitionistic case as follows: In a consistent analytic type theory L (i) with a recursive proof predicate, and (ii) with at least one model in which the numerals are standard, there is a proposition q which is neither provable nor refutable (hence undecidable). NB. Although $\neg q$ holds in every Boolean model in which the numerals are standard, $\neg q$ is not a theorem (whence L is incomplete). From this formulation of Gödel's incompleteness theorem follows: If the "usual" category of sets $\mathcal{S}$ is a Boolean local topos in which all numerals are standard, the set of propositions of L_0 which hold in $\mathcal{S}$ is not r.e. Hence $\mathcal{S}$ cannot be construed as the topos generated by an analytic type theory with a recursive proof predicate. And hence Gödel's proposition $\neg q$ holds in $\mathcal{S}$ but is not a theorem. Lambek and Scott then ask: How is Gödel's incompleteness to be interpreted, what is its philosophical meaning? According to them, Gödel himself interpreted it as showing that Formalism and Platonism are mutually incompatible philosophies of mathematics. For him, the ω-property must hold in the Platonic universe of mathematics. Lambek and Scott, however, maintain that this incompatibility vanishes if classical mathematics is abandoned for a moderate form of constructive mathematics, i.e. if the ω-property is replaced by what they call the ω^*-property, since a proposition in pure intuitionistic type theory L_0 is provable iff it holds in the free topos $\mathcal{F} = T(\mathsf{L}_0)$. Lambek and Scott suggest that the free topos should satisfy moderate adherents of diverse philosophical schools in FOM for various reasons: platonists, formalists, constructivists (or moderate intuitionists) and logicists. They refer to their point of view that the free topos should provide the new FOM as "constructive nominalism". However, there is a problem with constructive nominalism: Freyd showed how to construct a local topos in which all numerals are standard by assuming the existence of the "usual" category of sets $\mathcal{S}$. If a metamathematician is an intuitionist, she will believe $\mathcal{S}$ to be the free topos itself and will thereby reach a circularity. If a metamathematician is a platonist (believing anyway in classical logic), she will be able to prove the existence of classical model toposes (in which the terminal object is a generator and in which all numerals are standard). But will she be able to also "construct" one? Lambek and Scott conjecture that this might turn out to be impossible (at least if such toposes are required to possess certain reasonable properties).

From the discussion of provability, Lambek and Scott go on to the discussion of truth, and they compare different notions of truth. They take Brouwer's notion of truth to be provability in pure type theory L_0, hence truth in the free topos $\mathcal{F}$; they then relate Tarski's notion of meta-truth to the soundness of L_0 which (by Gödel's 2nd incompleteness theorem) is not provable in L_0. Finally they compare Tarski's to Gödel's notion of truth, which is truth in a classical Platonic universe $\mathcal{S}$. They observe that Tarski's notion implies soundness of L_1, where L_1 is pure classical type theory and, as before, the soundness of L_1 is not provable in L_1. According to Gödel's or Henkin's completeness theorem for classical type theory, there is no distinguished Boolean local topos as a candidate for the classical category of sets. Thus, one has to look at the totality of all models of classical type theory. Gödel's or Henkin's completeness theorem can be improved, based on the sheaf representation of toposes, to the following theorem: Every topos is equivalent to the topos of global sections of a sheaf of local toposes. Lambek and Scott note that the models of any type theory L are the points of a topological space and that the truth of a proposition of L varies continuously from point to point.

Jim Lambek and Phil Scott finally show that due to the fact that the free topos is local and has only standard numerals, a number of intuitionistic properties can be proved for pure intuitionistic type theory L_0: beyond the already mentioned ones of consistency, disjunction and existence, the extra ones include Troelstra's uniformity rule, Markov's rule and others. Lambek and Scott conclude that Gödel's incompleteness theorem does not show that provability falls short of capturing absolute truth in the Platonic universe; rather it shows (1) that other models of set theory are required than the alleged Platonic universe, and (2) that even the favorite candidate of a free topos depends on metamathematical assumptions which for various reasons are problematical. As they write: "metamathematics is an attempt by mathematicians to lift themselves up by their own bootstraps".

Section 4

In his contribution **Local Constructive Set Theory and Inductive Definitions** Peter Aczel explains Constructive Set Theory CST to be an open-ended set-theoretical framework for constructive mathematics which is not tailored to suit a particular brand of constructive mathematics, and which (by avoiding built-in choice principles) is also suitable for topos theory (i.e. mathematics carried out in an arbitrary topos with a natural numbers object).

Local Constructive Set Theory LCST is intended to be a local version of CST, i.e. a version which does not assume a single global universe of all the mathematical objects (which are in the range of the variables), but is formulated instead in a many-sorted language which has various forms of sort. LCST is some sort of a constructive counterpart to J.L. Bell's local set theory (which in turn is a certain kind of syntactic version of the notion of a topos).

Aczel asks the following motivating question: What interest is there in LCST? He answers:

(a) LCST allows for the formulation of predicative and generalized predicative versions of the notion of an elementary topos which might be appealing to some category theorists interested in constructive mathematics.
(b) LCST provides for a less complicated, more natural translation of the CST axiom systems (such as CZF or CZF^+) into constructive type theory CTT (as developed by Martin-Löf).
(c) In the presence of various competing foundational approaches to constructive mathematics such as the constructive type-theoretical, the constructive set-theoretical or the category-theoretical approach, it may be desirable to be able to carry over definitions and results from one to the other approaches. LCST serves this purpose nicely as it has a straightforward interpretation in (global) CST and a fairly straightforward direct interpretation in CTT.

Aczel's point of departure is the following: A set induction scheme of CZF (expressing that the universe V of sets is the smallest class s.t. every subset of the class is an element of the class) and the axiom REA, the Regular Extension Axiom, of CZF^+ play an essential role in the development of constructive mathematics. It is not difficult to state various inductive definition results of CST in LCST, but it is not easy to prove them in LCST, as the proofs in CST use the just mentioned set induction scheme and REA which both are global principles for which there seems to be no direct way of formulating a local version. Aczel's global goal is to get these inductive definition results in LCST.

Class terminology and notation are useful tools in classical axiomatic set theory, and Aczel claims that they are even more useful in CST (namely when many comprehension terms representing sets in classical set theory can only be taken to be classes in CST). In order to treat classes in a flexible way, a convenient way to go about it is by formulating a set theory in a suitable free logic. Aczel's approach to free logic differs from Beeson's by being more liberal. It is just a slight (free) variation of a Hilbert style axiomatization of intuitionistic predicate logic with equality.

Aczel's strategy to achieve his global goal consists of an introduction of new axiom systems CZFI and CZF^* which have axioms and schemes directly expressing the inductive definition results of CZF and CZF^+ (where CZF^+ = CZF+ REA. Let I refer to Inductive definition forms for sets and classes, BIS to Bounded Induction Scheme, and SSC to Strong Set Compactness property. Then CZFI = CZF^-+ I, and CZF^* = CZF + BIS + SSC.) CZFI and CZF^* both are extensions of the axiom system CZF^- which roughly results from dropping the set induction scheme in CZF. And they both have the crucial advantage of having local versions, that is, they are formal systems of LCST containing the just mentioned inductive definition results.

After having introduced all these formal systems, Aczel provides the free logic versions of them, that is the free logic version CZF^-_f of CZF^-, the free logic version CZF_f of CZF (which is at the same time a conservative extension of CZF), and the free logic version CZF^*_f of CZF^*.

In order to give a first example of a local set theory, Aczel turns to (a free logic version of) a local version LIZ of Intuitionistic Zermelo set theory IZ which is a fully impredicative formal system (each set having its powerset). Making this version of LIZ predicative amounts to constructing the local version $LCZF^-_f$ of CZF^-_f. By making use of the formal language of this free logic version of LIZ, Aczel achieves his global goal by defining the formal systems $LCZF^-_f$, $LCZF_f I$ and $LCZF^*_f$, which all are systems of LCST containing the inductive definition results mentioned earlier on.

Peter Aczel ends his contribution with an important application of inductive definitions in LCST: In constructive mathematics, well-founded trees are a particularly important tool, and inductive definitions can be used to generate them. Martin-Löf introduced the W-types of well-founded trees in his constructive type theory CTT. These types can fairly straightforwardly be represented in CST in a global way. Aczel then presents a local version of such a representation in the system $LCZF^*_f$.

Charles McCarty's starting point in his contribution **Proofs and Constructions** is his criticism of the use of "mathematical practice" (whatever that may be) in a normative way, that is e.g. to disqualify some other way of doing mathematics such as the intuitionistic one, for 2 reasons: (i) there is no univocal, self-justifying mathematical practice; (ii) if there were, more argument would be required to show that it serves as a norm of practicizing mathematics correctly.

Charles McCarty observes that mathematical practice and mathematical opinio communis are to be considered relative to history: What is common at one time in history, may be outdated, abandoned, disrespected at another. According to

McCarty, contemporary mathematicians would side far more with Brouwer's mathematical views than they would side with Hilbert's. Moreover, he maintains that the greatest mathematical minds in history were as revolutionary or reformist w.r.t. mathematics as were those mathematicians decried as such: they all changed the ways of thinking about mathematics.

McCarty sets off to put intuitionism on its feet: he criticizes intuitionism severely because of its offense of intuitionistic mathematics:

1. A. Heyting claims that intuitionistic mathematics consists of mental constructions. In connection with implication and negation, this claim can lead to blatant nonsense. It looks though like Heyting could be made sense of on the background of distinguishing between a so-called internal (from inside intuitionism) and a so-called external negation (from outside intuitionism). External negation is conceived of as being classical in its logic, i.e. obeying the Law of Excluded Middle LEM. McCarty dismisses this distinction as mere "moonshine" for the following reasons: (i) Nobody can change the meaning of logical operators by fiat. (ii) One can prove in intuitionistic mathematics that there is a unique operation of negation on the set of truth-values. In a first step, he proves that the idea that an external classical negation exists is mistaken. (Note that this proof rests on the intuitionistic Uniformity Principle UP.) And in a second step, he proves that there exists one and only one (internal) negation operator obeying the (intuitionistic) logical laws of negation. Finally, there is no rescue of the distinction of an internal and an external negation by means of the assumption that one can distinguish between "internal truth-values" and "external truth-values". McCarty claims that such a Quinean/Carnapian suggestion is based on a gross misunderstanding of intuitionistic mathematics.

   Charles McCarty subsequently draws attention to the view that presumably negation is not an exceptional case, but that intuitionism wholesale reforms logical and mathematical semantics. This is how classical philosophers like Carnap, Quine or perhaps Wittgenstein may view intuitionism (not how intuitionists themselves, and not Heyting in particular conceive of it). McCarty repudiates this view by observing that e.g. Heyting's assertions in his book *Intuitionism* are in no new language, but are simply new truths in an old language.
2. Now, it may be held that the distinction between internal and external negation is required to make sense of the Brouwerian weak counterexamples, which are apparent refutations of presumptive logical or mathematical principles on the grounds that they reduce to unacceptable instances of the quantified LEM.

   McCarty criticizes a certain sense of Brouwerian weak counterexamples as being blatantly fallacious: He points out that whether e.g. Goldbach's conjecture can be proved or refuted is of no relevance when it comes to determining the validity or lack thereof of a presumptive logical principle like LEM or a putative mathematical principle (which is an instance of the quantified LEM). It is the simple fallacy ad ignorantiam to infer that if an assertion is not proved, it isn't true (intuitionistically or other, where the negation here can't be internal, hence has to be external). McCarty concludes: "Unless reformulated, weak counterexamples are plain fallacies."

But unlike the distinction between an external and internal negation, the Brouwerian weak counterexamples can be rescued: This rescue is based on the observation that although $\neg\neg(p \vee \neg p)$ is a logically valid principle, the quantificational (1st or 2nd order analogies) $\neg\neg\forall\, x(P(x) \vee \neg P(x))$ and $\neg\neg\forall P\forall x(P(x) \vee \neg P(x))$ are not: By means of the Intuitionistic Church's Thesis ICT (relative consistent with Intuitionistic Zermelo-Fraenkel IZF) one can prove $\neg\forall P\forall x(P(x) \vee\neg P(x))$ which shows $\neg\neg\forall P\forall x(P(x) \vee \neg P(x))$ to be invalid. And one can show furthermore, that the initial, reductive part of the weak counterexample reasoning is sound: indeed $\neg\forall g : N \rightarrow \{0, 1\}(\forall n.g(n) = 0 \vee \exists n.g(n) \neq 0)$. (Note that negation this time is internal.)

3. Another intuitionistic erroneous view or myth is that when a proof exists, its existence and its significant proof-theoretical properties are totally obvious. The reasoning behind Brouwerian weak counterexamples apparently requires that when there is a proof of Goldbach's conjecture, one can immediately see that it is a proof of this conjecture and one is on its basis convinced of this conjecture. Heyting claimed that proofs or mathematical constructions generally ought to be or when expressed in a certain way are so immediate to the mind and its results so clear that no foundation for them are needed. McCarty puts this intuitionistic view straight too: He exposes Heyting as being overoptimistic, by discussing Kleene's example of a construction, a Heyting-style proof, that is a realizability witness for a specific instance of 1st order mathematical induction. And he shows that the facts relating to the existence and recursion-theoretical powers of this realizability witness can hardly be claimed to discern convincingly in a way which is immediate, clear and in need of no proof. And McCarty points out that reflection on such examples strongly suggests that the just mentioned common intuitionistic view is a myth.
4. Charles McCarty also suggests that Heyting's greatest error was to surmize that the special constructive meanings of statements made by intuitionists are defined by a special, priviledged relation between sentences (propositions) and proofs, the so-called BHK-interpretation of logical operators. These proofs have quite obviously to be intuitionistic. McCarty objects against this common intuitionistic view: (i) As in the case of negation, nobody has the power to stipulate the meaning of words expressing logical operators and to make it stick. (ii) The BHK-interpretation would be useless to any newcomer to intuitionism as she would have to know and master the new meaning before learning and coming to understand it. (iii) Point (ii) is the more forceful if it is assumed that this new intuitionistic meaning of the logical operators is radically new.
5. McCarty finally criticizes the intuitionistic conception of truth: A sentence is true iff it is provable intuitionistically, that is according to the BHK-interpretation. This view is wrong, since one can't prove sentences in an infinite regress. Sooner or later one hits on some axioms which are not provable anymore, they are only justifiable (in a apriori or aposteriori way). McCarty emphasizes that to say that the axioms of Intuitionistic Zermelo Fraenkel set theory IZF are true, but not provable obviously violates the intuitionistic conception of truth.

Now, there might be some sort of rescue to the preceding criticism if a distinction, crucial for intuitionism, is being made, namely the one between a proof and an (abstract) construction and if some erroneous intuitionistic beliefs about these constructions are discarded. According to these beliefs, these abstract constructions are ontologically and epistemically simple, and as far as truths about natural numbers are concerned, relatively lower-order. In the last part of his contribution, McCarty sets out to show that the opposite is the case. McCarty starts by defining the concept of Kleene-satisfiability by means of some sort of Kreisel-Troelstra-realizability for full impredicative 3rd order Heyting Arithmetic HA^3. He subsequently proves soundness of HA^3, Kleene-satisfiability of UP and the Fan Theorem FT, and finally he proves that ICT is not Kleene-satisfiable. He thus comes to the conclusion that ordinary Kleene-realizability does not validate Brouwerian analysis. Only when endorsing a higher-order version of realizability (as e.g. the one underlying Kleene-satisfiability) can intuitionists legitimately embrace a correct intuitionistic idea (of Brouwer's and Heyting's): namely, that the mathematical principles capturing facts of intuitionism can be deduced, with the aid of set theory, from a theory of (abstract) constructions. However, as can directly be understood from the definition of Kleene-satisfiability, the required fundamental properties of those (abstract) constructions are neither obvious and immediately graspable nor unshakably certain.

J. P. Mayberry begins his contribution **Euclidean Arithmetic: A Finitary Theory of Finite Sets** by claiming that conventional treatments of the foundations of arithmetic—classical, constructive, and finitary—rest on what he calls the "sorites fallacy", namely the belief that the notion of finite iteration is self-explanatory and therefore not in need of definition. The fallacy is manifested mathematically by the idea that both proof by induction and definition by recursion are self-evident, fundamental truths not requiring justification. This is true especially in conventional finitary and constructive foundations. From these observations he derives what is for him the major task for the foundations of arithmetic: to provide a precise, mathematical analysis and explanation of this notion of finite iteration.

Mayberry points out that Dedekind was fully aware of this fallacy and presented a set-theoretical solution to it in his theory of simply infinite systems, which, however, rests on strong infinitary assumptions. He proposes to follow Dedekind in giving a purely set-theoretical account of finite iteration (and therefore of natural number) but presented in a finitary theory of finite sets. He calls this theory "Euclidean arithmetic" because it can be seen to be an up-to-date version of ancient arithmetic, although it is also a finitary version of the intuitive idea of set that underlies the Zermelo-Fraenkel system.

In Euclidean arithmetic it is finite set not natural number that is taken as fundamental. Its fundamental operations and relations are those of the ZF system: pair set, power set, union, replacement, set-selection, membership, inclusion, and identity. Mayberry calls these operations and relations "global" since they apply to all finite sets and their members; "local" functions and relations are defined in the usual way to be sets of ordered pairs.

The theory is finitary not just in the sense that all sets are Dedekind finite, but also in that it conforms to what Mayberry calls Brouwer's Principle: A quantifier is subject to ordinary classical logic iff its domain of quantification is a set, which in Euclidean arithmetic is necessarily finite. Classical quantifiers with finite domains can be defined using the set-selection operator.

Mayberry describes two ways in which natural number arithmetic is embodied in Euclidean arithmetic:

1. Via the theory of "arithmetical" global functions and relations

A global function is called "arithmetical" if the cardinality of its value depends only on the cardinality of its arguments, and a global relation is "arithmetical" if its truth value depends only on the cardinalities of its arguments. Arithmetical global functions and relations corresponding to the basic functions and relations of conventional arithmetic can be defined, and the true equations of conventional arithmetic continue to hold when identity (of natural numbers) is replaced by cardinal equivalence (of sets). In fact, a formalised version of Euclidean arithmetic is mutually intertranslatable with the "weak arithmetic" $I\Delta_0 + \exp$.

2. Via a direct treatment of natural number systems in the manner of Dedekind

Thus given a unary global function, σ (as "successor function") and a set or individual a (as "initial element") there is a natural definition of a linear ordering L's being "generated from a by σ". Then such a pair σ, a is said to generate a natural number system if whenever σ generates the linear ordering $[a, \cdots, b]$ from a, then $\sigma(b) \notin \{a, \cdots, b\}$. The natural number system generated by σ from a is to be identified with the species (proper class), $\mathtt{N}$, of all linear orderings generated from a by σ; it corresponds to the species of initial segments in the corresponding simply infinite system as defined by Dedekind.

Setting $a = \emptyset$ and specifying σ to be $x \mapsto x \cup \{x\}$ we obtain what Mayberry calls the von Neuman natural number system. Similarly, specifying σ as $x \mapsto \{x\}$ or $x \mapsto x \cup \mathrm{P}(x)$ define the Zermelo and Cumulative Hierarchy systems.

From Dedekind's infinitary standpoint, all natural number systems can be shown to be isomorphic, but that is not the case in Euclidean arithmetic. There natural number systems come in different, even incomparable, lengths, and are closed under different arithmetical global functions. But to make this precise, constructive meanings have to be attached to propositions asserting that one natural number system, $\mathtt{M}$, is shorter than or equal to another, $\mathtt{N}$, or that a natural number system, $\mathtt{N}$, is closed under an arithmetical global function, φ.

Thus, e.g. to say that the natural number system $\mathtt{M}$ is "shorter than or equal in length to" the natural number system $\mathtt{N}$, $\mathtt{M} \preceq \mathtt{N}$, is to say that one can exhibit a "measure" for $\mathtt{M}$ in $\mathtt{N}$, i.e. a global function μ s.t. whenever L is a linear ordering lying in $\mathtt{M}$, $\mu(L)$ is a linear ordering lying in $\mathtt{N}$ of the same length as L. Under that constructive interpretation of $\mathtt{M} \preceq \mathtt{N}$, Mayberry takes its negation, $\mathtt{M} \npreceq \mathtt{N}$ to mean that for any μ it is impossible that μ should be a measure for $\mathtt{M}$ in $\mathtt{N}$.

Mayberry argues that the natural Euclidean notion of global function is that of a function defined in terms of the basic global functions of the theory. From a classical, infinitary standpoint this means that these global functions correspond to terms t in the natural formal language L for set theory (in which the basic set-theoretical operations are included in the primitive vocabulary), and this makes it possible e.g. to translate the foundational question whether a given natural number system N is closed under addition into the technical question whether an L-term t_+ with free variables a and b exists s.t. $\lambda xyt_+[x/a, y/b]$ defines addition in the standard infinitary model V_ω of the hereditarily finite pure sets. If N is the von Neumann, the Zermelo, or the Cumulative Hierarchy natural number system the answer, is NO. This is a classical, infinitary motivation to postulate the following: For any binary global function φ, it is impossible that for all linear orderings L_1 and L_2, $\varphi(L_1, L_2)$ lies in N and has the same cardinality as $L_1 +_C L_2$. It is also possible to provide classical, infinitary motivation to postulate that M $\not\preceq$ N both in the case in which M is the von Neumann system and N the Zermelo system, and the converse case in which M is the Zermelo system and N the von Neumann. Thus with these postulates added to Euclidean arithmetic, not only are there non-isomorphic natural number systems, but even pairs of natural number systems incomparable in length. Mayberry calls postulates of this kind "classically witnessed postulates", because there are classical, infinitary proofs that their claims are justified, and because they do not have any claim to being self-evident as the core axioms of Euclidean arithmetic do.

This doesn't square with our intuition, but our intuition is clearly extensional here. In Euclidean arithmetic natural number systems are, on the contrary, intensional in nature, so comparisons between them are determined by the logical and set-theoretical relations between their initial objects and, especially, their successor functions.

Mayberry finally points out that in Euclidean arithmetic there is an amazing variety of natural number systems with differing lengths and closure properties. Thus he defines the binary expansion N[2] of a natural number system N to be the class of linear orderings that are initial segments of the decimal numerals which are identified with finite sequences (L, f) where L is a linear ordering lying in N, and f : Field$(L) \to \{0, 1\}$. He then goes on to prove that N $\preceq$ N[2], that N is closed under addition iff N[2] is closed under multiplication, that N is closed under multiplication iff N[2] is closed under simple logarithmic exponentiation $(x, y \mapsto x^{log(y)})$, that N is closed under simple logarithmic exponentiation iff N[2] is closed under $(x, y \mapsto x^{log(y)^{log(log(y))}})$, and so on (all logarithms to the base 2).

If N is the von Neumann, the Zermelo, or the Cumulative Hierarchy system and one appeals to the classically witnessed postulate that it is not closed under addition, one obtains an unending sequence of natural number systems

$$\mathtt{N} \prec \mathtt{N}[2] \prec (\mathtt{N}[2])[2] \prec \cdots$$

each strictly longer than its predecessor and closed under faster growing arithmetical functions. But one never obtains a system closed under exponentiation, for the proposition that N[2] is closed under exponentiation is equivalent to the proposition that N itself is closed under exponentiation. Of course from the standpoint of Euclidean arithmetic all these systems are standard.

Mayberry concludes his account of Euclidean arithmetic with three observations:

i. All of this is in strong contrast to Dedekind's infinitary treatment of arithmetic. Indeed, Euclidean arithmetic can be augmented by a classically witnessed postulate that asserts that no natural number system is sufficiently long to count out every finite set.
ii. Mayberry remarks that he has expounded Euclidean arithmetic informally (although it could easily be presented as a formal system) because no formal system of the conventional sort can serve as foundation for arithmetic, the reason being simply that the conventional treatment of formal syntax rests on the very same sorites fallacy that underlies the conventional treatment of natural number arithmetic.
iii. Mayberry ends with a brief indication of how the method of classically witnessed postulates can be used to introduce the notion of a "long linear ordering" (which though finite includes an injective image of an entire natural number system as a proper initial segment). This makes possible a development of the Infinitesimal Calculus within Euclidean arithmetic.

Richard Tieszen starts his contribution **Intentionality, Intuition, and Proof in Mathematics** with the observation: Being directed towards the ordinary objects of mathematical research is quite different from being directed towards the consciousness of such objects. If one is directed towards the mathematical consciousness of and the mathematical thinking about such objects, one immediately comes across a central feature characteristic of many forms of this consciousness, namely intentionality. It is Tieszen's aim to bring the idea of intentionality of consciousness back into focus in considerations of FOM, and to show that the idea of intentionality of mathematical consciousness yields a philosophically rich and fruitful view of FOM.

'Intentionality' is meant to express "aboutness" or "directedness" of consciousness and Tieszen takes mathematical thinking as being intentional, i.e. directed towards various kinds of objects and states of affairs. He suggests to view mathematical sentences as expressions of the contents or intentions of acts of mathematical thinking. Furthermore, he claims that mathematicians are directed in their thinking towards certain objects or states of affairs by way of these contents or intentions, which he calls meaning-intentions—due to the phenomenological view that mathematicians as humans are meaning-bestowing beings. According to Tieszen, object-directedness is possible only due to the meaning-intentions of human acts of thinking (and not due to the existence of the objects). He closes these reflections on intentionality by noting that intentionality is ubiquitous in mathematics.

Tieszen continues to present a conception of intuition explained in terms of intentionality of human consciousness: Knowledge of objects or states of affairs requires evidence. Put in the language of intentionality: The meaning-intention by virtue of which someone is directed towards objects or states of affairs should be fulfilled or fulfillable. Intuition (i.e. meaning-fulfillment) as a dynamic fulfillment of a meaning-intention is essentially a sequence of mental acts in which someone comes to see something or to see that something is true.

Applying this conceptual apparatus to mathematics, Tieszen characterizes a mathematical proof as a fulfillment of a mathematical meaning-intention, i.e. an intuition and more specifically an intuition that. Tieszen draws attention to the fact that "there is a historical precedent for identifying only constructive proofs with fulfillments of mathematical meaning-intentions". Thus, a proof is a sequence of mental acts in which a mathematician comes to see that a sentence (proposition) in mathematics is true. And an open problem in mathematics can be conceived of as an expression of a mathematical meaning-intention which can be either fulfilled or frustrated. Tieszen adds a few characteristics of the notion of proof in this new sense:

1. A proof is one among different types of fulfillments of mathematical meaning-intentions.
2. The concept of proof can be understood according to the process/product distinction.
3. A proof (a mathematical intuition) is a filling of an empty mathematical meaning-intention.
4. Since "[g]oal-directedness is always involved in conscious knowledge-seeking", acquiring mathematical knowledge by means of proof is goal-directed.

Tieszen discusses this idea of mathematical proofs as fulfillments of mathematical meaning-intentions in more depth:

- Does this idea of mathematical proof also apply to so-called formal proofs, i.e. to proofs as defined w.r.t. a formal logical mathematical system? The answer is fairly straightforward: If formal (or mechanical) proofs involve only the mechanical manipulation of meaningless syntax then the sentence to be formally proved is about nothing. But then the formal proof cannot be said to be the fulfillment of a mathematical meaning-intention expressed by this sentence, since no meaning-intention is expressed by this sentence. Hence a formal proof cannot count as a mathematical intuition. Tieszen points out that this is just where Brouwer's worries about formalism come in: In the shift from the directedness towards mathematical objects to directedness towards language, there is a move away not only from mathematics, but also from mathematical intuition. And he also points out that according to Gödel "[t]he incompleteness theorems only show that something was lost in translating the concept of proof as "that which provides evidence" into a purely formalistic (and hence, relative) concept".
- Do mathematical sentences express a meaning-intention if they are neither sentences of constructive mathematics (with a so to speak natural meaning-intention) nor sentences of formalist mathematics (with no meaning-intention)? According to Tieszen they do and they acquire it in the following way: They acquire it through the history of regions of mathematics, "through the sedimentation of concepts and results, false starts and readjustments, applications, ..., and so on", by what Tieszen calls a "meaning-history".

- Turning to the axiomatic method in mathematics, it appears that the intuitive evidence provided by a proof can only be as strong as the intuitive evidence associated with any axioms or assumptions on which the proof is based. Here Tieszen indicates some of the issues that distinguish constructive from non-constructive proofs in axiomatic systems and notices that accounting for the more general (not merely constructive) features of the idea of mathematical proofs as intuition is, in several respects, an open field of research.

After having considered the cases where the concepts of fulfillment and intuition are construed constructively and where they are construed classically, Tieszen turns to the case in which there (provably) cannot be a fulfillment of a meaning-intention whatsoever; that is, there can be no intuition that the proposition is true, and much less intuition of the intended object(s) the proposition is about. Tieszen calls such a proposition a "frustrated meaning-intention" (in contrast with the propositions with a proof called "fulfilled meaning-intentions"). A sentence with a frustrated meaning-intention will lead to a contradiction. As a consequence, negation is itself based in intuition. Tieszen introduces again a historical dimension into his considerations: Mathematical propositions cannot simply be classified apriori as frustrated or as fulfilled (fulfillable) meaning-intentions. Tieszen also holds that intuitions in mathematics can be corrected by subsequent intuitions in mathematics.

In the sequel Tieszen reflects on further aspects of proofs as fulfillments of mathematical meaning-intentions:

(a) If there is more than one proof of a mathematical proposition, there are multiple but different intuitions, i.e. fulfillments of the same mathematical meaning-intention. Tieszen notes how this is related to the objectivity of the intuitions.
(b) Proofs (fulfillments) can exceed their meaning-intention, that is they can provide more information than is required to prove the proposition in question. Proofs can also support a different proposition from the one they were intended to be proofs of. There is then a mismatch between proofs and their meaning-intentions.
(c) Sometimes proofs use only concepts and results lying inside the "scope" of the expressed meaning-intention; and sometimes proofs use (some) concepts or results lying outside the "scope" of the expressed meaning-intention. Tieszen calls the former "internal proof" and the latter "external proofs". He observes that proofs in axiomatic systems are always internal, if the axiomatic method is used in a strict way.
(d) Tieszen considers the possibility of mistaken proofs (intuitions) and claims that while intuitions are foundational, in the sense that they are required for knowledge and can only be corrected by subsequent intuitions, they are not foundational in the strong epistemic sense of being infallible.

Tieszen dedicates the last part of his contribution to the conception of constructive proof. This conception historically originates with Brouwer's intuitionism. O. Becker was the first to identify proofs with fulfilled mathematical meaning-intentions, and Heyting used Becker's idea to interpret the intuitionistic logical operators, which is nowadays called the BHK-interpretation of the constructive logical operators. Tieszen formulates the BHK-interpretation in terms of meaning-intention and fulfillment, and he remarks that this interpretation does not by itself force a constructive interpretation of logic, it does so only under the appropriate understanding of the notions of construction, fulfillment or intuition. The intuitionistic conception of proof has in contrast to other constructivist and non-constructivist conceptions been of special relevance to the philosophy of mind and the study of mathematical consciousness, as no other movement or position in the foundations of mathematics FOM has considered features of human consciousness and human mental processes as essential as intuitionism has.

Tieszen's final conclusion is that intentionality in mathematics is quite obviously ubiquitous, whereas intuition is not. However, for intuition in mathematics not to be ubiquitous by no means means that it is inexistent.

Whilst several other contributions compare and contrast set theory with toposes as foundations for mathematics FOM, the message of Awodey's paper is that these are logically equivalent. On this basis, Paul Taylor rejects them all, on the grounds that they capture theories of the discrete—dust that needs to be glued back together in order to reconstruct familiar objects such as the real line.

In his contribution, **Foundations *for* Computable Topology**, Taylor argues instead that foundations should take direct account of natural continuous mathematical objects, and proposes a methodology for doing this.

Besides the nature of the basic objects of mathematics, he also disputes the logical strength that is appropriate. The traditional view based on set theory seeks to allow arbitrary collections, getting into the kinds of difficulties that Feferman discusses in his contribution: we forever need bigger and bigger collections. According to Taylor, we run into the same problem looking at things from a logical point of view, because Gödel's Incompleteness Theorem says that we need ever stronger hypotheses in order to prove consistency. Each individual mathematician therefore stops at a particular level of complexity on pragmatic grounds, namely what that scholar needs or understands. The result is that algebraists and logicians use systems of different strengths, whilst everybody argues about the axiom of choice AC.

In contrast to this tower of babel, as Taylor puts it, the logical situation for computation is very simple. Alan Turing and Alonso Church independently proposed their own axiomatisations of computability, but found that they were equivalent. A huge variety of machine architectures and programming languages have since been developed, but we routinely translate between them using compilers and emulators. There is no disagreement about logical strength. Besides this, for Taylor, going back before the invention of set theory, the tradition of mathematics was quite explicitly computational. Computations and not collections are the natural FOM.

During the reign of set theory it also became customary to develop mathematics in an axiomatic fashion, but the axioms and theorems stayed in the textbooks. Meanwhile, people started using machines to do computation, but by writing programs in FORTRAN whose mathematical basis (if there was any) had no relationship to Richard Dedekind's axiomatisation of the continuum.

Taylor observes that modern technology provides the ability to perform computation that would have been inconceivable before. In particular, it is now possible to compute directly with the axioms. Functional programmers have now been doing this with integer arithmetic for three decades. Once they competed to write the slowest program for 5!, but now their languages are as fast as the machine-oriented ones and are used in heavy-duty industrial tasks. For computation with real numbers, there are several similar mathematically based approaches in the type theory community that are beginning to show signs of practical success.

Taylor considers that the foundations *for* a particular mathematical discipline should be re-cast in a mould designed for its own theorems and computations. He proposes a methodology for doing this that is based on two very general but technical principles:

1. Category theory is used on a day-to-day basis by many kinds of mathematicians, but for reasons that have nothing to do with its foundational claims. They are that the other categorical sub-disciplines (many of which pre-date the invention of toposes) provide tools that very closely match the phenomena that arise elsewhere in mathematics. Chief amongst these is the notion of an adjunction. This is also found in many other forms such as free constructions, universal properties, etc. It is very common for mathematicians to formulate the main theorems that characterise their discipline as adjunctions.
2. There is a quite general equivalence between categorical and symbolic (type-theoretical) formulations. Awodey uses this in his Step 2 in the specific case of structure like that in set theory, but the correspondence is much more general than this. The pattern is that the introduction, elimination, beta and eta-rules for a particular logical symbol correspond precisely to an adjunction. This was observed by Bill Lawvere, first in the case of the λ-calculus and cartesian closed categories and then for the quantifiers as adjoints to substitution.

Taylor's contribution here has been to go in the opposite direction, from the categorical to the type-theoretical side. The methodology that he proposes in his paper begins by making a judicious selection of the characteristic theorems of a subject, expressed as adjunctions. Under the formal correspondence, these become the formal rules of a new symbolic logic *for* that subject.

Then, since this logic contains only what is needed, it is often directly amenable to computation.

The bulk of Taylor's paper goes on to develop the mathematics of a new system of computable topology. The theorems that he selects are

1. (An abstract categorical idea that was inspired by) Marshall Stone's duality between topology and algebra,
2. The classification property of the Sierpiński space Σ, i.e. that any open subspace $U \subset X$ is the inverse image of the point $\top \in \Sigma$ along a unique continuous function $\phi : X \to \Sigma$ and any closed subspace $C \subset X$ is the inverse image of the point $\bot \in \Sigma$ along a unique continuous function $\phi : X \to \Sigma$, and

3. Scott continuity, which is related to the "finite open sub-cover" definition of compactness.

The first two of these theorems have analogues in set theory, where open subspaces of topological spaces play the role of general subsets of sets.

Stone duality for sets is known classically as the Lindenbaum–Tarski theorem. This says that the category of complete atomic boolean algebras and their homomorphisms is dual to the category of sets. This breaks into two parts:

1. Sobriety: any function $f : X \to Y$ that induces an isomorphism $f^{-1} : \mathcal{P}(Y) \cong \mathcal{P}(X)$ is itself a bijection, and
2. Spatiality: every complete atomic boolean algebra is $\mathcal{P}(X)$ for some set X (its subset of atoms).

The idea of Stone duality is that "spaces" and (continuous) functions are exactly determined by their algebras and homomorphisms. Abstract Stone duality says that the carriers of the algebras are themselves "spaces" of the same kind. Taylor has developed a "type theory" that exploits this situation.

The second theorem (the classification property) gives rise to a logic on the Sierpiński space Σ that is similar to higher order predicate calculus. However, it is weaker in that

1. It has no negation,
2. Universal quantification $\forall$ is only allowed over compact spaces, and
3. Existential quantification $\exists$ is only allowed over a new kind of space, called overt.

Overtness brings together numerous ideas from topology, logic and computation, such as open maps, having a countable dense subspace and recursive enumerability.

These ideas are consonant with Taylor's thesis that the usual logic of mathematics is too powerful:

1. Overtness is invisible in classical topology because in the overly strong logic one can prove that all spaces are overt, whilst
2. He shows at the end of his paper that the difference between topology and set theory is exactly that arbitrary sets admit negation and quantification, i.e. they are all discrete, compact and overt.[19]

[19] I'd like to thank some of the contributors for their help in writing the analytical summary of this last contribution.

References

Mostowski, A. (1965) *Thirty Years of Foundational Studies. Lectures on the Development of Mathematical Logic and the Study of the Foundations of Mathematics in 1930–1964*, vol. XVII, Helsinki: Acta Philosophica Fennica.

Parsons, C. (2006) (art.) Foundations of Mathematics, in Borchert, D.M., ed. *Encyclopedia of Philosophy*, vol. 6, Detroit: Thomson Gale, second edition, 20–57.

Wang, H. (1958) Eighty Years of Foundational Studies, *Dialectica* **12**, 466–497.

Part I
Senses of 'Foundations of Mathematics'

Foundational Frameworks

Geoffrey Hellman[1]

1 Introduction: Questions of Justification and Rational Reconstruction (Between Hermeneutics and Cultural Revolution)

A good place to begin is with a delineation of the *sorts* of questions that foundational frameworks for mathematics, as we conceive them, can usefully address and which motivate their development in the first place. This will in turn guide us in framing some important *desiderata* that such frameworks should try to meet.

Although the well-known distinction between questions of discovery and questions of justification has been criticized as oversimplified, distorting the complex reality of science and mathematics as practiced, we take it to be nevertheless fundamentally sound. It is true that justifying a method or a principle often involves citing successful practice which may well include elements and aspects of discovery, perhaps inevitably so in connection e.g. with the "fruitfulness" of theories or methods. And it is also true that discovery may be stimulated or driven by questions of justification (as, for example, was the rigorization of analysis via the method of limits in the nineteenth century). But far from undermining the distinction, such considerations rather reveal that the two "contexts" or types of questions are interrelated in interesting ways. Indeed, it would be difficult to describe such interrelations without implicitly acknowledging it! Even in mathematics, where practice involves proof more heavily (and generally of a more conclusive kind) than anywhere else, the distinction still stands between asking how a certain theorem came to be believed and asking how it is justified. For us, a foundational framework is not, at least in the first instance, addressing questions of discovery, even if one would be happy to learn that it shed light on processes of discovery or helped lead to new knowledge.

[1] Thanks to Roy Cook for helpful comments on an earlier draft.

G. Sommaruga (ed.), *Foundational Theories of Classical and Constructive Mathematics*,
The Western Ontario Series in Philosophy of Science 76,
DOI 10.1007/978-94-007-0431-2_1,

There lurks, however, an ambiguity in questions of the form, "How is [such and such a claim] justified?", an ambiguity marking a second (but overlapping) crucial distinction: On the one hand, one can ask how a claim is or has actually been justified in practice, in which case we are asking about methods and standards actually employed or invoked by relevant individuals or groups typically engaged in the practice. On the other hand, one can ask how a claim *can* or *could in principle* be justified (or "*is* justified" in the sense of having a status according to some (perhaps implicit) standard, as opposed to the sense of an activity or practice, corresponding to the grammatical distinction between an adjective and a participle). Here one is asking about a kind of justification that often goes beyond actual practice, according to an "ideal" standard or at least a "better than usual" one in certain respects. A familiar example (indeed of both distinctions) would be the contrast between actual theorem-formulating and proving in an historical episode (e.g. the case of the (Descartes-) Euler conjecture (with the $F - E + V = 2$ formula) on (topologically) simple polyhedra, as Lakatos (1976) famously described) and a formal statement and correct proof in an axiomatic framework (say, a modern descendant of Hilbert's system of geometry), where what is at stake is an ideal standard of clarity and formal rigor (made precise through twentieth century logical standards such as "that the set of numerical codes of theorems be recursively enumerable", etc.).[2] Or the contrast might be that between good textbook proofs of theorems in analysis, which seldom cite set existence (e.g. comprehension) principles at all, and proofs in subsystems of second-order arithmetic, the staple of reverse mathematics, detailing the sufficiency and necessity of relatively weak comprehension axioms[3]; or proofs which tacitly rely on the law of excluded middle (applying, e.g., Weak König or sequential compactness (equivalent to full König)) but which can be replaced with a constructive proof (possibly with a reformulation of the result regarded as "slight" in the context), etc. In these and many other instances, it is the question "How *can* this be (well-) justified?" that takes center stage, however interesting and important for various purposes the question of actual justification-in-practice may be. Thus the programs of reverse mathematics (Friedman, Simpson et al.), predicative analysis (Feferman et al.), constructive analysis and algebra (Bishop, Bridges, Richman et al.), and others clearly highlight and exploit the distinction. In figuring out "what really rests

[2] For a critical assessment of Lakatos' program of a purported "logic of mathematical discovery", see Feferman (1981). As he points out (criticism (v), p. 317), Lakatos offers no theoretical criterion for what counts as "improvement in a proof" but relies on some shared understanding of that phrase and even seems to acknowledge that mature mathematical theories achieve high standards that no longer yield to counterexamples or other challenges. Feferman's points that such plateaus are indeed achieved, and that some starting concepts for developing proofs are extremely clear and never the source of criticism of proofs (e.g. 'natural number'—of course efforts to ground that concept in others have given rise to criticism) seem quite accurate as pertaining to actual mathematical practice. From the standpoint of epistemology concerned with *justifiability-in-principle*, the Lakatosian rationale for rejecting logical foundations of mathematics is all the more dubious.

[3] See Simpson (1991) for a detailed, masterful treatment. Memorable results include, e.g., the equivalence (over the weak base theory, RCA_0) of the separable Hahn-Banach theorem with the Weak König lemma, similarly for the Weierstrass approximation theorem; the equivalence (over RCA_0) of the extreme value theorem with Arithmetic Comprehension, likewise for sequential compactness (Bolzano-Weierstrass theorem); and many others.

on what?", such logico-mathematical investigations are also part of mathematical epistemology, functioning as "rational reconstructions" of bodies of mathematical practice even when they also involve creative mathematics.

Similar remarks can be made regarding less directly proof-theoretic (and more ontologically or semantically motivated) undertakings such as nominalistic reconstructions, articulations of structuralism, reworkings of logicist systems, and so forth. A major goal of the nominalist investigations, for instance, is to undercut the claim of indispensability of reference to mathematical *abstracta*—a claim that has helped give objects-platonism the unwarranted status of *default* interpretation of mathematical practice—by providing worked-out alternatives that *can in principle* serve the needs of the sciences (and, in several cases, the needs of pure mathematics or much of it as well). But even if suitably idealized humans (not a conceit of science fiction) could effectively deploy such systems, the more important point is that they (purport to) provide a way of *reflecting upon* and *understanding* mathematics as practiced that avoids various problems confronting objects-platonism (*whether or not* mathematics as practiced "really implies" that position), may well allow for a better account (in the making) of human inquiry in general, and may better achieve certain more specific goals (including meeting some of the *desiderata* to be listed in a moment) motivated by practice but not explicitly addressed within it. Whether we classify these reconstructions as FoM or PoM (foundations or philosophy of mathematics) or as appealing to "scientific" vs. "metascientific" standards, is not very important. In any case, in their own distinctive ways they all pass easily and safely through a(n apocryphally) *wide* channel between the Scylla of "hermeneutics"—claiming to tell us what working mathematics "really means" (whatever *that* really means)—and the Charybdis of "cultural revolution"—advocating the replacement (even if without public shaming of platonists) of actual mathematical language and theories-in-use with those of the favored system(s) or scheme.[4]

With these preliminaries in mind, let us now turn to some key *desiderata* of a foundational framework for mathematics.

[4] Cf. e.g. Burgess and Rosen (1997), Hellman (1998), also (2001), Chihara (2005), and Rosen and Burgess (2005). As should be clear to those who have followed this debate, I remain unpersuaded by Burgess' and Rosen's more recent articulations of their position and critique of nominalist reconstruction programs. While this is not the place to give a full response, let me cite one of several objections I would make: they now speak of "*revolutionary* nominalists" as proposing "amendments to [mathematics], not exegesis but emendations". (Burgess, 2004, p. 23) But amendments can be friendly or hostile in varying ways and degrees, and to continue to use the term "revolutionary" in this blanket way serves to obscure the course of the bold sailors' ships (at least to the narrators if not to the sailors themselves).

Moreover, why pick on nominalists? Why not apply the same (false) dichotomous reasoning to reverse mathematics, predicative analysis, Bishop constructivism, etc.? After all, most mathematical practice doesn't share their justificatory standards either. (For instance, when was the last time you saw a textbook proof of a theorem citing its reliance on an impredicative comprehension axiom, or on the law of excluded middle?)

2 Desiderata

Without any claim of completeness, we may list the following:

1. Articulate standards of proof:
   a. Indicate background logic, axioms;
   b. Provide a hierarchy of systems of increasing strength (e.g. consistency strength, strength of set-existence principles as in reverse mathematics, etc.) in which the vast bulk of known mathematics can be reconstructed;
   c. Indicate the strongest justification for key axioms or postulates of the systems of 1.b.
2. Articulate means of expressing mathematical structures of interest and interrelations among them:
   a. Indicate machinery for "axioms" as defining-conditions on structures (à la abstract algebra), e.g. 2nd-order logic or a fragment, a suitable set theory, etc.
   b. Indicate means of expressing interrelations among mathematical structures (the various morphisms, functors, etc.);
   c. As under 1, seek to do "more with less".
3. In 1. and 2., clearly identify the logical and mathematical primitives; the better these are antecedently understood the better, *ceteris paribus.*
4. Preserve past gains, e.g. set-theoretic reduction of classical analysis (deriving least-upper-bound principle from Boolean closure, à la Dedekind); many examples from modern proof theory, e.g. forward directions of reverse mathematics (Simpson, 1991); mathematical induction from bounded separation and weak 2nd-order comprehension (Feferman-Hellman); etc. Provide the strongest justification available for assumption of infinitely many objects and infinite totalities.
5. Accommodate multiple approaches (pluralism), e.g. multiple conceptions of *continua*, including intuitionist, predicativist, non-standard analytic, smooth infinitesimal, etc. Exhibit interrelations among theories and models, e.g. compatibility or incompatibility, incomparability (in a specified sense), embeddability (up to isomorphism), etc.
6. Provide for extendability of universes of discourse for mathematics especially in connection with indefinitely extensible concepts such as 'set', 'ordinal', 'cardinal', 'large category', etc.

We remark that, as *desiderata*, some of these might not be regarded as *requirements.* Although 1(a) and (b) and 2(a) and (b) should surely be required, 1(c), for instance, is not usually thought of as a component of a foundational framework, but rather as an epistemological investigation about it.[5] And 6 may be regarded as a special problem concerning indefinitely extensible concepts and modality, beyond the

[5] Mathematics texts, however, may offer reasons for extending a system with new axioms. A good example is Drake (1974) on large cardinal extensions of ZFC.

usual scope of "foundations". In any case, what matters is not how we choose to use words but that a foundational framework at least allow for a satisfactory resolution of the problems and paradoxes associated with "proper classes" or "incompleteable totalities", etc.[6]

Point 5, accommodating multiple approaches and conceptions, deserves special comment. Traditionally, this has been uncharacteristic of much foundational work, which naturally concentrated on developing a given approach or scheme and which, equally naturally, tended to proclaim its own superiority. Perhaps it is symptomatic of progress that now we can see that, at least in some important cases, conflicts have been exaggerated or misdescribed and that advantages of a given approach are often accompanied by advantages of another on a different score, and that we really need to take a broader view, accommodating a variety of perspectives, recognizing trade-offs, and preserving gains while minimizing losses. As has been recognized, the complexity and variety of mathematical practice may well require multiple approaches even at the foundational level.[7]

3 Implications: Set Theory and Category Theory

It would take us too far afield to try to survey the full gamut of foundational programs in light of these desiderata. For present purposes, we will confine ourselves to a few remarks about set theory and category theory *considered as foundational frameworks*. We recognize, of course, that these are both part of ongoing mathematics proper, and, considered as such, are beyond the scope of this assessment.

Suitably supplemented, the prevailing set theory, ZFC, and many variants, provide a major, well-known success story regarding most of the first three *desiderata* listed above. Of course, the axioms themselves don't tell us how 1 (b) is satisfied, but it is routine to recast the various systems studied in proof theory (such as subsystems of second-order arithmetic, systems of predicative analysis, etc.) in purely set-theoretic terms. Indeed, a tiny fragment of ZFC suffices, so little relative to its full power that it can be misleading to describe "ordinary mathematics" as really "set-theoretic", except in its pervasive use of set-theoretic language and operations. (For this reason, it may be preferable to describe set theory *qua* mathematics as "extraordinary mathematics". Obviously, "ordinary", as used by Friedman, Simpson et al. and as intended here, leaves plenty of space for as much ingenuity, creativity, and insight as there is.) Similarly, special work is required in order to develop 2 (c), regarding efficiency of expressive power. But, in light of the hierarchies studied in descriptive set theory and related work, a great deal is known and can be taken intact as contributing to this goal.

---

[6] In Zermelo's potent language: provide a resolution of the conflicting tendencies of human thought, those of "creative progress" and the striving for "all-embracing completeness" (Zermelo, 1930, p. 47).

[7] Cf. Feferman (1977), Hellman and Bell (2006).

1 (c), it seems to me, is still a work in progress, not surprisingly given its "philosophical" character. "Strength" of justification refers to some combination of epistemic security of the justifying assumptions and their epistemic independence of the axiom or axioms in question (non-circularity), and reasonable minds can differ on these matters. Indeed, the very target of justification depends on background assumptions about the foundational system. Consider the Axiom (Scheme) of Replacement of ZFC: On a face-value, traditional platonist reading, the target is the instances of the scheme taken as purportedly true statements about functions and sets: in effect, the range (forward image) of any (1st-order expressible) function on a set is itself a set. Here axioms are treated as assertions about a presumed reality, à la Frege, and the epistemological question is, "What reasons can be given for believing these statements to be true?". On this view, the purely *mathematical* motivation of wanting enough closure for an elegant theory of transfinite ordinal arithmetic is hardly compelling, if relevant at all! Why should set-theoretic reality conform to our desires, especially desires for "elegance" reflecting our own standards? (For that matter, why should even the most elementary axioms of set theory be true? Maybe, for instance, some pairs simply don't exist!) Similar embarrassing questions arise for other attempts at justification, e.g. that the universe of sets should be truly large relative to any of its members. On the other hand, on a structuralist view of set theory, the target shifts: the axioms are not read as assertions at all but rather as conditions implicitly defining the sort of structure or structures we are interested in studying, as is commonplace elsewhere in mathematics. (This is the Hilbertian view of "axioms", also called "algebraic", "structural", or "abstract" (the latter a confusing label in philosophical contexts, I would say).) From this perspective, the question is not why we should believe this statement, but rather, "What reasons can be given for believing that the overall system, including the axiom(s) in question and presumably incorporating Robinson arithmetic, is consistent and, moreover (if we seek to respect intended interpretations), coherent, or satisfiable (in a sense of logical possibility)?"[8] Although such questions are indeed challenging and, as we have learned from Gödel, cannot be answered in the ideal Hilbertian manner via a formalizable, finitistic consistency proof, they are arguably more relevant to mathematical practice than the objects-platonist ones they supplant. And there is at least the prospect of supporting the working hypotheses of consistency and coherence through successful mathematical practice, even where such practice is scarcely relevant to the question of whether "there really are enough sets" for actual truth of an axiom such as Replacement.[9]

Moving to *desideratum* 5, set theory—again identified with ZF or ZFC—is capable of modelling any consistent first-order theory whatever, at least if we believe

[8] On the role and importance of "coherence", see Shapiro (1997).

[9] Thus, a structuralist with respect to ZFC side-steps the kind of question raised by Boolos in his (1998), whether there "really are" as many sets as are guaranteed by ZFC, e.g., via Hartogs' theorem (which follows from Replacement), the least fixed point of the $\aleph$ function. What matters for mathematical practice is not this but whether there is any inconsistency or incoherence in the statement that there are (or whether ZFC is at least consistent and, moreover that a structure satisfying *second-order* ZFC is a logical possibility).

the Gödel strong completeness theorem for first-order logic.[10] In this sense, many theories indeed are "respected", including many set theories other than ZFC itself (that is on the assumption of their consistency, surely a prerequisite of "respectability").[11] But is this more than what *noblesse oblige*? On the Fregean view of axioms, a foundational framework based on ZFC gives it a privileged status as providing the truth about sets, collectively forming a unique cumulative hierarchy. While various extensions of ZFC, e.g. by large cardinal axioms, may add to that truth, other axiom systems inconsistent with ZFC (e.g. violating Choice or Regularity or instances of Separation, etc.) would be regarded as mistaken, whatever other virtues they might have; similarly regarding alternatives to the classical continuum which add non-classical axioms leading to formal negations of classical theorems (such as intuitionistic analysis with special continuity principles for choice sequences implying e.g. the Brouwer uniform continuity theorem, or smooth infinitesimal analysis with the Koch-Lawvere axiom governing (smooth) maps from nilsquare infinitesimals to the reals). It is one thing to provide for set-theoretic models of such alternative theories; it is another to recognize those theories as having as good a claim to "truth" as the favored classical set theory. Although such truth would not pertain to the very same cumulative hierarchy of sets, other structures would need to be recognized as answering to those theories without having to be identified as a set within the privileged hierarchy. Finally, on this point, many theories or structural descriptions, compatible with ZFC or not, postulating "large" domains cannot be faithfully modelled as ZFC sets (in essentially the same way that the reals cannot be faithfully modelled as a countable set). For instance, any "large category", such as the category of all groups or of all topological spaces, etc., from the perspective of ZFC, contains items identified as sets of arbitrarily large ordinal rank and so cannot be identified with any set at all. If they are instead identified as proper classes (in an extension of ZFC, such as NBG (conservative) or Morse-Kelley (non-conservative), then that will be transcended if we then consider a category of such categories or of all the functors relating them, etc. Similar problems arise for various alternative set theories. The assumed fixed-universe ontology of ZFC thus becomes a liability when it comes to respecting multiple frameworks.

That same background ontology also leads to special problems highlighted in *desideratum* 6, providing for extendability of universes of discourse for mathematics. As I have treated various aspects of this in detail in several places,[12] a summation of the import of those investigations should suffice here. In a nutshell, while the problem of providing for extendability without exceptions seems intractable for any traditional platonist view of mathematics, which sanctions the notion of "absolutely all mathematical objects", it is especially intractable for set theory on the fixed hierarchy view we've been discussing. As in the case of *desideratum* 5, there

---

[10] In fact, this metatheorem is provably equivalent to the Weak König lemma (over RCA_0), i.e. a tiny bit of set theory suffices. See Simpson (1991, pp. 139, ff).

[11] For information on natural ways of passing from e.g. toposes to set-models (of suitable set theories) and vice versa, as well as modelling relations in both directions between sets or categories and models of type theories, see Awodey ("From Sets to Types, to Categories, to Sets", this volume).

[12] See Hellman (1989, 2002, 2005).

is a sense in which ZFC, say, respects extendability, viz. by treating models of theories (assumed consistent) as sets and recognizing sets and models (of the same or different theories) of arbitrarily higher cardinality (e.g. by the upward Löwenheim-Skolem theorem). But again this can involve distortion of intended meanings, and moreover it leaves unresolved the problem raised by the presumed fixed, universal, hence maximal background of sets and ordinals. As I have argued, the problem persists even if we refuse to speak of *any* kind of collection or totality of "all sets" or "all ordinals" or "all cardinals". Even if we adopt a logic of plurals as governing these locutions, we sin against the injunction not to limit "creative progress" if we refuse to allow mathematical thought to introduce new language recognizing still further objects (at least as *possibilities*) behaving just as the ones already assumed (something that *can be done*). "Absolutely all sets or ordinals" is not really better off than "the class or totality of absolutely all sets or ordinals": both are at best a convenient piece of mythology.

Turning to category and topos theory (CTT), let us begin at the end: Regarding *desiderata* 5 and 6, this framework represents a significant advance beyond set theory, brought out nicely through Bell's "many toposes" perspective (1986, 1988). Toposes satisfying distinctive axioms, each with its own internal logic, serve as universes of discourse for substantial bodies of mathematics reflecting a variety of interests, methods, and commitments, e.g. classical set-theoretic, constructive, smooth-analytic and differential-geometric, etc. None is privileged as an all-encompassing background. And there is not the motivation as in set theory to take seriously "absolutely all sets" or "absolutely all ordinals", etc. Since a plurality of universes is characteristic of this approach, there is no particular reason to embrace "absolutely all toposes" or "absolutely all categories" either. One can describe or posit a (meta)category of categories without thinking of it as having "absolutely all categories" as objects, and likewise for functor categories.

Interestingly, it is in how CTT grapples with parts of *desiderata* 1–3 that the most serious problems arise (and, of course, these are the most clearly traditionally "foundational" ones).

It would appear that CTT, understood as a body of mathematical (and metamathematical) practice, does not point clearly to a single system of Fregean-style axioms the way set-theoretic practice points to ZFC as a core, standard system.[13] This may be in part because of the way in which category theory arose, motivated by problems within algebraic topology and geometry, not impelled by paradoxes threatening fundamentals as Zermelo's set theory was. Further, category theory has become integrated with mathematical practice as an organizing tool with widely applicable machinery (the language of '*objects*' (usually mathematical *structures*) and '*arrows*' or '*morphisms*' under a '*composition*' operation) well adapted to structuralist ideas, e.g. that structure matters only "up to isomorphism", that one can

[13] Here we are referring to set theory as mathematics in its own right, such as the study of large cardinals, of determinacy, of the continuum problem or constraints, and so forth. Although the *language* of sets and elements is ubiquitous throughout mathematics and can serve to define functions, operations, and relations, the powerful axioms of set theory, such as Power Set and Replacement, are not needed for the vast bulk of mathematics, i.e. it is in this sense "extraordinary mathematics".

abstract from the particular items belonging to a collection, and so forth. Clearly, the "axioms" presented for describing what a category is or what a topos is are Hilbertian, defining-conditions on types of structures, not assertions. The primitive notions function *schematically*, taking on definite meanings in particular applications and contexts (cf. Awodey, 1996, 2004).

Category-theoretic axiom systems have been proposed, however, with a foundational role in mind (e.g. Lawvere, Mac Lane, McLarty) and, as McLarty (2004) has stressed, these should not be read schematically but as (presumably true) claims about mathematical reality, as much as ZFC is. The main systems relevant here are the first-order ones, ETCS and CCAF (the elementary theory of the category of sets, and the category of categories as a foundation). I have taken account of this in my (2006a), making points which we develop a bit further here:

(1) While ETCS (or ETCS + R, with a CT version of Replacement added to obtain a mathematical equivalent of ZFC) is surely interesting in its own right as articulating a structuralist-functionalist description of sets forming an iterative hierarchy, when considered in the role of a foundational framework, it inherits some of the same problems affecting ZFC, especially regarding *desiderata* 5 and 6. Presumably other versions of set theory would be presented with different (substantive) axiom systems, *either* treated as satisfied by structures as objects internal to the discrete topos posited as the subject matter of ETCS + R, in which case, that topos is given privileged status; *or* those versions would be taken as describing different self-standing toposes, in which case ETCS + R is not the relevant, comprehensive foundational framework that would be adequate. Presumably such a framework would provide Fregean-style axioms on the mathematical existence of many (discrete and other) toposes and of course it would then have to be judged on its merits, including how well it meets the *desiderata* 1–6 considered in this paper. Perhaps the framework would be (an extension of) CCAF, the substantive theory of a category of categories and functors McLarty has helped formulate. Perhaps that framework would also help with *desideratum* 6 by providing CTT versions of extendability principles generalizing modal-structural theories of sets.

(2) In connection with CCAF, I wrote:

> ...it is clear that these axioms (as in McLarty, 1991) are not employing the CT primitives ('object', 'morphism','domain', 'codomain', 'composition') schematically, as in the algebraic defining conditions, but with intended meanings presumably supporting at least plausible truth of the axioms. The objects are categories, the morphisms are functors between categories, etc. ...when we speak of the "objects" and "arrows" of a metacategory of categories as *categories* and *functors*, respectively, what we really mean is "structures (or at least "interrelated things") satisfying the algebraic axioms of CT", i.e. we are using "*satisfaction*" which is normally understood set-theoretically. That is not to say that there are no alternative ways of understanding "*satisfaction*"; second-order logic or a surrogate such as the combination of mereology and (monadic) plural quantification of modal-structuralism would also suffice. But clearly there is some dependence on a background that explicates satisfaction of sentences by structures, and this background is not "category theory" itself, either as a schematic system of definitions or as a substantive [first-order theory] of a metacategory of categories. But this need for a background theory explicating satisfaction was precisely the conclusion we came to in our (2003) paper, reinforcing the well-known critique of Feferman from (1977), which exposed a reliance on general notions of "collection"

and "operation". It was precisely to demonstrate that this in itself does not leave CT structuralism dependent on a background set theory that I proffered a membership-free theory of large domains as an alternative. Although the reaction, "Thanks, but no thanks!", frankly did not entirely surprise me, it will also not be surprising if a perception of dependence on a background set theory persists.

Thus, with respect to our *desiderata,* CCAF is not yet adequate, at least on the assumption that '*category*' and '*functor*' are not to be taken as primitive (in view of *desideratum* 3). While, to be sure, mathematicians are fully conversant in these terms, that is because they have understood their definitions as technical terms of art or at least got themselves into a position to do so. Presumably that would be by understanding a variety of examples, e.g. of familiar spaces, topological, metric, etc., or algebraic structures along with the relevant types of morphisms. That is, what is required is an understanding of "structure-preserving map", and surely that derives from familiarity with Bourbaki-style mathematics and the ability to generalize from it. And this, in turn, involves generalizing on collections, operations and relations on a domain, bringing us full circle again back to a theory dealing directly with these notions, such as membership-based set theory, higher-order logic, or an equally powerful substitute such as mereology + plural quantification.[14]

Why not appeal instead to ETCS? While formally this theory can recover a theory of operations and relations—and McLarty explains that it already *is* a theory of the very same hierarchy (or hierarchies) of sets studied in membership-based set theory—can we really understand ETCS as doing all this without a prior, independent understanding of "collection of elements", "operation taking tuples of elements to elements", etc.? Treating sets as "points" connected to and from other points by "arrows"is fine, but even so elementary a definition as "generalized element" (as an arrow from a terminal point to a point) can be seen to work only when we realize that in "the" Set Topos, any singleton is terminal; we are then thinking of points as collections of elements.

But there is a further, perhaps more decisive reason why appeal to a first-order theory such as ETCS (+ R) would not suffice. Even if the case were made that it could be well-enough understood in its own terms to support an account of collections, operations, and relations for foundational purposes as effective as ZFC's account, that would not extend to mathematics beyond the reach of those theories, such as that involved in a general treatment of categories and toposes, including those too large to be faithfully modelled as sets. Stepping back, recall that the CTT framework, including theories such as CCAF, held the promise of improving on membership-based set-theoretic foundations precisely in its capacity to meet

[14] Have we departed from our stance in the introduction of focussing on how assumptions *can* or *could* be justified rather than how they actually *are* justified? Not really: we are suggesting that, without the kind of preparation in familiar mathematics that we all get, it is unlikely at best that we would gain an adequate grasp of "structure-preserving map" for first-order theories of a category of categories like CCAF to serve the justificatory role of a foundation. That this may well turn on empirical facts about our central nervous system doesn't make it either irrelevant or false. As we said above (p. 2), we are interested in the real possibilities of justification by (mathematically capable) humans, agreeing to this extent with Errett Bishop (1967) when he said, "If God has mathematics that needs to be done, let him do it himself!"

desiderata 5 and 6, i.e. *without* embedding everything within a single, fixed, background universe of sets. But if we rely on a categorical set theory for an account of mathematical structure, that promise is severely compromised as such an account is too restricted. This is the fundamental dilemma confronting categorical foundations. If *desiderata* 2 and 3 are met by appealing to a first-order set theory, even if (for the sake of argument) it is granted that a categorical set theory would serve, then membership-based set theory's limitations in connection with *desiderata* 5 and 6 will not have been overcome but merely reproduced. If 2 and 3 are not met in that way, then we do not yet know how they will be met. It appears that we have not yet been presented with a really satisfactory, self-standing CT foundational framework.

4 Modal-Structural Mathematics and Foundations

There are various ways of building up modal-structural (MS) mathematics, developed in Hellman (1989, 1996), and (2005), the last providing a compact, semi-formal summary. All have in common a translation pattern sending any ordinary mathematical sentence to a sentence (in a preferred nominalist language) asserting what would hold in any structure of appropriate type (satisfying relevant axioms or conditions) that there might be. This is called the "hypothetical component" of an MS interpretation of the mathematics in question. In addition, there is a corresponding "categorical" assertion that structures of the relevant type are (logico-mathematically) possible. From a foundational perspective, there is room for improvement on earlier presentations, however, in two respects: (1) The possibility of infinitely many things is simply postulated (in the manner of logicism, set theory, and predicative mathematics), instead of being derived from more fundamental axioms. (2) The possible existence of a progression (i.e. objects satisfying the (second-order) Peano-Dedekind axioms, PA^2) is derived from the axiom of infinity in a archtypically impredicative fashion, relying on a nominalistic version of second-order comprehension to obtain the minimal closure of an initial object under an operation (behaving as successor, though described nominalistically). This just mimics the usual "top-down" procedure (taking the intersection of all inductive classes). A better strategy is, first, to derive a suitable axiom of infinity from natural, general principles, rectifying (1), and then to build up a model of the elementary theory of finite-sets and classes (EFSC) (Feferman and Hellman (1995, 2000)), thereby rectifying (2). EFSC is a theory with a (concretely interpretable) pairing operation, an urelement under pairing, an empty finite set, closure under adjunction with a single individual, bounded separation (i.e. applied to finite sets), and weak second-order class comprehension (i.e., classes of individuals—interpretable as wholes of atoms[15] or via plurals—defined by formulas with bound individual and finite-set

[15] A mereological *atom* is an individual with no proper parts. The Atomicity axiom states that every individual has an atom as a part. Other axioms guarantee Boolean closure, adjusted for lack of a null individual. (x is a *proper part* of y just in case x is part of y and $x \neq y$. Also it is convenient to define 'x *overlaps* y' as "x and y have a common part".)

variables only), but *no* axiom of finite-set induction. EFSC derives the existence of an **N**-structure, i.e. satisfying the axioms of PA^2 (with second-order induction), uniquely up to isomorphism.[16]

All this can be accomplished in what we call our *Core System*, which consists of the following:

1. Classical first-order logic along with logic of plurals (nearly equivalently, monadic second-order quantifier logic);
2. Atomic mereology with variables for finite wholes ("fusions" or "sums" of atoms), with a "comprehension" axiom scheme guaranteeing fusions of any instantiated predicate;
3. S5 quantified modal logic (without Barcan or converse Barcan formulas);
4. An axiom of extendability of domains or structures:

$$\Box\forall M \Diamond \exists M'[M' \succ M], \tag{EP}$$

   where M, M' are plural variables and $\succ$ means "properly extends".[17]
5. An axiom scheme of reflection of the form:

$$S_{PT} \rightarrow \Diamond \exists M[M \text{ models } S], \tag{MSR}$$

   where S_{PT} designates the Putnam-translate of S, typically a first- or second-order sentence of the language of set theory.[18] The motivation behind this is that the mathematical possibilities are "indescribable", in the sense that what S

Atomic mereology, i.e. with the Atomicity axiom, is a convenience for the MS framework, and could be bypassed, allowing atomless individuals (called *gunk*). What matters is the possibility of infinitely many pairwise non-overlapping individuals, to obtain the effect of steps 4. and 5. below; and with gunk that can always be arranged. E.g. for infinity, as in 4., take as a "next" individual any given one g (of *some* (plural) assumed proper parts, G, of the gunk) combined with a proper part of the "complement" of g.

[16] The idea is to build up an **N**-structure "from below" as a "union" of finite initial segments. Note that, whereas in the context of the system EFSC, the notion of "finite" was taken as antecedently given, in the present context of mereology with (monadic) logic of plurals, it is explicitly definable: for instance, use the formulation given in note 18, below, to define "the atoms of w (resp. of X) are *infinite*", and define '*finite*' as 'not infinite'. What EFSC shows is that mathematical induction (for a successor structure) is derivable predicatively *relative to the notion of 'finite set'*. That carries over in the MS framework, even if we choose to define 'finite' as just indicated, rather than take it as given. Whether there is a gain in the use of mereology and plurals for this sort of definition *vis-à-vis* the usual set-theoretic definitions is a question meriting further attention.

In any case, this recovery of induction provides an answer to Poincaré's objection that it cannot be usefully logically derived. Without assumption of any definite infinite structure of finite sets, a particular infinite structure of the natural numbers is derived. (For a more detailed defense, see Feferman-Hellman (2000).)

[17] In practice, versions of EP may contain further details specifying just which theories the structures model, how the relations and functions of the structures are related, and so forth. In particular, we can for instance adopt as a special case that "Any objects M there might be could be 'minimally' extended, i.e. with a single additional atom, a." This will be useful in proving Theorem 1 below.

[18] To illustrate, the Putnam translate of a (set-theoretic) sentence with unbounded quantifiers, say, of form $\forall X \exists y \Phi$ is

says about them, as spelled out in the Putnam translate, fails in that S already holds in a small segment thereof, i.e. relativized to a single model (of the set-theory or proto-set-theory in question). Interestingly, the combination of EP and MSR imply "Possibly there are infinitely many objects, M".[19]

Theorem 1 *Let T be the theory consisting of just the extralogical axiom,* $\exists x[x = x]$. *Then possibly there is an infinite model of T.*

Proof Let S be the sentence, "For any plurality M_i there is a minimally greater M_j (in the sense of the previous footnote)". Then observe that S_{PT} is equivalent to the minimalist version of EP of the previous footnote (with the innocuous further condition that the pluralities model T). So, by MSR, possibly there is a plurality M modelling T & S, whence the M are infinitely many. ∎

With this result, *ordered pairing* of arbitrary individuals can be introduced (à la Burgess et al., 1991) allowing reduction of n-ary relations to unary, so that logic of plurals of 1. has the expressive scope of full second-order logic. Then we adopt

6. An axiom scheme of "extensional comprehension", providing necessary existence of any relation specified by a modal-free second-order formula,

$$\Box\exists R\forall x_1 \ldots \forall x_n[R(x_1 \ldots x_n) \leftrightarrow \Phi], \qquad \text{(Logical Comp)}$$

where Φ lacks free 'R' and is modal free.

Theorem 2 *Possibly there is a countable model of EFSC.*

Proof This follows from the first theorem, mimicking well-known set-theoretic techniques in the theory of mereology + logic of plurals, including 5. ∎

(While 6. permits arbitrary impredicative instances for later developments, only those predicative relative to finite sets (wholes) are used for obtaining an EFSC structure.)

$$\Box\forall M\forall X[X \subseteq |M| \rightarrow$$
$$\Diamond\exists M'\exists y(y \in |M'| \ \& \ M' > M \ \& \ \Phi)],$$

where Φ has no unbounded quantifiers. It is clear how this generalizes to arbitrary first- and second-order set-theoretic sentences in prenex form.

[19] A sufficient condition for a plurality of individuals X to be *infinitely many* is this: a is one of the X and for any y which is one of the X, there is a unique atom b such that $b + y$ is also one of the X. This was the definition used in postulating an axiom of infinity in earlier presentations of modal-structural mathematics (as, e.g., in my (1996)).

Note that, once the possibility of infinitely many X is established, it will be routine to introduce ordered pairing and therewith a reduction of polyadic to monadic second-order logic, whence the notions of "Dedekind-infinite" and "Cantor-Frege infinite" (in terms of a natural-numbers structure) can be explicitly defined. Either of these notions can then be used in standard ways, replacing the above (merely sufficient) condition for "the X are infinitely many".

In this Core System, the main subsystems of second-order arithmetic studied in reverse mathematics are interpretable (in the sense of Tarski) (most in a predicative fragment), as indeed is full classical analysis and even classical third-order arithmetic.[20]

Regarding our first two *desiderata,* this Core System already encompasses (deductively and expressively) vast portions of mathematics as practiced, and its justification is no less secure than that of the principles EP and MSR, along with the comprehension principles of mereology and the (monadic) logic of plurals.

The last two *desiderata* on pluralism and extendability are perhaps modal-structuralism's strongest suits. Extendability principles regarding large structures are readily formulated as modal principles, the simplest being of the form, "For any structure of such-and-such type there might be, there might also be a proper extension (satisfying the same or different conditions)". "Extended extendability principles" take the form, "For any ordinal sequence of structures (of such-and-such types) that there might be (the ordinals coming from a hypothetical model of a suitable theory, e.g. ZFC or ETCS + R or an extension), there might also be a common proper extension...". The possibility of a sequence or collection or just the compossibility of "all possible models or structures" is never recognized (see the extensional Logical Comp, which lacks necessity operators attached to the universal quantifiers), and this is as it should be, as we have explained (see e.g. Hellman (2005, pp. 553ff)). This is, of course, motivated by the aim of enforcing unrestricted extendability; but it also stems from the ordinary idea of sets as refications of "collecting" operations or activities, together with the obvious fact that you can only collect what exists and *could* only collect (in given circumstances) what would *then* exist (i.e. in *those* circumstances).

Regarding pluralism, clearly the Core System can be extended in many different ways and directions, e.g. postulating (possibility of) toposes and other categories or metacategories of various kinds, set-theoretic models of various kinds, *sui generis* topological or geometric spaces of many kinds, etc. Moreover, all this can be done without taking any one kind as privileged either as a background for others or in some other way. (And, of course, privileges may be granted wherever they are well-earned, respecting the merits of others!) Moreover, if honest toil is really preferable to theft, the possibility of the relevant kinds of structures need not simply be postulated but may be derived in an extension of the Core System by *large domains*, described in the language of the Core System.[21] Multiplicity of logical systems governing structures is also recognized (as in topos theory and even in set theory via representation of satisfaction relations internal to a structure that may

[20] If a stronger axiom of infinity is adopted postulating the possibility of a continuum of atoms, then fourth-order arithmetic becomes available. This particular extension of the Core System marks a boundary on the power of traditional nominalism where points of physical space, time, or space-time are admitted as objects.

[21] See Hellman (2003), where we sketch the description of domains of atomic individuals analogous to set-domains of strongly inaccessible height, relative to which set-models, categories, toposes, etc. can be constructed.

differ from the usual, classical background).[22] Of course, a choice of background logic for a foundational framework itself must be made, although even here variety is possible, indeed even required, e.g. for the distinct purposes of *desiderata* 1 and 2, deduction vs. expression, where the latter typically demands the greater power of second- or higher-order logic (whereas proofs would naturally be given in an axiomatic fragment or in first-order systems). There is nothing to prevent a modal-structural framework from demanding constructive proofs wherever they can be provided, including where possible proofs of (possible) existence of relevant types of structures, although the scope and limits of such an approach remain to be determined. In any case, it may well be that a classical background is suitable for demonstrating or at least motivating the possibility of structures for theories built on intuitionistic logic. Smooth infinitesimal analysis provides an interesting case in point: the theory requires restriction to intuitionistic logic for survival[23], but that theory does not lend itself to a thoroughgoing constructive interpretation, and we wouldn't expect to be able to provide a constructive proof of nil-square infinitesimals $\neq 0$, or of a model recognizing such things. (SIA proves $\neg\forall x(x^2 = 0 \rightarrow x = 0)$ but, on pain of contradiction, cannot prove $\exists x(x^2 = 0 \wedge x \neq 0)$.)

The question of understanding the relevant sense of possibility, the matter of *desideratum* 3, has been addressed to some extent in (2005). Although the notion does not occur explicitly in ordinary mathematics, it is arguably implicit in many contexts, as when one describes intuitive geometrical constructions to demonstrate the satisfiability (hence consistency) of non-Euclidean geometries, or ways that some topological condition may fail, or instances of algebraic structures via specified transformations of an assumed space, etc. If (*per impossibile*, I realize) pure set-theoretic modelling could not reproduce such examples, surely we would fault the set theory as a standard rather than conclude that the mathematical conditions in question are unsatisfiable or incoherent or impossible *tout court*. No doubt the language of sets is more precise than ordinary mathematical discourse, but if the notion of possibility is up to the task required—and for the vast bulk of mathematics, the above axiom of infinity (in the context described) suffices—then the MS framework can survive on its own, postponing the burdens of a strong set theory such as ZF or even Z for a later stage.[24]

Another relevant consideration, especially in the context of foundations, is the epistemic status sought for the framework's basic postulates. In the case of mathematical existence of infinite structures, it is (and always has been) too much to demand the security of finitistic operations. Even the term "knowledge" is too strong, except perhaps in an everyday practical sense. No one seriously believes that classical analysis is inconsistent, much less that Peano arithmetic is, but we have learned

---

[22] See my (2006b) for an example involving smooth infinitesimal analysis.

[23] See Bell (1998); also Hellman (2006b).

[24] It is worth noting that, in motivating various axioms of set theory itself, independent of other entrenched ones, such as Choice, Replacement, or a large cardinal axiom implying existence of a non-constructible set, etc., set theorists (who *are* mathematicians) do speak of allowing the universe of sets to be as extensive and rich as possible. Can this be understood entirely as "formally consistent"?

(from Gödel et al.) not to expect finitistic consistency proofs of these theories, much less proofs of the mathematical existence of standard models. If foundations in its entirety required the mathematical certainty of elementary computability, it would simply not get very far. Instead, we adopt certain starting points as reasonable working assumptions suitable to the goals of the enterprise. From the perspective of, say, strict finitism, our (Ax $\Diamond$ ∞) is indeed extravagant, especially if the goal is to secure just mathematics used as an instrument of computation in scientific applications. Such an approach may be satisfied with a formalist or instrumentalist conception of mathematics generally. (The early nominalist efforts of Goodman and Quine (1947) adopted this stance.) For a structuralist, however, interested in respecting the truth in some objective sense of most of classical as well as constructive mathematics as practiced, stronger assumptions are unavoidable. If we are also skeptical of actual infinites of objects not part of spacetime and of the various arguments that we must accept them, we are then motivated to seek an alternative. Clearly, that has been and continues to be our own stance.

Finally, if the approach to foundations sketched here is viable, then, while strong foundationalism, seeking to ground all of mathematics on certain or self-evident assumptions, is indeed a lost cause, a modest, well-tempered foundationalism is alive and well. While it cannot pretend to achieve the beauties of Bach, it can provide a measure of epistemic order, a balance of unity and diversity, and perhaps even insight into the nature of our subject.

References

Awodey, S. (1996) "Structuralism in Mathematics and Logic", *Philosophia Mathematica* **4**, 209–237.

Awodey, S. (2004) "An Answer to Hellman's Question: 'Does Category Theory Provide a Framework for Mathematical Structuralism?' ", *Philosophia Mathematica* **12**, 54–64.

Bell, J.L. (1986) "From Absolute to Local Mathematics", *Synthese* **69**, 409–426.

Bell, J.L. (1988) *Toposes and Local Set Theories*, Oxford: Oxford University Press.

Bell, J.L. (1998) *A Primer of Infinitesimal Analysis*, Cambridge: Cambridge University Press.

Bishop, E. (1967) *Foundations of Constructive Analysis*, New York: McGraw.

Boolos, G. (1998) "Must We Believe in Set Theory?", in Boolos, G., *Logic, Logic, and Logic*, Cambridge, MA: Harvard University Press, pp. 120–132.

Burgess, J.P. (2004) "Mathematics and *Bleak House*", *Philosophia Mathematica* **12**, 18–36.

Burgess, J.P., A. Hazen, and D. Lewis (1991) "Appendix on Pairing", in Lewis, D., *Parts of Classes*, Oxford: Blackwell, pp. 121–149.

Burgess, J.P. and Rosen, G. (1997) *A Subject with No Object: Strategies for Nominalistic Interpretation of Mathematics*, New York: Oxford University Press.

Chihara, C. (2005) "Nominalism", in S. Shapiro, ed. *The Oxford Handbook of Philosophy of Mathematics and Logic*, Oxford: Oxford University Press, pp. 484–514.

Drake, F. (1974) *Set Theory: An Introduction to Large Cardinals*, Amsterdam: North-Holland.

Feferman, S. (1977) "Categorical Foundations and Foundations of Category Theory," in R. E. Butts and J. Hintikka, eds., *Logic, Foundations of Mathematics and Computability Theory,* Dordrecht: Reidel, pp. 149–169.

Feferman, S. (1981) "The Logic of Mathematical Discovery vs. the Logical Structure of Mathematics," P. D. Asquith and I. Hacking, eds. *PSA 1978* Vol. **2**, East Lansing, MI: The Philosophy of Science Association, pp. 309–327.

Feferman, S. and Hellman, G. (1995) "Predicative Foundations of Arithmetic", *Journal of Philosophical Logic* **24**, 1–17.

Feferman, S. and Hellman, G. (2000) "Challenges to Predicative Foundations of Arithmetic", in G. Sher and R. Tieszen, eds., *Logic and Intuition*, Cambridge: Cambridge University Press, pp. 317–338.

Goodman, N. and Quine, W.V.O. (1947) "Steps Toward a Constructive Nominalism", *Journal of Symbolic Logic* **12**, 97–122.

Hellman, G. (1989) *Mathematics without Numbers: Towards a Modal-Structural Interpretation*, Oxford: Oxford University Press.

Hellman, G. (1996) "Structuralism without Structures", *Philosophia Mathematica* **4**, 100–123.

Hellman, G. (1998) "Maoist Mathematics?", Review of Burgess and Rosen, *A Subject with No Object*, *Philosophia Mathematica* **6**, 357–358.

Hellman, G. (2001) "On Nominalism", *Philosophy and Phenomenological Research* **62**, 3, 691–705.

Hellman, G. (2002) "Maximality vs. Extendability: Reflections on Structuralism and Set Theory", in D. B. Malament, ed. *Reading Natural Philosophy: Essays in the History and Philosophy of Science and Mathematics*, La Salle, IL: Open Court, pp. 335–361.

Hellman, G. (2003) "Does Category Theory Provide a Framework for Mathematical Structuralism?", *Philosophia Mathematica* **11,** 129–157.

Hellman, G. (2005) "Structuralism", in S. Shapiro, ed. *The Oxford Handbook of Philosophy of Mathematics and Logic*, Oxford: Oxford University Press, pp. 536–562.

Hellman, G. (2006a) "What is Categorical Structuralism?", in J. van Benthem, G. Heinzmann, M. Rebuschi and H. Visser, eds. *The Age of Alternative Logics. Assessing philosophy of logic and mathematics today,* Dordrecht: Kluwer, pp. 151–161.

Hellman, G. (2006b) "Mathematical Pluralism: The Case of Smooth Infinitesimal Analysis", *Journal of Philosophical Logic* **35**, 621–651.

Hellman, G. and Bell J.L. (2006) "Pluralism and the Foundations of Mathematics", in C.K.Waters, H. Longino, and S. Kellert, eds. *Scientific Pluralism*, Minnesota Studies in Philosophy of Science, Vol. **19**, Minneapolis, MN: University of Minnesota Press.

Lakatos, I. (1976) *Proofs and Refutations*, Cambridge: Cambridge University Press.

McLarty, C. (1991) "Axiomatizing a Category of Categories", *Journal of Symbolic Logic* **56**, 1243–1260.

McLarty, C. (2004) "Exploring Categorical Structuralism", *Philosophia Mathematica* **12**, 37–53.

Rosen, G. and J.P. Burgess (2005) "Nominalism Reconsidered", in S. Shapiro, ed. *The Oxford Handbook of Philosophy of Mathematics and Logic*, Oxford: Oxford University Press, pp. 515–535.

Shapiro, S. (1997) *Philosophy of Mathematics: Structure and Ontology*, New York: Oxford University Press.

Simpson, S.G. (1991) *Subsystems of Second Order Arithmetic*, Berlin: Springer-Verlag.

Zermelo, E. (1930) "Über Grenzzahlen und Mengenbereiche: Neue Untersuchungen über die Grundlagen der Mengenlehre", *Fundamenta Mathematicae* **16**, 29–47.

The Problem of Mathematical Objects

Bob Hale

In seeking a foundation for mathematics, one may be looking for what may be called a foundation in the *logical* sense: a single, unified set of principles—perhaps unified by their jointly constituting an acceptable axiomatization of some concept or concepts plausibly taken as fundamental—from which all, or at least a very large part of, mathematics can be derived. In this sense, some version of set theory is plausibly taken as a foundational. But one may also be interested in an *epistemological* foundation—roughly, an account which explains how we can know standard mathematical theories to be true, or at least justifiably believe them. A foundation in this sense could not be provided by any *mathematical* theory, however powerful and general, by itself. Indeed, the more general and powerful a mathematical theory is, the more problematic it must be, from an epistemological point of view. What is called for is a philosophical account of how we know, or what entitles us to accept, the mathematical theories we do accept. Since such an account cannot very well be attempted without adopting some view about the nature of the entities of which the mathematical theories treat, this is likely to involve broadly metaphysical questions as well as epistemological ones. It is not certain either that we are right to demand such a foundation, or that one can be given, but I shall proceed on the assumption that it is reasonable to seek one.

If fundamental mathematical theories such as arithmetic and analysis are taken at face value, any attempt to provide such a foundation must confront the problem of mathematical objects—the problem of explaining how a belief in the existence of an infinity of natural numbers, an uncountable infinity of real numbers, etc., is to be justified. Of course, as already noted, these theories may be derived within a suitable theory of sets, but then we simply replace the problem of justifying belief in numbers of various kinds with the problem—unlikely to be easier—of justifying belief in the existence of the universe of set theory.

I am going to discuss only one small, but fundamental, part of the problem—whether we can be justified in believing that there is a denumerable infinity of natural numbers—or, more generally, an infinity of objects of any kind. I shall consider two broad approaches to this problem—what I shall call *object-based* and

G. Sommaruga (ed.), *Foundational Theories of Classical and Constructive Mathematics*, The Western Ontario Series in Philosophy of Science 76,
DOI 10.1007/978-94-007-0431-2_2,

property-based approaches. By a property-based approach I mean any approach which argues *indirectly* for an infinity of objects, in the sense that our access to an infinite sequence of objects is seen as dependent on an underlying infinity of properties. By an object-based approach I mean any attempt to argue *directly* that we can have access to, or knowledge of, an at least potentially infinite sequence of objects. These descriptions are rather vague, but I think there are some rather clear instances of the approaches I wish to contrast. The clearest—and in my view, most promising—attempt to develop the object-based approach forms part of Charles Parsons's investigation of the popular but normally rather opaque and ill-explained idea of mathematical intuition. Another example is Dedekind's notorious argument that the realm of his own thoughts—which comprises the possible objects of his thought—is infinite. Perhaps the most obvious example of the property-based approach is to be found in Frege's attempt to prove the infinity of the natural numbers.

1 Parsons on Mathematical Intuition[1]

1.1 Intuition of and Intuition That

Central to Parsons's account is a distinction between objectual and propositional intuition—between *intuition of objects* and *intuition that p*, for some *p*. The former notion is to play the fundamental role—much as perceptual knowledge that *p* (e.g. that there is a duck on the pond) depends upon perception of the duck, so intuitive propositional mathematical knowledge is to depend, at bottom, on intuition of certain objects.

1.2 Pure Abstract and Quasi-concrete Objects

But the numbers themselves are not—in Parsons's view—among the objects of intuition. He distinguishes between pure abstract objects and quasi-concrete ones. Quasi-concrete objects are, roughly, objects which are closely connected with concrete objects, or 'belong to' such objects—e.g. shapes, edges, and, importantly for Parsons, linguistic expressions considered as types, as opposed to their concrete tokens.

Pure sets and numbers are not so related to concrete objects, are usually taken to exist independently of concrete objects, and would presumably all rank as pure abstract objects for Parsons.

The objects of intuition are restricted, for Parsons, to the quasi-concrete. This restriction is, I think, crucial. So long as it is in place, it is at least very plausible to claim that we may stand in a direct cognitive relation—analogous to, and perhaps

[1] Parsons (1980). This discussion of Parsons's view draws on Hale and Wright (2002).

often mediated by, ordinary sense perception—to the objects of intuition. But to claim that we may enjoy such a relation to abstract objects quite generally, and so to pure sets and other pure abstract objects, which have no essential connection with concrete objects, would be to cast the notion of intuition back into the depths of obscurity from which Parsons has done much to rescue it.

1.3 The Language of Stroke Strings

Parsons's detailed account of intuition works with the 'language' whose sole primitive is the single stroke '|', and whose well-formed expressions are just the strings: |, | |, | | |, | | | |, ... , for strings of any finite length. As he observes, this sequence of stroke-strings is isomorphic to the natural numbers, if we take | as 0 and the operation of adding a stroke on the right of any given string as the successor operation.

In this case, we can literally *perceive token* stroke-strings—the objects of *intuition* are stroke-string *types*. In the simplest cases, intuiting a type-string consists in seeing a token and seeing it as a token of its type. So intuition requires possession of the concept of a type, and a grasp of some notion of type-identity. Not all intuition of types is directly parasitic upon perception of tokens, in Parsons's account, as we'll see.

Intuition of type-strings can, Parsons claims, ground propositional knowledge concerning the system of type strings. For example, we can—he claims—know intuitively, on the basis of intuition *of* type-strings *that* | | | is the successor of | |. More ambitiously, he claims[2] that intuition can give us knowledge of analogues of the elementary Dedekind-Peano axioms:

$DP1_s$ | is a stroke-string

$DP2_s$ | is not the successor of any stroke-string

$DP3_s$ Every stroke-string has a successor which is also a stroke string

$DP4_s$ Different stroke-strings have different successors

Parsons takes $DP3_s$ to be equivalent to the claim that each stroke-string can be extended by one more, and regards it as 'the weakest expression of the idea that our "language" is potentially infinite' (Parsons, 1980, p. 105).

Obviously, if Parsons is right that these axioms can be known by intuition, then we can have intuitive knowledge that there are potentially infinitely many objects. But is he right?

Part of the difficulty here is that $DP2_s$—$DP4_s$, unlike $DP1_s$, are *general*. How can intuition of objects—which, though types, are still *particular*—yield knowledge of general truths? Parsons agrees that $DP2_s$—$DP4_s$ cannot be known by intuition founded on actual *perception*. But he thinks we can solve the problem by extending the notion of intuition to cover also intuition of types based upon *imagined* tokens:

[2] Parsons does not explicitly formulate the claim as stated here. I am following James Page's (1993) exposition. Parsons takes no exception to this claim in his reply to Page.

> But if we imagine any string of strokes, it is immediately apparent that a new stroke can be added. One might imagine the string as a Gestalt, present all at once: then, since it is a figure with a surrounding ground, there is space for an additional stroke ... Alternatively, we can think of the string as constructed step by step, so that the essential thing is now succession in time, and what is then evident is that at any stage one can take a further step. (Parsons, 1980, p. 106)

As Parsons himself emphasizes, the crucial thing—if the required generality is to be achieved—is that one has to imagine an *arbitrary string of strokes*. But doesn't this run straight into an analogue of Locke's problem of the abstract general triangle? Just as we have trouble imagining a triangle which is 'neither oblique nor rectangle, neither equilateral, nor equicrural, nor scalenon; but all and none of these at once' (Locke *Essay* Bk IV chapter 7, section 9), so we shall have trouble imagining a string of strokes which is neither one stroke long, nor two, nor three, ..., nor any other number, but which has, nevertheless, some definite length.

Parsons thinks we can get around the problem, in one of two ways:

> [by] imagining vaguely, that is imagining a string of strokes without imagining its internal structure clearly enough so that one is imagining a string of n strokes for some particular n, or taking as a paradigm a string (which now might be perceived rather than imagined) of a particular number of strokes, in which case one must be able to see the irrelevance of this internal structure, so that in fact it plays the same role as the vague imagining (ibid)

I don't think either suggestion works. The trouble with vaguely imagining a string extended by one stroke is that it is unclear how this is supposed to give us knowledge that *any* string can be extended by one, rather than just knowledge that *some string or other* can be so extended. The difficulty with the second suggestion is to give a satisfactory explanation of what, in this kind of case, *seeing the irrelevance* of the internal structure of the imagined or perceived string is supposed to consist in. Consider, by way of contrast, the relatively clear case where we do a geometrical proof using a particular figure, e.g. a particular right triangle in the proof of Pythagoras's Theorem—we can see that the other angles, and the lengths of the sides, is irrelevant, because we know that no appeal to them is involved in the reasoning. But there is no reasoning in the intuitive case—this is why it's so obscure how the notion that internal structure is irrelevant is to be cashed in.

I think there is a more general, and more fundamental difficulty, which concerns what is taken to be required for the *existence of type-strings*.

There are three possible positions:

(i) A type-string exists (if and) only if there exists at least one token of that type
(ii) The existence of type-strings is entirely independent of their actual or possible concrete or imaginative instantiation
(iii) A type-string exists (if and) only if there *could be* a token of that type

On option (i), there will be infinitely many type-strings only if there are infinitely many tokens. More generally, knowing that there are infinitely many quasi-concrete objects will require knowing that there are infinitely many concrete objects. This is obviously no good. We don't have any such knowledge. And anyway, knowledge of pure arithmetic should not depend upon it, even if we did.

Adopting (ii) amounts to rampant Platonism. Quite apart from any other objections which may be raised against that, Parsons couldn't buy into it without abandoning his claim that types are quasi-concrete, and so abandoning all hope of having intuitive knowledge of them.

That leaves (iii). Defending this would require making good the claim that there is at least a potential infinity of concrete objects (token-strings). The problem here is not that this isn't believable enough—it is rather to see how Parsons could *justify* the belief. His claim would be that given any perceived or imagined token-string, a single-stroke extension of it is *imaginable*. That is, it must be imaginable that there be such a token-string, even if none actually exists.

The problem now is with the modality in imagin*able*. Does it involve rigidifying on our *actual imaginative capacities*, or not? If so, we are in trouble, since our actual capacities are limited, so it will be just false that given any perceived or imagined token-string, an extension can be imagined—this fails, beyond the limit. But if we appeal instead to *possible extensions* of our imaginative capacities, we have to be able to say what extensions are admissible—and now the difficulty is to see how answer that question without appealing (in the case in hand) to some antecedent conception of the range of forms, or types, there are for token-strings to instantiate! But that means that intuition is useless as a means of getting to know facts about that.

I don't think the problems I've described here are peculiar to Parsons's particular approach—they seem likely, rather, to afflict any attempt to base a knowledge of infinity in apprehension of possibilities supposedly grounded in imagination. Since I cannot see how an object-based approach could get anywhere without appealing to some such grasp of possibilities, I think that that approach cannot work. Can a property-based approach do better? I shall now try to argue that it can.

2 Frege's Proof

Frege sketched, in *Grundlagen* §§82-3, a proof of the proposition that the number of numbers less than or equal to n immediately succeeds n. Given other propositions he has already derived from what is now generally called Hume's principle—

HP $$\forall F \forall G(NxFx = NxGx \leftrightarrow F \approx G)$$

(where '$NxFx$' and '$F \approx G$' abbreviate, respectively, 'the number of Fs' and 'the Fs and the Gs correspond one-to-one'), this suffices, if acceptable, to establish that the sequence of finite numbers is infinite.[3]

Frege's attempted proof exemplifies what I am calling the property-based approach. For Frege, numbers are fundamentally and essentially numbers belonging to concepts, so that a number exists only if there is a concept to which it belongs.

[3] Frege derives HP from his explicit definition of $NxFx$ as the extension of the concept 'equinumerous to the concept F'.

In particular, to establish, for an arbitrary finite number n, that n has an immediate successor, one must be able to exhibit a concept to which the number $n+1$ belongs. Although Frege had not, in *Grundlagen*, settled into using the term 'concept' (i.e. 'Begriff') to denote the reference, as opposed to the sense, of a predicate, I think we can reasonably interpret him as holding that numbers belong to *properties*—in the basic case, to first-level properties, so that a statement of number such as 'Nx x is a moon of Jupiter = 4' ascribes the second-level property of *having 4 instances* to the first-level property of *being a moon of Jupiter*.

I shall not go through the proof, since the issues I want to discuss don't depend upon its detail. No one now doubts that the Dedekind-Peano axioms can be derived, in second-order logic, from Hume's principle—i.e. that what George Boolos called *Frege's Theorem* is indeed a theorem. In particular, we know that the existence of successors for all finite numbers can be established, given Hume's principle, in pretty much the way Frege proposed.[4] What remains open to question is whether Frege's Theorem has the philosophical significance some of us claim for it. In particular, the idea that Frege (as good as) proved that there are infinitely many finite numbers has met with some substantial objections. I can discuss only what I take to be the most serious of them here.[5]

3 Dummett's Objections

In his Edwards *Encyclopaedia* article on Frege,[6] Michael Dummett diagnoses what he takes to be the fundamental error in Frege's theory of the natural numbers, including his attempted proof of their infinity, as follows:

> ... the assumption that certain class terms can be formed and treated as having reference is an existential assumption and can hardly be grounded on logic alone. The minimal assumption Frege needed in *Grundlagen* is that there is a mapping by which each concept F is mapped onto a class of concepts containing just those concepts G such that there are just as many Fs as Gs. When classes are taken, as Frege took them, to be objects—that is, to be in the domain over which the concepts in question are defined—this is tantamount to the assumption that there are at least denumerably many objects. From this point of view it does not much matter whether the numerical operator is taken as primitive or as defined in terms of classes: the existential assumption is the same. Now, admittedly, if numbers (or

[4] See Boolos and Heck (1998)

[5] I should mention one line of objection I'm not going to discuss in any detail. This is an old objection, going back to Poincaré, and discussed subsequently by Charles Parsons and others, to the effect that any attempt to carry out anything like the logicist programme must make use of mathematical induction, and so must be viciously circular. Here I can only say that I have always had difficulty seeing the force of this objection. It is certainly true, in particular, that Frege's proof of the successor theorem is inductive. But Frege thinks he is entitled to employ mathematical induction by his definition of the natural numbers:

$Nat(x) \leftrightarrow x = 0 \vee S^*(0, x)$

where S^* is the ancestral of the successor relation. This amounts to defining the natural numbers so that induction holds, but it does not assume the existence of any objects except 0.

[6] Dummett (1967).

> classes) are taken to be objects, then it is reasonable to assert that there are infinitely many objects (far more reasonable than Russell's "axiom of infinity' which asserts that there are infinitely many individuals—objects which are neither classes nor numbers—and is probably not even true). But then the recognition of the truth of the statement that there are infinitely many objects cannot be held to precede a grasp of the notion of number, which is required for an understanding of the domain over which the individual variables are taken as ranging.[7]

As Dummett observes, the notion of class and the identification of numbers with certain classes play no essential part in the objection he is presenting here. The argument loses none of whatever force it possesses if recast so as to eliminate talk of classes in favour of just numbers—that is, if recast as:

> ...the assumption that certain *numerical* terms can be formed and treated as having reference is an existential assumption ... The minimal assumption Frege needed ... is that there is a mapping by which each concept *F* is mapped onto *an object in such a way that concepts F and G are mapped onto the same object just in case there are just as many Fs as Gs*. When *numbers* are taken ...

Thus if the objection is good, it must tell equally against more recent attempts to re-instate Frege's programme by basing it on Hume's principle, understood as an implicit definition of the number operator.[8] I want to consider Dummett's objection as directed against this position.

There are two separable objections suggested by the passage I've quoted. The first four sentences seem to express the objection that Frege cannot proceed without an assumption that there exist infinitely many objects, and that this assumption cannot be grounded in logic alone. But the last two sentences seem rather to be claiming that Frege's procedure involves a vicious circularity, in consequence of the fact that the numbers themselves are taken—as they must be taken, if Frege's proof of infinity is to work—as belonging to the domain over which the individual quantifiers range. It is not clear whether Dummett sees himself as making two separable objections—in fact, I doubt it—but they should be separated anyway, because they raise quite different issues. I'll begin with the second—circularity—objection.

There is something very odd about this objection. Dummett concedes that the assertion that there are infinitely many objects is reasonable, if numbers are (taken to be) objects, but then complains that if they are so taken, there can be no recognition of that fact—by Frege's route—without prior grasp of the concept of number, and that prior grasp of the concept of number is required for understanding the individual quantifiers, if numbers are taken as lying within their range. But Frege would surely have *agreed* that there can be no recognition of the existence of infinitely many objects without a prior grasp of the concept of number—at least, he would certainly agree that one cannot follow his proof, and so come in that way to see that there are infinitely many objects, unless one already has the concept of (finite) number. To think that that is an objection is, on the face of it, simply to misunderstand how Frege thought we could come to know that there are infinitely many objects. There

[7] Dummett (1967 p. 236), and Dummett (1978, p. 114).

[8] See Hale and Wright (2001), and references therein to earlier work.

is no reason to suppose that Frege thought that we could come to know that *except* by seeing that there are infinitely many numbers—i.e. that he thought we could independently establish the existence of an infinity of objects of some other kind. So what is the intended objection?

There are two possibilities, between which Dummett's objection hovers indecisively. First, he may be claiming that Frege's *definition* of number (or equally, Hume's principle put forward as replacement for it) is viciously circular because one has to possess the concept of *number* in order to understand the individual quantifiers involved in the definition, once their range is taken to include the numbers themselves. Second, he may be claiming that Frege's *argument* that each finite number has a successor, and hence that there are infinitely many objects, is viciously circular, because it rests on the assumption that there are infinitely many objects. In this case, the alleged vicious circularity is *epistemological*, rather than *definitional*.

In so far as the passage I've quoted offers any support for the first claim—that Frege's definition and its neo-Fregean replacement are viciously circular—it lies in the appeal to the idea that when the embedded first-order quantifiers are understood as ranging *impredicatively* over a domain including the numbers themselves, one must already grasp the concept of *number* if one is to understand those quantifiers. However, whilst we should agree that one cannot fully understand quantifiers without know what their bound variables range over, this cannot plausibly be taken to require that for any specific kind of object lying in their range, we must be in possession of the concept of that kind of objects. Possession of the concept of *aardvark*, for example, is not a prerequisite for understanding quantification over all terrestrial animals. To be sure, it might be claimed, first, that one must have some concept covering all the objects in the range of a first-order quantifier, and second, that in the case of *number* there is no more general and inclusive concept standing to it as *terrestrial animal* stands to *aardvark*, so that one cannot understand quantification over numbers without having the concept *number*. The first part of this claim is plausible, provided it is not understood as implying that *only* restricted quantification is intelligible. If it is so understood, I think we should reject it, since I can see no compelling reason why one cannot quantify unrestrictedly over all objects whatever.[9] And if one can do so, one can understand quantifiers which range over—amongst other things—the numbers, without having the concept of number.

Dummett seems fairly clearly to think otherwise—that quantifiers must be understood to range over some definite domain, and that one cannot coherently quantify over a domain including all objects whatever. Earlier in the article, he writes:

> ...the paradoxes of set theory reveal that it is impossible coherently to interpret bound variables as ranging simultaneously over all objects which could be comprised within a domain over which bound variables could coherently be interpreted as ranging.[10]

---

[9] I will not dispute the first claim if all it means is that one needs a concept under which fall all and only the entities over which one intends to quantify. For unrestricted first-order quantification, that means one must have a general concept of *object*—but I think we do have such a concept. Something is an object just in case it falls under some sortal concept, where a sortal concept is one having an associated criterion of identity as well as what Dummett calls a criterion of application.

[10] Dummett (1967 p. 230), Dummett (1978 p. 99)

Since what the paradoxes of set theory may be taken to reveal is that there cannot be a class or totality of all objects whatever (given that any class or totality is itself an object) it seems that Dummett's thought here must be that interpreting bound variables as ranging *simultaneously over all objects* involves thinking of all objects as forming or being comprised in a class, or collection, or totality. Any objects whatever may be taken to lie within a domain of quantification, but no single domain can encompass all of them, so there can be no unrestricted quantification over all objects. But this line of thought is obviously resistable—why should quantifying absolutely unrestrictedly over all objects require taking the objects quantified over to form a class or totality?[11]

If the complaint is seen rather as charging Frege's proof with *epistemological* circularity, it still needs to be distinguished from the objection presented in the first part of the quoted passage, that Frege is relying upon a non-logical assumption. But I think that essentially the same reply may be made to both objections. As far as the alleged assumption that numerical terms have reference is concerned, the immediate reply should be that there *is no such assumption*, at least when Hume's principle is put forward as an implicit definition. To suppose that there is is to miss the point that Hume's principle is the second-order universal closure of the *biconditional* schema:

$$\mathrm{NxFx} = \mathrm{NxGx} \leftrightarrow \mathrm{F} \approx \mathrm{G}$$

so that the truth of its instances need not be taken, and should not be taken, as requiring the existence of referents for their left-hand side numerical terms. This is a perfectly general point about abstraction principles—that is, principles of the shape:

$$\forall\alpha\forall\beta[\S(\alpha) = \S(\beta) \leftrightarrow \mathrm{Eq}(\alpha,\beta)]$$

where Eq is an equivalence relation on entities of the type of α and β, and if the abstraction is good, § is a function from entities of that type into objects.

The effect of laying down such a principle as an implicit definition of the abstractive operator is to ensure that any identity of the type occurring on the left-hand side of the biconditional coincides in truth-value with the corresponding right-hand component. But the biconditionals instantiating the abstraction principle are to be so understood that their truth is consistent with their ingredient abstract terms lacking

[11] I think—though I am not at all confident about attributing this line of thought to him at the time of writing his *Encyclopaedia* article—that Dummett would claim that when bound variables are not taken as ranging over a definite domain, but understood as ranging unrestrictedly, over what he later called an indefinitely extensible domain, we are no longer able to employ classical logic, and must replace it by intuitionist logic. This would allow the intelligibility of unrestricted quantification, but only at the cost of classical logic. But anyone, including Frege, unwilling to sacrifice classical logic must—or so Dummett would claim—forswear absolutely unrestricted quantification and accept that we may only coherently quantify over definite domains. I cannot discuss this claim here, save to observe that even if it can be made good, it would not affect the cogency of Frege's argument for the infinity of the finite numbers, which does not appear to depend on any distinctively classical inferences.

reference.[12] The truth of the left-hand side identities—and hence the existence of the relevant abstract object—may only be inferred with the aid of the corresponding right-hand side statements as supplementary premises. Of course, in the particular case of Hume's principle, there are instances whose truth is a matter of logic, since the identity-map ensures that $F \approx F$, for any concept F—so that the existence of NxFx is guaranteed, for a suitable[13] concept F. But the fact that something can be *proved by means of* a principle is not in general taken to show that one is, in asserting the principle, simply *assuming* its truth.[14]

4 Dummett's Objection Refurbished, and a Related Objection from Charles Parsons

To the reply just given to Dummett's objection, it may be countered that even if there is no *explicit* assumption to the effect that terms formed by means of the number operator have reference, or that there are infinitely many objects, there is nevertheless an assumption *implicit* in the use of Hume's principle as basis for arithmetic which is tantamount to assuming an infinity of objects. For Hume's principle quantifies over properties, and these properties must—if the existence of numbers is to be an objective, mind-independent matter—be conceived as existing independently of the abstraction itself. To clarify the point: it may of course be maintained that we in some sense create or construct the concept of each of the requisite properties—in this case, properties in the sequence: *finite number* $\leq n$. But the properties themselves must be conceived as existing independently of our concept-forming activity, as 'there' all along. However, the assumption of sufficiently many such mind-independent properties—that is, at least countably infinitely many of them—is an existential assumption every bit as problematic as the assumption of the existence of infinitely many objects.

No-one, as far as I know, has pressed the existential assumption objection in quite this form. But Charles Parsons, in an essay which appeared a little before Dummett published his encyclopaedia article, makes an objection to Frege along rather closely related lines[15]:

> As a concession to Frege, I have accepted the claim of at least some higher-order predicate calculi to be purely logical systems.... The justification for not assimilating higher-order logic to set theory would have to be an ontological theory like Frege's theory of concepts as fundamentally different from objects, because "unsaturated". But even then there are dis-

[12] Obviously this presupposes that the underlying logic is free—since atomic sentences cannot be true unless their ingredient singular terms refer, some modest restrictions on the quantifier rules are needed.

[13] A suitable F must be sortal, but arguably further restrictions are required—for example, to exclude indefinitely extensible concepts, such as *ordinal* and *cardinal*, *set*, and *object* itself.

[14] For further discussion of this, and the related objection that a stipulation of Hume's principle has no advantage over a direct stipulation of the Dedekind-Peano axioms, see Hale and Wright (2009).

[15] Parsons (1965). The quotation is from section VII of the essay.

tinctions among higher-order logics which are comparable to the differences in strength of set theories. Higher-order logics have existential commitments. Consider the full second-order predicate calculus, in which we can define concepts by quantification over *all* concepts. If a formula is interpreted so that the first-order variables range over a class D of objects, then in interpreting the second-order variables we must assume a well-defined domain of concepts applying to objects in D which, if it is not literally the domain of *all* concepts over D, is comprehensive enough to be closed under quantification. Both formally and epistemologically, this presupposition is comparable to the assumption which gives rise to both the power and the difficulty of set theory, that the class of all subclasses of a given class exists. Thus it seems that even if Frege's theory of concepts is accepted, higher-order logic is more comparable to set theory than to first-order logic.

Parsons's objection, then, is that even if we take higher-order variables to range over Fregean concepts—incomplete or 'unsaturated' entities, fundamentally different from objects, and so from sets—it remains that higher-order logic involves existential commitments just as problematic as those of set theory. And if we think of Fregean concepts as properties—as being the referents of predicates rather than their senses—then the objection may be seen telling against what I am calling the property-based approach.

The first objection would be devastating, if it were true that the assumption of the existence of infinitely many properties (of the type: *finite number* $\leq n$) is as problematic as the assumption of infinitely many objects. But it seems to me that whether it is problematic depends upon how one conceives of properties. Of course, the objector is right to insist that properties must be conceived as objective and so mind-independent. But so conceiving of them does not settle whether properties are to be conceived in purely extensional terms, or in some other way. The assumption *would* be as problematic, if properties were thought of as individuated *extensionally*. For then, there would only be enough *properties* if there were enough *objects*. In general, properties P and Q would be distinct only if there were an object having one of them but not the other. Of course, for any finite number of properties, only a smaller finite number of objects is needed—but for infinitely many properties, we need infinitely many objects. For similar reasons, Parsons's complaint appears entirely justified, as directed against any view on which the values of higher-order variables are, whilst not identified with sets, taken to be individuated extensionally. So in particular, it seems a fair enough complaint against Frege, since he did conceive of his concepts in a purely extensional fashion. More generally, it appears that any property-based approach is a waste of time—or at least can enjoy no ontological or epistemological advantage over an object-based approach—if it treats properties in a purely extensional way.

However, it seems to me that one does not have to conceive of properties in a purely extensional way, and that a proponent of the neo-Fregean abstractionist approach should not do so. To put it very roughly, if one thinks of properties—the values of bound higher-order variables—as individuated purely extensionally, then however much one emphasizes the supposed ontological *differences* between properties and sets, one will not be able to get away from the fact that properties behave *just like sets* at least in important respects. And one will be very hard put to it to make out that there is really any philosophical advantage in a foundation which assumes higher-order logic—one might as well start with set theory. This was, near enough, Parsons's point. But if one thinks instead of properties as individuated *non-extensionally*, there is at least some chance of philosophical advantage. In particular, the assumption that there are infinitely many properties—say, of the Fregean form: *finite number* $\leq n$, for finite n—does not amount to the assumption that there are infinitely many objects, and may be significantly weaker, and so epistemologically less problematic—perhaps weak enough for it to form part of a foundation for arithmetic.

One central task, on this approach, is to make clearer just what the non-extensional conception of properties comes to. Rejecting purely extensional individuation of properties is consistent with different—more or less demanding—positive conditions for the identity and distinctness of properties, and indeed, for the existence of properties. Some of the choices which confront us here seem likely to have significant implications for how much mathematics can be recovered on an abstractionist approach. Here I can say only a little more about these questions.

On what is very naturally described as an *Aristotelian* conception of properties, properties exist, or have being, only through their instances—there are *no uninstantiated properties*. The first point I need to make is that the abstractionist is, fairly obviously, committed to a *non-Aristotelian* conception of properties. If we focus on the argument for the infinity of the natural numbers, it is clear that it can't get started at all unless it is accepted that there is at least one uninstantiated property—if we are to get zero as the number of Fs, for some suitable (sortal) property F, then we require a property which no objects possess. Indeed, if we are to *define* zero in this way, we surely need a property which is *necessarily* uninstantiated—so we are committed to rejecting even a weakened form of Aristotelianism, which does not require *actual* instantiation, only *possible* instantiation.

The Fregean argument for the infinity of the natural numbers appeals to properties of a specific type—for each n, we require the property of being a natural number less than or equal to n. But this does not generate any *additional* demand for uninstantiated properties—0 is by definition a natural number, so that the existence of 0 ensures that the property of being a natural number ≤ 0 is instantiated. And the existence of the natural numbers from 0 up to and including n likewise ensures that the property of being a natural number $\leq n+1$ is instantiated. Indeed, the Fregean can, and in my view should, think that the relevant properties are all *necessarily* instantiated—but this necessity derives not from any general Aristotelian requirement on properties; it derives rather from the necessary existence of the natural numbers themselves.

To explain why I think it is reasonable to adopt a non-Aristotelian conception of properties, it will help to draw attention to another distinction, between what can be called *sparse* and *abundant* conceptions of properties. Roughly speaking, on an abundant conception, given any well-formed and meaningful first-level predicate or open sentence in one free individual variable, there is a corresponding first-level property, or property of objects. This is a very undemanding conception of properties—according to it, there are not only such simple properties as being round and being white, but also various kinds of complex properties, including negative, conjunctive and disjunctive properties (such as not being white, being white and spherical, and being red or white), and relational properties (such as being less massive than Nelson's Column, being less than 1 km distant from an aardvark, etc.). In sharp contrast with this, proponents of a sparse conception wish to insist on much more stringent conditions for something to count as a genuine property—they may, for example, wish to restrict attention to properties which play some role in causal explanations of the behaviour of objects which possess them, or to properties which cannot change simply as a result of the change or loss of properties of objects other than objects which possess them (as I might lose the property of being less than 1 km away from an aardvark, even though I do not move, because all the aardvarks in my vicinity head off to the nearest waterhole or ant heap).

There may be good reasons to be interested in some sparse notion or notions of properties. But however good such reasons may be, they cannot, in my view, add up to reasons for denying that a less demanding abundant conception is possible or legitimate. Indeed, in my view, the best—and I would argue, only—way to understand a sparse conception is as a restriction of the more generous abundant conception.

One might propose what could be called a *weakly non-Aristotelian* conception, according to which a property is just a *way things might be*—so that a first-level property is a way objects might be, a second-level property a way first-level properties might be, and so on. This stands opposed to the strongly Aristotelian conception, according to which a property is a way some things *are*. Although this appears to me a perfectly intelligible conception, it is, as I've already remarked, insufficient for the abstractionist's purposes. The abstractionist should *define* 0 by reference to some property which is *necessarily* uninstantiated—it seems clearly unsatisfactory to define 0 by reference to some property which is merely *contingently* uninstantiated. And that requires a *strongly non-Aristotelian* conception which admits properties which are *ways things couldn't be*, corresponding to predicates which are necessarily unsatisfiable. But once one embraces the abundant conception, there is no good reason to deny that there are such properties. In fact, necessarily uninstantiated properties will be unavoidable, if the class of properties is, so to speak, closed under conjunction and negation. For then where P is any property, not-P and hence P-and-not-P will also be properties, and the last will be necessarily uninstantiated. But closure under negation, conjunction, etc., is guaranteed, if any well-formed and meaningful open sentence stands for a property. A property is, to put it somewhat loosely, a *way things might or might not be*—more accurately, a first-level property is a *condition which objects do or don't satisfy, either contingently or as a matter of necessity*.

To sum up: I have argued that of the two approaches to the problem of proving the existence of an infinity of objects I contrasted, the property-based approach is best placed to succeed. Specifically, I've defended Frege's attempted proof of the infinity of the natural numbers against what seems to me to be the most serious objection to it—the charge that it is in some way viciously circular. As against Dummett, I argued that a version of Frege's argument based on Hume's principle involves neither a vicious circle in *definition*, nor an *epistemological* circularity—contrary to what Dummett claims, there need be no *assumption* of the existence of a denumerable infinity of objects. I then turned to a strengthened version of Dummett's objection, which claims that the Fregean argument avoids the assumption of an infinity of *objects* only by relying instead upon the equally problematic assumption of an infinity of *properties*. I argued that this, along with Charles Parsons's closely related objection to Frege's reliance on second-order logic on the ground that involves existential assumptions comparable to those of set theory, can be answered, if we understand properties in a *non-extensional* sense and adopt a very modest conception—the *abundant* conception—of what is required for their existence.

References

Boolos, G. and Heck Jnr, R. (1998) " Die Grundlagen der Arithmetik §§83-3", in M. Schirn, ed. *Philosophy of Mathematics Today*, Oxford: Clarendon Press.

Dummett, M. (1967) "Frege, Gottlob", in P. Edwards, ed. *The Encyclopaedia of Philosophy*, New York: Macmillan; reprinted as "Frege's Philosophy", in M. Dummett (1978), *Truth and Other Enigmas*, London: Duckworth.

Hale, B. and Wright, C. (2001) *The Reason's Proper Study: Essays towards a Neo-Fregean Philosophy of Mathematics*, Oxford: Clarendon Press.

Hale, B. and Wright, C. (2002) "Benacerraf's Dilemma Revisited", *European Journal of Philosophy* **10**, 101–129.

Hale, B. and Wright, C. (2009) "Focus Restored – Comments on John MacFarlane", in O. Linnebo, ed. special issue of *Synthese* on "Bad Company" **170**, 457–482.

Page, J. (1993) "Parsons on Mathematical Intuition", *Mind* **102**, 223–232.

Parsons, Ch. (1965) "Frege's Theory of Number", in M. Black, ed. *Philosophy in America*, London: Allen & Unwin; reprinted in Ch. Parsons (1983), *Mathematics in Philosophy*, New York: Cornell University Press.

Parsons, Ch. (1980) "Mathematical Intuition", *Proceedings of the Aristotelian Society* **80**, 145–168; reprinted in W.D. Hart (1996), ed. *The Philosophy of Mathematics*, Oxford: Oxford University Press – page references are to this reprint.

Set Theory as a Foundation

Penelope Maddy

The view of set theory as a foundation for mathematics emerged early in the thinking of the originators of the theory and is now a pillar of contemporary orthodoxy. As such, it is enshrined in the opening pages of most recent textbooks; to take a few illustrative examples:

> All branches of mathematics are developed, consciously or unconsciously, in set theory. (Levy, 1979, p. 3)

> Set theory is the foundation of mathematics. All mathematical concepts are defined in terms of the primitive notions of set and membership ... From [the] axioms, all known mathematics may be derived. (Kunen, 1980, p. xi)

> [M]athematical objects (such as numbers and differentiable functions) can be defined to be certain sets. And the theorems of mathematics (such as the fundamental theorem of calculus) then can be viewed as statements about sets. Furthermore, these theorems will be provable from our axioms. Hence, our axioms provide a sufficient collection of assumptions for the development of the whole of mathematics – a remarkable fact. (Enderton, 1977, p. 10–11)

From its Cantorian beginnings through its modern flowerings, set theory has also raised problems of its own, like any other branch of mathematics, but its larger, foundational role has been and remains conspicuous and distinctive.[1]

The initial stages of this foundational project have already been sketched: The finite ordinals begin with the empty set, $\emptyset$, and continue at each stage by taking the set of previous ordinals, so that $\emptyset$ is followed by $\{\emptyset\}$, $\{\emptyset,\{\emptyset\}\}$, $\{\emptyset,\{\emptyset\},\{\emptyset,\{\emptyset\}\}\}$, and so on . (The first infinite ordinal, ω, is again the set of all smaller ordinals.) We can define an operation—'S' for successor—on these finite ordinals by $Sn = n \cup \{n\}$, a binary relation—'<' for less than—by $n < m$ iff $n \in$ m, and all the usual facts about the natural numbers, successor and less than will be provable. We can identify $\emptyset$ as 0, $\{\emptyset\}$ as 1, $\{\emptyset,\{\emptyset\}\}$ as 2, and so on, and define binary operations + and $\times$ by

[1] There are those who propose category theory as an alternative to set theoretic foundations, but at least for now, this has not changed the fact that set theory is so viewed.

G. Sommaruga (ed.), *Foundational Theories of Classical and Constructive Mathematics*, The Western Ontario Series in Philosophy of Science 76,
DOI 10.1007/978-94-007-0431-2_3,

recursion,[2] and again all the usual facts about these particular numbers and these arithmetic operations become theorems of set theory. Even mathematical induction becomes a theorem.[3]

Given these set theoretic versions of the natural numbers, integers can be identified as ordered pairs, the various operations and relations on integers defined in terms of the prior relations on natural numbers, and again all the usual theorems proved.[4] Similarly, ordered pairs of integers can serve as rational numbers or ratios.[5] The real numbers are more difficult, but here Dedekind cuts do the job.[6] From there, curves, complex numbers, algebraic structures, geometric and topological spaces, functions, functionals, functors, and all the vast menagerie of modern mathematics can be represented within set theory, and its standard theorems proved from ZFC.

This is a truly 'remarkable fact', but mathematicians and philosophers have long debated precisely what it shows. The strongest reading of Frege's project would see him as discovering or uncovering the true identity of the natural numbers; from this point of view, the current set theoretic versions of numbers, functions, spaces, etc. would show us what numbers, functions, spaces, etc. really are.[7] Benacerraf (1965) has argued that this interpretation is implausible, beginning with the set theoretic account of natural numbers, because many different identifications seem equally good. For example, von Neumann (1923) made the identification described above—0 with $\emptyset$, 1 with $\{\emptyset\}$, 2 with $\{\emptyset,\{\emptyset\}\}$, and so on—and it works perfectly well, while Zermelo, in fact, made another identification—0 with $\emptyset$, 1 with $\{\emptyset\}$, 2 with $\{\{\emptyset\}\}$, and so on—that works just as well.[8] Once this is noted, we realize that many other identifications would also work, in fact infinitely many. There may be technical reasons to prefer one to another,[9] but nothing deep enough to motivate a metaphysical argu-

---

[2] That is, $n + 0 = n$ and $n + Sm = S(n + m)$ for addition, $n \times 0 = 0$ and $n \times Sm = (n \times m) + n$ for multiplication.

[3] For details, see Enderton (1977, chapter 4). This formulation is preferable to Frege's because the system in which it is framed, ZFC, is not prone to Russell's (or any other known) paradox. In the cumulative hierarchy corresponding to ZFC, Frege's candidates for the numbers do not exist. This is because, for example, there are new three-element sets formed at every stage of the hierarchy, so there is no stage of the hierarchy at which the set of all three-element sets could be formed.

[4] The integers are the positive and negative whole numbers. The underlying idea is to let $<n,m>$, where n and m are natural numbers, represent the integer $n - m$. So e.g. the integer $<n,m>$ is less than the integer $<n',m'>$ iff $n + m'$ is less than $n' + m$ as natural numbers. See Enderton (1977, p. 90–101), for the niceties.

[5] Here the idea is to let $<a,b>$, where a and b are integers, represent the fraction a/b. For details, see Enderton (1977, p. 101–11).

[6] See Dedekind (1872). A Dedekind cut is a pair (A,B) of sets of rationals such that A and B are non-empty and disjoint, every element of A is less than every element of B, and every rational is in either A or B. So, for example, if A is the set of all negative rationals and all positive rationals whose square is less than 2, and B is the rest of the rationals, then (A,B) is the cut corresponding to $\sqrt{2}$. For more, see Enderton (1977, p. 111–120).

[7] This would be a metaphysical reading, as it purports to reveal the true nature of numbers, etc.

[8] See Zermelo (1908). On this account, other adjustments must also be made; e.g. the successor of n is $\{n\}$, not $n\cup\{n\}$.

[9] Von Neumann's version is in fact preferred because it carries over directly to the transfinite ordinals.

ment that one rather than the other uncovers the true identity of the natural numbers. And the other identifications, of integers, rationals, reals, functions, etc., all share this type of arbitrariness.[10]

In fact, one (admittedly controversial) reading of the Fregean text suggests that even Frege did not take himself to be discovering the underlying nature of the natural numbers themselves. In a much-discussed footnote to his definition of 'the number of *F*s' as 'the extension of the concept "equinumerous with *F*" ', he writes, 'I believe that for "extension of the concept" we could write simply "concept" ' (Frege, 1884, §68). He raises some potential objections to this move, remarks that he thinks they could be met, then sets them aside because they 'would take us too far afield for present purposes' (ibid.). Now if our 'present purpose' is revealing the true nature of the natural numbers, then the choice between these two answers—an extension and a concept—is central, surely not an expendable detour. Later, in his summary, Frege writes:

> In this definition the sense of 'extension of a concept' is assumed to be known. This way of getting around the difficulty [the Julius Caesar problem] cannot be expected to meet with universal approval, and many will prefer other methods of removing the doubt in question. I attach no decisive importance even to bringing in the extensions of concepts at all. (Ibid. §107)

It's clear that Frege wants very much to find some logical surrogate for the natural numbers, but perhaps he is not concerned to find a unique such surrogate. If so, he could hardly have taken his task to be the discovery of precisely which logical objects the numbers are.[11]

In any case, from this point of view, the job of set theoretic foundations is not to reveal the true identities of the various mathematical objects. Another possibility, championed by Quine,[12] is that the set theoretic versions of various mathematicalia provide an 'ontological reduction', that is, they show us that we can legitimately replace a world view that countenances both natural numbers, integers, rationals, reals, etc., on the one hand, and sets on the other, with a more streamlined world view that countenances only the sets. The motivation for such a replacement is twofold: first, the observation that natural science is generally chary of new entities,[13] and secondly, the conviction that abstracta tend to generate philosophical problems, for example problems of identity and (perhaps) epistemology.[14] Under the circumstances, a scientifically-minded philosopher would naturally prefer a more austere ontology, and the set theoretic reduction of mathematics is one way of achieving this.

But whatever the merits of ontological reduction, I think this is not what is at issue in mathematical discussions of set theoretic foundations. Consider Moschovakis,

[10] We noted earlier that Cantor and Dedekind gave distinct but equally workable accounts of the real numbers.

[11] See Benacerraf (1981) for this interpretation of Frege, and Burge (1984) for the other side of the story. Wilson (1992) comes at the issue from another angle.

[12] See Quine (1964) and (1969a).

[13] I come back to this issue in part II of my book Maddy (1997).

[14] See, for example, Quine (1948) or (1960), chapter 7. Without a set theoretic reduction, we would 'face the old abstract objects . . . in all their primeval disorder' (Quine, 1960, p. 267).

in another recent textbook; he begins by considering the relation between geometric objects and their numerical counterparts in analytic geometry:

> A typical example of the method we will adopt is the 'identification' of the ... geometric line ... with the set ... of real numbers. ... What is the precise meaning of this 'identification'? *Certainly not that points are real numbers.* ... What we mean by the 'identification' ... is that the correspondence ... gives a **faithful representation** of [the line] in [the reals] which allows us to give arithmetic definitions for all the useful geometric notions and to study the mathematical properties of [the line] **as if points were real numbers**. (Moschovakis, 1994, p. 33–34)[15]

Such 'identifications' occur on a much broader scale in set theoretic foundations:

> In the same way, we will discover within the universe of sets *faithful representations* of all the mathematical objects we need, and we will study set theory on the basis of the lean axiomatic system of Zermelo **as if all mathematical objects were sets**. The delicate problem in specific cases is to formulate precisely the correct definition of 'faithful representation' and to prove that one such exists. (Ibid. p. 34)

So the job of set theoretic foundations is to isolate the mathematically relevant features of a mathematical object and to find a set theoretic surrogate with those features.

Notice that both our earlier notions of set theoretic reduction are explicitly rejected on this account. The identification of the natural numbers with, say, the finite von Neumann ordinals is not claimed to reveal their true nature, but simply to provide a satisfactory set theoretic surrogate; thus, no problem arises from the observation that more than one satisfactory set theoretic surrogate can be found. Likewise, the identification is not understood as a prelude to the repudiation or elimination of the original mathematical objects; though they may well drift off into irrelevancy for the most part, no such strong ontological conclusion is drawn. But if neither metaphysical insight nor ontological economy is forthcoming, what is gained by the exercise?

The answer to this question lies in mathematical rather than philosophical benefits. The force of set theoretic foundations is to bring (surrogates for) all mathematical objects and (instantiations of) all mathematical structures into one arena—the universe of sets—which allows the relations and interactions between them to be clearly displayed and investigated. Furthermore, the set theoretic axioms developed in this process are so broad and fundamental that they do more than reproduce the existing mathematics; they have strong consequences for existing fields and produce a mathematical theory that is immensely fruitful in its own right. Finally, perhaps most fundamentally, this single, unified arena for mathematics provides a court of final appeal for questions of mathematical existence and proof: if you want to know if there is a mathematical object of a certain sort, you ask (ultimately) if there is a set theoretic surrogate of that sort; if you want to know if a given statement is provable or disprovable, you mean (ultimately), from the axioms of the theory of sets.

---

[15] Here a point on the line corresponds to the real number that measures the distance to that point from an arbitrarily chosen origin using an arbitrarily chosen unit of length. For plane geometry, a point corresponds to an ordered pair of numbers determined in the familiar style of Cartesian coordinates.

Moschovakis's example—the identification of geometric points with real numbers—was among the first and most dramatic examples of the power of set theoretic foundations. Chapter I.1 of Maddy (1997) shows how Cantor's work on trigonometric series eventually required a precise account of the real numbers, but the need went much deeper than this. The Greeks were first to worry over the paradoxes of continuous structures: the Pythagoreans discovered that the points on a line were too numerous to be labelled by rational numbers,[16] and Zeno's familiar paradoxes raised further worries about the coherence of our understanding of space and time. For centuries, these troubles were avoided by a deliberate separation between geometry and arithmetic, until Descartes introduced analytic geometry and Newton and Leibniz developed the calculus (in the late seventeenth century). This spectacularly successful method was nevertheless beset by unclarities and outright contradictions; many of these shortcomings were brought out with devastating *élan* by Bishop Berkeley, who ridiculed infinitesimals as 'the Ghosts of departed Quantities' (Berkeley, 1734, p. 199). Only after the work of Cauchy, Weierstrass, and others in the late nineteenth century were infinitesimals eliminated in favour of the notion of a limit, and even then the fundamental theorems about limits could not be proved because the reals themselves were not well understood.[17]

What mathematicians had, throughout this long development, was an intuitive idea of continuity, of the structure of a geometric line; this was enough to tell them that the rational points on the line left gaps, and that gaps of any kind were inconsistent with the notion that the line is continuous. What they lacked was a clear and consistent characterization of continuous structure. And this, finally, is what the set theoretic reductions of Cantor and Dedekind provided: each described a set with an ordering that contained no gaps. In this case, set theory provided a clear and compelling surrogate for what had heretofore been a troubling source of fundamental confusion and concern. These set theoretic reals are constructed out of rational numbers, which are constructed out of pairs of integers, which are constructed out of pairs of natural numbers; Frege saw himself as completing the process of foundation by providing an account of the natural numbers themselves. And, as we've seen, that account now also takes the form of a set theoretic surrogate. In this way, set theoretic foundations ended centuries of mathematical disarray.

Since then, the set theoretic point of view has allowed existence questions to be clearly posed and answered. For example, Lebesgue analysed the idea of the 'length' of a point set and produced his notion of Lebesgue measure; then the question—is there a point set with no determinate length?—could be well posed, and in the context of ZFC, it could be answered, 'Yes, there is'. Such conclusive resolutions can be found in various branches of mathematics, including algebra, topology, and analysis. Other cases turn out to be beyond the reach of our current axioms; such examples are

---

[16] e.g. by erecting an isosceles right triangle on the unit length of a line, then using a compass to lay out the length of the hypotenuse onto the line, we can generate a point that corresponds to no rational number. (If it did, i.e. if a/b is in lowest terms and $(a/b)^2 = 2$, then $a^2 = 2b^2$, so a is even. But then $a = 2c$, for some c, so $b^2 = 2c^2$, and b is also even. This contradicts the assumption that a/b is in lowest terms.)

[17] See Boyer (1949) for more.

the subject of chapter I.4 in Maddy (1997). In these cases, the set theoretic perspective at least puts a stop to doomed efforts at proof. Beyond that, further investigation often uncovers the sensitivity of such existence questions to experimental hypotheses of higher set theory, that is, to new axiom candidates, which allows still deeper interconnections to be drawn and holds out the promise of solution if and when a final consensus is reached on these hypotheses. How such a consensus might be rationally achieved is the central theme of Maddy (1997).

In sum, then, despite the lack of metaphysical or ontological payoff, I think it is fair to attribute considerable mathematical success to set theoretic foundations of the sort Moschovakis describes: mathematical objects and structures are identified with or instantiated by set theoretic surrogates and the classical theorems about them proved from the axioms of set theory. Mathematics is profoundly unified by this approach; the interconnections between its branches are highlighted; classical theorems are traced to a single source; effective methods can be transferred from one branch to another; the full power of the most basic set theoretic principles can be brought to play on heretofore unsolvable problems; new conjectures can be evaluated for feasibility of proof; and ever stronger axiomatic systems hold the promise of ever more fruitful consequences. As the desired mathematical payoffs can be achieved by this modest version of set theoretic foundations, I will assume no more than this in what follows.

So far, we've seen that the mathematical fruits of set theoretic foundations can be achieved without the various metaphysical and ontological claims that have worried many commentators. So far, so good. But we have yet to touch on the various related epistemological themes. Once again, I think the mathematical benefits can be preserved without any strong (and controversial) additional theses; to suggest why this is so, let me examine a few of the characteristic concerns.

Given that Zermelo introduced axiomatic set theory in part to avoid the nagging paradoxes, it is not surprising that set theoretic foundations are sometimes understood as the project of installing mathematics on a consistent basis. Zermelo himself, before presenting his axioms, laments:

> I have not yet even been able to prove rigorously that my axioms are consistent, though this is certainly very essential; instead I have had to confine myself to pointing out now and then that the antinomies discovered so far vanish one and all if the principles here proposed are adopted as a basis. (Zermelo, 1908, p. 200–201)

Poincaré makes the same point in the form of an objection: 'We have put a fence around the herd to protect it from the wolves but we do not know whether some wolves were not already within the fence' (quoted in Kline (1972, p. 1186)). If our axiomatization is to protect us from contradiction, we must be sure that it harbours no inconsistencies of its own.

The trouble, it later turned out, was not simply that Zermelo and others were unable to provide a consistency proof, but that a meaningful consistency proof is not possible. In his (1931), Gödel showed that any consistent theory strong enough to reproduce arithmetic—and we've seen that ZFC can do this—cannot prove its own consistency. So, if classical mathematics is identified with set theory—as set theoretic foundations would have it—then the consistency of set theory cannot be proved by any method of classical mathematics. The modern form of this objection, making explicit use of Gödel's second incompleteness theorem,[18] can be found in the writings of MacLane, a prominent contemporary critic of set theoretic foundations:

> Now in one sense a foundation is a security blanket: If you meticulously follow the rules laid down, no paradoxes or contradictions will arise. In reality there is now no guarantee of this sort of security; we have at hand no proof that the axioms ZFC for set theory will never yield a contradiction, while Gödel's second theorem tells us that such a consistency proof cannot be conducted within ZFC. (MacLane, 1986, p. 406)

Since Gödel, as MacLane notes, Zermelo's hope of establishing the consistency of his axioms has been effectively dashed.

So, if founding mathematics on set theory is supposed to provide it with a foundation in MacLane's sense of a 'security' blanket, keeping us secure from Poincaré's 'wolves', that is, from contradiction, then Gödel's second theorem presents a formidable objection.[19] Much as we might like to have a guarantee of consistency, we now know this is not forthcoming; the question is whether or not a 'foundation' can provide anything of value without this. Here I think our previous discussion points to a positive answer: the mathematical benefits provided by set theoretic foundations do not depend on set theory being provably consistent. We can admit that freedom from contradiction is an obvious desideratum for any mathematical theory, especially one destined for a role as central as the one we've proposed for set theory; we can admit that new axiom candidates should be viewed with suspicion (at least) to the extent that they seem likely to introduce contradictions; but for all this, we can still hold that the unification provided by set theoretic foundations is mathematically valuable despite the lack of a consistency proof.

Suppose, then, that we adopt a version of set theoretic foundations that does not claim to make mathematics safe from contradiction. We might still wonder about the relative certainty of set theory and the mathematical theories represented in it. Founding arithmetic, for example, on set theory is sometimes thought to base the relatively certain on the relatively uncertain, or, in Quine's phrase, 'a case of *obscurum per obscurius*'.[20] But again, this criticism only has force if our foundation aims at establishing what it founds with a higher degree of certainty. And again, the mathematical achievements of set theoretic foundations survive without this.

In fact, the idea that to found a theory is to base it on something more certain was rejected early on, by Russell himself:

> There is an apparent absurdity in proceeding, as one does in the logical theory of arithmetic, through many rather recondite propositions of symbolic logic, to the 'proof' of such truisms as 2 + 2 = 4: for it is plain that the conclusion is more certain than the premises, and the supposed proof therefore seems futile. (Russell, 1907, p. 272)

---

[18] Gödel's first incompleteness theorem says that for any sufficiently strong theory, there is a sentence it cannot prove or disprove. The second incompleteness theorem, cited here, follows from the first. See Enderton (1972) for a textbook treatment.

[19] Detlefsen (in his (1986)) defends Hilbert's programme for establishing the consistency of classical mathematics by arguing that a criticism based on Gödel's second theorem need not be viewed as conclusive. So far as I know, no one has succeeded in exploiting the loopholes Detlefsen identifies.

[20] Quine (1969a, p. 43). For this and related objections, see Steiner (1975, chapter 2).

But this absurdity is only apparent, Russell argues. When we examine various inferences we discover that

> [t]he word 'premise' has two quite different senses: there is what we may call the 'empirical premise', which is the proposition or propositions from which we actually are led to believe the proposition in question; and there is what we will call the 'logical premise', which is some logically simpler proposition or propositions from which, by a valid deduction, the proposition in question can be obtained. (Ibid. pp. 272–273)

So, for example, we might come to believe the 2 + 2 = 4 on the basis of various observations involving stones, sheep, or ginger snaps, various observations that support the proposition inductively; these are the empirical premisses from which the proposition can be inferred. On the other hand, when we prove 2+2=4 from the axioms of pure logic, these axioms are logically simpler propositions, involving fewer notions; they are logical premisses. Of course, what is simple in one context may be complex in another; the logical premiss of one argument may be proved from premisses still simpler in another argument.

The mistake philosophers make, according to Russell, is supposing that the scale of logical simplicity coincides with the scale of obviousness or certainty:

> The propositions that are easiest to apprehend are somewhere in the middle [of the logical scale], neither very simple nor very complex. Generally speaking, they become simpler as civilization advances. Thus *we* probably find it easier to think of fishing than of trout-fishing or salmon-fishing; but I am told that savages are apt to have a verb for trout-fishing and another for salmon-fishing, but no verb for fishing. (Ibid. p. 273)

So, in our example, the observations that two stones plus two stones make four stones, and that two sheep plus two sheep make four sheep, etc.—these observations are less simple than 2 + 2 = 4, and 2 + 2 = 4 is itself less simple than the axioms of pure logic.

> The proposition 2 + 2 = 4 itself strikes us now as obvious; and if we were asked to prove that 2 sheep + 2 sheep = 4 sheep, we should be inclined to deduce it from 2 + 2 = 4. But the proposition '2 sheep + 2 sheep = 4 sheep' was probably known to shepherds thousands of years before the proposition 2 + 2 = 4 was discovered; and when 2 + 2 = 4 was first discovered, it was probably inferred from the case of sheep and other concrete cases. (Ibid. p. 272)

So the highest degree of obviousness occurs toward the centre of the scale of logical simplicity, and its position can vary with time.

Now when a mathematician proves a theorem, the proposition proved is generally so complex that the premisses of the argument are both more obvious and more logically simple. 'Thus in mathematics, except in the earliest parts, the propositions from which a given proposition is deduced generally give the reason why we believe the given proposition' (i.e. logical and empirical premisses coincide: ibid. p. 273). But the cases that concern us, cases like the proof that 2 + 2 = 4, are different:

> in dealing with the principles of mathematics, this relation is reversed. Our propositions [e.g. the principles of pure logic] are too simple to be easy, and thus their consequences [like 2 + 2 = 4] are generally easier than they are. Hence we tend to believe the premises because we can see that their consequences are true, instead of believing the consequences because we know the premises to be true. (Ibid. pp. 273–274)

In other words, the more certain is derived from the less certain, thus boosting our confidence in the latter.

Similar approaches to axiomatics appear in the work of Zermelo, Gödel, and many of their successors, as we will see, but they are not without their critics. For example, Tiles argues that to defend an axiom in terms of its consequences is inappropriate

> to the conception of set theory as providing a logical *foundation* for mathematics. To claim this status for set theory it is necessary to claim an independent and intrinsic justification for the assertion of set-theoretic axioms. It would be circular indeed to justify the logical foundations by appeal to their logical consequences, i.e. by appeal to the propositions for which they are going to provide the foundation. (Tiles, 1989, p. 208)

But, as we've seen, this style of objection only applies to a foundation in the epistemic sense, one intended to provide a 'secure given starting point' (ibid.). Set theoretic foundations need not be so intended, and their mathematical benefits remain even if such epistemic benefits are foresworn.

This, of course, is what Russell does; he explicitly renounces the epistemic goal of founding mathematics on something more certain than the statements of mid-level mathematics. Having done so, he must see some other goal as sufficiently attractive to motivate the project, and he does:

> The advantage of obtaining simple logical premises in place of empirical premises is partly that it gives a greater chance of isolating a possible pervading element of falsehood, partly that it organises our knowledge, and partly that the logical premises have, as a rule, many more consequences than the empirical premises, and thus lead to the discovery of many things which could not otherwise be known. The law of gravitation, for example, leads to many consequences which could not be discovered merely from the apparent motions of the heavenly bodies, which are our empirical premises. And so in arithmetic, taking the ordinary propositions of arithmetic as our empirical premises, we are led to a set of logical premises from which we can deduce Cantor's theory of the transfinite. (Russell, 1997, p. 275)

Of course, Russell's claim to have deduced Cantor's theory from pure logic is subject to dispute—the axioms of Infinity, Choice, and Reducibility used in *Principia Mathematica* do not seem, on the face of them, to be logical—but the general tone of this passage agrees with our assessment: the true benefits of set theoretic foundations are not epistemic but mathematical (e.g. uncovering inconsistencies, organizing knowledge, leading to theories of greater power and fruitfulness).

Before leaving the subject of set theoretic foundations, I would like to consider one last, generally epistemological objection, an objection less precise than those of Poincaré or Tiles, but one that I think is deeply felt by many of the mathematicians who count themselves as strong opponents. The most explicit statement of this objection that I've been able to find occurs in a debate between MacLane, as the opponent, and Mathias, as the defender.[21] In this passage, Mathias describes what he takes to be the concern underlying MacLane's unease:

> Set theory is so rich a theory that it has been claimed for much of this century to be the foundation of mathematics. In ontological terms this claim is not unreasonable; but MacLane

[21] See MacLane (1992) and Mathias (1992).

> resists. I would guess that his reason is not so much that he objects to the ontology of set theory but that he finds the set-theoretic cast of mind oppressive and feels that other modes of thought are more appropriate to the mathematics he wishes to do. (Mathias, 1992, p. 115)

I think it cannot be denied that mathematicians from various branches of the subject—algebraists, analysts, number theorists, geometers—have different characteristic modes of thought, and that the subject would be crippled if this variety were somehow curtailed. Mathias appreciates this:

> One of the remarkable things about mathematics is that I can formulate a problem, be unable to solve it, pass it to you; you solve it; and then I can make use of your solution. There is a unity here: we benefit from each other's efforts. ... But if I pause to ask *why* you have succeeded where I have failed to solve a problem, I find myself faced with the baffling fact that you have thought of the problem in a very different way from me: and if I look around the whole spectrum of mathematical activity the huge variety of styles of thought becomes even more evident.
>
> Is it desirable to press mathematicians all to think in the same way? I say not.... Uniformity is not desirable and an attempt to attain it, by (say) manipulating the funding agencies, will have unhealthy consequences. (Ibid. p. 113)

If set theoretic foundations were understood to entail that all mathematics is set theory, in the sense that all mathematicians might as well be set theorists, restricting themselves to the methods characteristic of set theory, then it would surely represent an unhealthy push towards uniformity. This, Mathias suggests, is (part of) what's bothering MacLane, and we should admit, along with Mathias, that this would be something worth denouncing.

But set theoretic foundations need not be so understood. For the sake of comparison, consider the claim that everything studied in natural science is physical; it doesn't follow from this that botanists, geologists, and astronomers should all become physicists, should all restrict their methods to those characteristic of physics. Again, to say that all objects of mathematical study have set theoretic surrogates is not to say that they should all be studied using only set theoretic methods. Mathias makes this same point in reply to MacLane: 'The purpose of foundational work in mathematics is to promote the unity [as opposed to the uniformity] of mathematics; the larger hope is to establish an ontology within which all can work in their different ways' (ibid. p. 114). So, once again, set theoretic foundations can do their job without insisting that all legitimate methods are included in the usual methods of set theory.

In what follows, then, I will assume that set theory provides a foundation for mathematics in the modest sense delimited here: for all mathematical objects and structures, there are set theoretic surrogates and instantiations, and the set theoretic versions of all classical mathematical theorems can be proved from the standard axioms for the theory of sets (ZFC).[22] This includes no claim about the real identity of mathematical objects, no claim to have reduced ontology, no claim to have founded mathematics on something provably free from contradiction or more certain, and

[22] Even MacLane allows this much: 'The rich multiplicity of mathematical objects and the proofs of theorems about them can be set out formally with absolute precision on a remarkably parsimonious base' (1986, p. 358). He is referring to ZFC.

no claim that all mathematical methods can be replaced by set theoretic methods.[23] For all that, set theoretic foundations still play a strong unifying role: vague structures are made more precise, old theorems are given new proofs and unified with other theorems that previously seemed quite distinct, similar hypotheses are traced at the basis of disparate mathematical fields,[24] existence questions are given explicit meaning, unprovable conjectures can be identified, new hypotheses can settle old open questions, and so on. That set theory plays this role is central to modern mathematics, that it is able to play this role is perhaps the most remarkable outcome of the search for foundations. No metaphysics, ontology, or epistemology is needed to sweeten this pot!

References

Benacerraf, P. (1965) "What numbers could not be", reprinted in P. Benacerraf and H. Putnam, eds. *Philosophy of Mathematics*, Second edition, Cambridge: Cambridge University Press, 1983.

Benacerraf, P. (1981) "The last logicist", *Midwest Studies in Philosophy*, **6**, 17–35, repr. in Demopoulos (1995), 41–67.

Berkeley, G. (1734) The Analyst, reprinted in *'De Motu' and 'The Analyst'*, edited and translated by D. Jesseph, Dordrecht: Kluwer, 1992, 157–221.

Boyer, C. (1949) *The History of the Calculus and its Conceptual Development*, New York: Dover, 1959.

Burge,T. (1984) "Frege on extensions of concepts, from 1884 to 1903", *Philosophical Review*, **93**, 3–34.

Dedekind, R. (1872) "Continuity and irrational numbers", reprinted in Dedekind (1901), 1–27.

Dedekind, R. (1901) *Essays on the Theory of Numbers*, translated W.W. Beman, La Salle, Ill: Open Court, 1948.

Detlefsen, M. (1986) *Hilbert's Program*, Dordrecht: Reidel.

Demopoulos, W., ed. (1995) *Frege's Philosophy of Mathematics*, Cambridge MA: Harvard University Press.

Enderton, H. (1972) *A Mathematical Introduction to Logic*, New York: Academic Press.

Enderton, H. (1977) *Elements of Set Theory*, New York: Academic Press.

Kline, M. (1972) *Mathematical Thought from Ancient to Modern Times*, New York: Oxford University Press.

Kunen, K. (1980) *Set Theory: An Introduction to Independence Proofs*, Amsterdam: North Holland.

Levy, A. (1979) *Basic Set Theory*, Berlin: Springer.

MacLane, S. (1986) *Mathematics: Form and Function*, New York: Springer.

MacLane, S. (1992) "Is Mathias an ontologist?", in H. Judah, W. Just, and H. Woodin, eds., *Set Theory of the Continuum*, Berlin: Springer, 119–122.

Maddy, P. (1997) *Naturalism in Mathematics*, Oxford: Oxford University Press.

Mathias, A. R.D. (1992) "What is MacLane missing?" in H. Judah, W. Just, and H. Woodin, eds., *Set Theory of the Continuum*, Berlin: Springer, 113–118.

Moore, G. H. (1982) *Zermelo's Axiom of Choice*, New York: Springer.

[23] It needn't even include the claim that set theory is the only theory that could serve as this sort of foundation.

[24] A striking example: forms of the Axiom of Choice turn up in the fundamental assumptions of algebra, topology, analysis, and logic, as well as set theory. See Moore (1982) for a description of how these interconnections were discovered, and their impact on the various fields.

Moschovakis, Y. (1994) *Notes on Set Theory*, New York: Springer.
Quine, W.V. (1948) "On what there is", reprinted in Quine (1980), 1–19.
Quine, W.V. (1960) *Word and Object*, Cambridge, MA: MIT Press.
Quine, W.V. (1964) "Ontological reduction and the world of numbers", reprinted in Quine (1976), 212–120.
Quine, W.V. (1969a) "Ontological relativity", reprinted in Quine (1969b), 26–68.
Quine, W.V. (1969b) *Ontological Relativity and Other Essays*, New York: Columbia University Press.
Quine, W.V. (1976) *The Ways of Paradox and Other Essays*, revised edition, Cambridge MA: Harvard University Press.
Quine, W.V. (1980) *From a Logical Point of View*, Second edition, Cambridge MA: Harvard University Press.
Russell, B. (1907) "The regressive method of discovering the premises of mathematics", reprinted in Russell, *Essays in Analysis*, edited by D. Lackey, New York: George Braziller, 272–283.
Steiner, M. (1975) *Mathematical Knowledge*, Ithaca, NY: Cornell University Press.
Tiles, M. (1989) *The Philosophy of Set Theory*, Oxford: Basil Blackwell.
Van Heijenoort, J., ed. (1967) *From Frege to Gödel*, Cambridge, MA: Harvard University Press.
Von Neumann, J. (1923) "On the introduction of transfinite numbers", reprinted in van Heijenoort (1967), 346–354.
Wilson, M. (1992) "Frege: the royal road from geometry", *Nous*, **26**, 149–180, reprinted (with postscript) in Demopoulos (1995), 108–159.
Zermelo, E. (1908) "A new proof of the possibility of a well-ordering", reprinted in van Heijenoort (1967), 183–198.

Foundations: Structures, Sets, and Categories

Stewart Shapiro

Recent years have seen a wealth of discussion on the topic of the foundations of mathematics, and the extent to which category theory, set theory, or some other framework serves, or can serve, as *a* foundation, or *the* foundation of some, most, or all of mathematics. Of course, adjudications of these matters depend on what, exactly, a foundation is, and what it is for, and it depends on what mathematics is. It is like a game of Jeopardy. We are given some answers: set theory, category theory, abstraction principles, etc., and we have to figure out what the questions are. Most of the participants in this debate are at least fairly clear about what *their* questions are, but it seems that the participants do not have the same questions in mind. And some of the questions have disputable presuppositions concerning the nature of mathematics. My purpose here is to survey some of the terrain. The goal is to clarify the discussion, and perhaps to advance parts of it, without plumping for one or the other view.

1 Ontology, Maybe Even Metaphysics

In one sense, the purpose of a foundation of mathematics is to describe, or otherwise provide for, the *objects* of mathematics. From this perspective, an advocate of set-theoretic foundations would claim, or propose, that the objects of mathematics—perhaps all of mathematics—are sets. An advocate of categorical foundations, in this sense of "foundations", would hold that the objects of mathematics are categorical objects and/or arrows, perhaps in a single topos; an advocate of Scottish neo-logicism would hold that the objects of mathematics are introduced via certain abstraction principles; an ante rem structuralist would claim that the objects of mathematics are places in structures; and similarly for any other ontological foundational program presented.

Presumably, an advocate of any of these ontological foundations would agree that there are mathematical objects. A nominalist, who denies that mathematics has

G. Sommaruga (ed.), *Foundational Theories of Classical and Constructive Mathematics*, The Western Ontario Series in Philosophy of Science 76,
DOI 10.1007/978-94-007-0431-2_4,

a distinctive subject matter, would find the whole project of describing the ontology of mathematics wrong-headed. On that view, there is nothing to describe. Presumably, the rejection of mathematical ontology would apply to the proposed foundation itself, assuming that the foundation is itself a mathematical theory. The nominalist would provide a (fictional, modal, etc.) interpretation or reinterpretation of that very theory. So we can excuse nominalists from this part of the discussion. Perhaps a nominalist would claim that if mathematics had a subject matter, it would be ..., but it is hard to see the point of that.

There are different senses to the ontological question. A *metaphysical* ontological foundationalist would claim that the proposed foundation provides *the* (ultimate) subject matter of mathematics, *the* underlying metaphysical nature of mathematics. The foundation tells us what mathematics is *about*, appearances to the contrary notwithstanding. According to this view, natural numbers, complex numbers, analytic functions, quaternions, points in Euclidean space, really are sets, or arrows, or places in structures, or abstracts, etc., depending on what the proposed foundation is.

One version of this view, in line with traditional metaphysics, is that the foundation says what the essence of mathematical objects has always been, or at least what has become the essence of mathematical objects from a certain point in history. The thesis is thus a *descriptive* one. The metaphysician takes himself to be delimiting the underlying fabric of reality. To the extent that mathematics is part of this reality, it must be included, and the foundation describes it.

I confess some perplexity with theses like this one. How does the philosopher get at the underlying, deep structure of mathematical reality? How do we determine the extent to which mathematics is part of the ultimate fabric of the universe? What would count as an argument in favor of, say, set theory as against category theory or Scottish neo-logicism? We can look at what mathematicians say and do, of course, but advocates of all of the foundational projects do that. Few of them accuse mathematicians of making mistakes. The goal of the foundational enterprise, from this ontological perspective, is to interpret the practice, in metaphysical terms, trying to figure out somehow, what it is all about. To paraphrase Carnap, what are the rules for this metaphysical game?

Instead of this descriptive, metaphysical orientation, an ontological foundationalist might adopt the holistic spirit of the ship of Neurath. She makes a *proposal* that we take the subject matter of mathematics to be sets, arrows, abstracts, etc. The thesis is that this will make for a tighter ship. The claim is that the mathematical/scientific enterprise will function better, on some specified grounds for what counts as "better", if we unify the ontology of mathematics in the way indicated. One stated criterion for this enterprise is simplicity and uniformity of ontology. Our ontologist might follow Quine and ask why we need to have sets, points, numbers, functions, etc. when sets alone will do. And they might further follow Quine and ask why we need non-constructible sets, thus making a proposal to set theorists themselves.

A third orientation to ontological foundations is a claim that the proposed foundation provides direct *mathematical* benefits. Consider this passage from a chapter of Yiannis Moschovakis (1994) text, entitled "Are sets all there is":

> [Consider] the "identification" of the ... geometric line ... with the set ... of real numbers, via the correspondence which "identifies" each point ... with its coordinate ... What is the precise meaning of this "identification"? *Certainly not that points are real numbers.* Men have always had direct geometric intuitions about points which have nothing to do with their coordinates ... What we mean by the "identification" ... is that the correspondence ... gives a **faithful representation** of [the line] in [the real numbers] which allows us to give arithmetic definitions for all the useful geometric notions and to study the mathematical properties of [the line] **as if points were real numbers** ... [W]e ... discover within the universe of sets *faithful representations* of all the mathematical objects we need, and we will study set theory ... **as if all mathematical objects were sets**. The delicate problem in specific cases is to formulate precisely the correct definition of "faithful representation" and to prove that one such exists. (Moschovakis, 1994, pp. 33–34)

The idea, then, is that the proposed ontological foundation provides *surrogates* for some or perhaps all mathematical objects. This is accomplished by isolating "the mathematically relevant features of a mathematical object", as Penelope Maddy (1997, p. 26) puts it. We learn things about the originals by studying the surrogates, and, to follow Moschovakis, we do not care much about whether the original mathematical objects or structures are identical with the proposed surrogates.

This ontological, *mathematical* perspective comes with no metaphysical claims and it comes with no claims concerning how the ship of Neurath should be reorganized. In what I take to be a similar spirit, Mark Wilson (1981) once remarked that "any notion that the reals should not be identified with sets represents as great a misunderstanding of mathematical ontology as the claim that they should." It is the surrogate-relation that is doing the mathematical work.

Maddy nicely articulates some mathematical benefits of this sort of ontological foundation:

> The force of set-theoretic foundations is to bring (surrogates for) all mathematical objects and (instantiations of) all mathematical structures into one arena—the universe of sets—which allows the relations and interactions between them to be clearly displayed and investigated ... perhaps most fundamentally, this single, unified arena for mathematics provides a court of final appeal for questions of mathematical existence ... if you want to know if there is a mathematical object of a certain sort, you ask (ultimately) if there is a set-theoretic surrogate of that sort.
>
> ... vague structures are made more precise, old theorems are given new proofs and unified with other theorems that previously seemed distinct, similar hypotheses are traced at the basis of disparate mathematical fields, existence questions are given explicit meaning, unprovable conjectures can be identified, new hypotheses can settle old open questions, and so on. (Maddy, 1997, p. 26, 34f)

To continue to wax Carnapian, note that questions of the existence of certain mathematical objects are not limited to philosophers interested in metaphysics and ontology or even to naturalistic philosophers concerned with tidying up the ship of Neurath. Sometimes mathematicians ask questions about the existence of a type of mathematical object "internally", so to speak—although we need not take a stance on the relationship between those questions and the more philosophical ones. To return to a matter I have invoked elsewhere, Wilson (1993, pp. 208–209) illustrates the historical development and acceptance of a space-time with an "affine" structure on the temporal slices:

> ... the acceptance of ... non-traditional structures poses a delicate problem for philosophy of mathematics, viz., how can the novel structures be brought under the umbrella of *safe* mathematics? Certainly, we rightly feel, after sufficient doodles have been deposited on coffee shop napkins, that we understand the intended structure ... We would hope that "any coherent structure we can dream up is worthy of mathematical study ... " The rub comes when we try to determine whether a proposed structure is "coherent" or not. Raw "intuition" cannot always be trusted; even the great Riemann accepted structures as coherent that later turned out to be impossible. *Existence principles* beyond "it seems okay to me" are needed to decide whether a proposed novel structure is genuinely coherent ... [L]ate nineteenth century mathematicians recognized that ... existence principles ... need to piggyback eventually upon some accepted range of more traditional mathematical structure, such as the ontological frames of arithmetic or Euclidean geometry. In ... our century, set theory has become the canonical backdrop to which questions of structural existence are referred.

This nicely illustrates, or at least nicely dovetails with, one of key mathematical benefits that, according to Maddy, comes from the set-theoretic foundation. For mathematicians of our time—if not philosophers—a set-theoretic proof of satisfiability resolves questions of existence, as they arise.

Advocates of other ontological foundations can, of course, make similar claims on behalf of their own systems. Colin McLarty (2004) follows Saunders Mac Lane in arguing that a set theory formulated in categorical terms, such as ETCS or CCAF, will work even better as a canonical backdrop for mathematical ontology than the more standard Zermelo-Fraenkel set theory.

At this point, the philosopher might ask what it is about mathematics that allows the mathematician to get so much mileage out of *surrogates*. Why, after all, can we learn so much about a branch of mathematics simply by studying surrogates for the objects studied by that branch? And why does the existence of a faithful surrogate in the iterative hierarchy resolve questions of existence? Prima facie, this does not happen as often in other scientific and everyday inquiries. After all, how much can we learn about, say, the human eye, by studying a surrogate about it (whatever that would be). How can we establish the existence of a psychological mechanism by finding a surrogate for it? I have addressed these questions elsewhere (Shapiro, 2004), taking the phenomenon to be evidence in favor of structuralism, and so will not go any further down that line here.

Although Steve Awodey (2004) opposes, or at least downplays, ontological foundations, he provides a nice summary of it. The ontological perspective, of whatever stripe, is a "bottom-up" approach that tries to "build up specific 'mathematical objects' within a particular 'foundational system' in such a way that

1. there are enough such objects to represent the various kinds of numbers, as well as spaces, groups, manifolds, etc. of everyday mathematics, and
2. there are enough laws, rules, and axioms to warrant all of the usual inferences and arguments made in mathematics about these things ..." (Awodey, 2004, pp. 55–56)

2 Epistemology: What We Know and How We (Can) Know

A second, but rather different, sense of "foundation" concerns the epistemic basis, or the *justification*, of mathematical propositions. The idea is that the grounds for true, or at least knowable, mathematical propositions are to be found in the proposed foundation.

One strong version of this perspective is a claim that no one knows any mathematical propositions until she has derived them from the proposed foundation. Our set-theorist would claim that no one knew that, say, every natural number has a successor until that proposition was rendered in the language of set theory and derived from the axioms thereof. And similarly for the other foundations: ante rem structuralism, category theory, Scottish neo-logicism, etc.

This strong take strikes me as absurd—so absurd that I presume that no one held such a view. Surely, the mathematical and scientific community had considerable mathematical knowledge long before the foundational work began. And even a contemporary mathematician who is hostile to, oblivious of, or just disinterested in foundational studies, continues to have considerable mathematical knowledge.

Articulating a more sensible, epistemological sense of foundations is not exactly straightforward. The perspective has a natural home in Gottlob Frege's logicism since that was, at least in a large part, an epistemological project. Robin Jeshion (2001, pp. 939–940) summarizes various aims that have been attributed to Frege's project by different scholars. It will prove useful to consider these epistemological goals with respect to other proposed foundations:

> **Mathematical Rationale:** Frege's motives are mathematical. He desired to prove some theorems. He believed that whatever admits of proof ought to be proved. And he thought that the propositions of arithmetic, hitherto unproved, admit of proof. So they should be proved.
>
> **Logico-Cartesian Rationale:** Frege is a reformer who aims to perfect arithmetical knowledge. He thought that the logical source alone is capable of producing knowledge possessing absolute self-evidence, certainty, and clarity. He also thought that actual arithmetical knowledge is marred by doubt, uncertainty, and unclarity. He aimed to establish logicism because he thought doing so was necessary for demonstrating the epistemological superiority of arithmetical knowledge.
>
> **Knowledge-of-Sources Rationale**. Frege desired to discern the philosophical status of our arithmetical knowledge i.e., to determine whether arithmetic is analytic or synthetic, a priori or a posteriori. He aimed to establish logicism because he thought that proving the propositions of arithmetic was necessary for determining the epistemological source of our arithmetical knowledge.

Frege (1844, §2) wrote that "it is in the nature of mathematics to prefer proof, where proof is possible", noting that "Euclid gives proofs of many things which anyone would concede him without question". He then tells us why it is that mathematicians "prefer proof, where proof is possible":

> The aim of proof is, in fact, not merely to place the truth of the proposition beyond all doubt, but also to afford us insight into the dependence of truths upon one another. After we have convinced ourselves that a boulder is unmoveable, ... there remains the further question, what is it that supports it so securely?

So from the perspective of the first, epistemological-mathematical rationale, our set theorist might note that in arithmetic, the principle of mathematical induction is usually taken as an axiom, and thus not proved (except in the trivial sense of a one-line proof). In contrast, the set-theoretic "translation" of the principle of arithmetic induction, via von Neumann ordinals, is a theorem. We have thus accomplished something here. It is this set-theoretic derivation that "supports" induction "so securely". The Scottish neo-logicism might make an analogous claim that we can prove the selfsame induction principle from Hume's principle, and similarly for the other foundations.

Maddy (1997, pp. 26) goes a bit further than this, suggesting that the set-theoretic foundation helps adjudicate questions in practice concerning *what counts as a proof*:

> [The] single, unified arena for mathematics provides a court of final appeal for questions of mathematical ... proof: ... if you want to know if a given statement is provable or disprovable, you mean (ultimately), from the axioms of the theory of sets.

Presumably, a similar claim can be made on behalf of the other foundations. This rationale dovetails a bit with the mathematical version of the ontological rationale broached in the previous section. In proving mathematical results from the foundation, the mathematician also uncovers "the relations and interactions between" various structures, as Maddy put it. This, it would seem, helps to show us why the boulder cannot be moved.

As with the ontological issues above, there is a nagging problem of *identifying* the mathematics that is supposedly proved from the foundation. To claim that the set-theoretic derivation in question is, in fact, a proof of the induction principle of arithmetic, the theorist must hold that there is some crucial content in common between the set-theoretic theorem and the original induction principle it "translates". We saw above that the set-theorist is not looking to identify mathematical objects, such as the natural numbers, but only to find surrogates of them. So how can we prove something about the natural numbers by proving something about their surrogates?

Analogous questions apply to the other foundational programs. The Scottish neo-logicist must show that the conclusion of the relevant derivation from Hume's principle is, in fact, the induction principle *of arithmetic*, and not some nice-looking imposter. And, if the category theorist adopts this version of epistemological foundations, she must also indicate why the deliverances of the foundation are, in fact, proofs of the original theorems. Again, I will not pursue these questions here, noting that the plausibility of the claims shows something about the nature of mathematics (again, see Shapiro (2004)).

From the present orientation, it seems, the different foundations are not rivals to each other. Consider, for example, the set-theoretic and the neo-logicist proofs of the induction principle. It is commonplace that the very same proposition can have several proofs. In fact, much mathematical activity consists of finding new proofs of established theorems. So can't we hold that *both* foundations provide proofs of the induction principle? An advocate of one of them need not challenge the legitimacy of the other.

Let us turn to what Jeshion calls the Logico-Cartesian Rationale. As the connection with Descartes suggests, the idea here is that Frege thought that, at the time, arithmetical knowledge was not as secure as one would like. Advocates of this reading have Frege holding that arithmetical knowledge is marred by doubt and unclarity. He was out to show how arithmetic has (or can have) the self-evidence and certainty it deserves. It is an instance of the rationalist quest to put everything in terms of clear and distinct ideas.

As the saying goes, that was then, this is now. It seems to me that this Logico-Cartesian Rationale is a non-starter on the set-theoretic foundation. Can anyone claim that the axioms of Zermelo-Fraenkel set theory are *more* clear and distinct, more self-evident, or in any way more secure than the Dedekind-Peano postulates in arithmetic? Arithmetic commits one to a denumerably infinite ontology. By contrast, the ontological commitments of set theory are staggering. Suppose that someone does harbor doubts about the Dedekind-Peano postulates, or the other basic propositions of arithmetic. How can these doubts be allayed by showing how to derive the axioms of arithmetic from the axioms of set theory?

To a lesser extent, the same goes for the Scottish neo-logicist program. Can someone claim that deriving the Dedekind-Peano postulates from Hume's principle removes any uncertainty or brings more clarity and distinctness to arithmetic? The discussion surrounding the Bad Company issue indicates that there are doubts about the acceptance of neo-logicist abstraction principles. Can the same be said for arithmetic itself? Of course, some do doubt arithmetic. There are skeptics, strict finitists and nominalists among us. But will the neo-logicist derivations help them? Notice, moreover, that Hume's principle is strictly *stronger* than ordinary second-order arithmetic. Second-order arithmetic is actually equivalent to a version of Hume's principle restricted to finite concepts.

Even if it is possible for category theory to play this logico-Cartesian role, with respect to arithmetic, analysis, and the like, this is certainly not the motivation of its founders or its contemporary advocates. They do not share the premise that there is something uncertain or doubtful or otherwise epistemically lacking with arithmetic, analysis, etc. So let us move on.

The third of Jeshion's rationales is what she calls "Knowledge-of-Sources". Although she lists it as a rival interpretation, it is actually close to the one she adopts herself. Frege believed that propositions have objective dependence relations to one another. It is not a matter of how some person or other comes to discover or believe a given proposition, or even of how some person or other comes to know the proposition. Rather, it is a matter of what *grounds* this truth. Frege (1884, §3) writes:

> ... we are concerned here not with the way in which [the laws of number] are discovered but with the kind of ground on which their proof rests; or in Leibniz's words, "the question here is not one of the history of our discoveries, which is different in different men, but of the connection and natural order of truths, which is always the same" (Frege (1884, §17), Leibniz, *Nouveaux Essais*, IV, §9)

Frege's accounts of both the analytic-synthetic and the a priori-a posteriori distinctions are formulated in terms of these dependency relations:

> [T]hese distinctions between a priori and a posteriori, synthetic and analytic, concern, as I see it, not the content of the judgement but the justification for making the judgement ... When ... a proposition is called a posteriori or analytic in my sense, this is not a judgement about the conditions, psychological, physiological, and physical, which have made it possible to form the content of the proposition in our consciousness; nor is it a judgement about the way in which some other man has come ... to believe it true; rather it is a judgement about the ultimate ground upon which rests the justification for holding it to be true.
>
> ... The problem becomes ... that of finding the proof of the proposition, and of following it up right back to the primitive truths. If, in carrying out this process, we come only on general logical laws and on definitions, then the truth is an analytic one ... If, however, it is impossible to give the proof without making use of truths which are not of a general logical nature, but belong to the sphere of some general science, then the proposition is a synthetic one. For a truth to be a posteriori, it must be impossible to construct a proof of it without including an appeal to facts, i.e., to truths which cannot be proved and are not general ... But if, on the contrary, its proof can be derived exclusively from general laws, which themselves neither need nor admit of proof, then the truth is a priori. (Frege, 1884, § 3)

So from this perspective, the point of deriving axioms of arithmetic from the foundation is to determine the epistemic pedigree of the propositions—whether they are analytic or synthetic, a priori or a posteriori. If we see that the Dedekind-Peano axioms can be derived from general logical laws and definitions, then we see that the axioms and theorems of arithmetic are analytic. This was Frege's goal, in opposition to Kant's thesis that arithmetic is synthetic a priori.

It seems to me that despite Frege's use of terms like "proof" and "justification", his relation of dependence is as much metaphysical as it is epistemic. As he notes himself, the relation in question has nothing to do with how people come to believe propositions. Indeed, for most of us, the belief in the axioms of arithmetic did not go via the proposed founding definitions. Frege's framework thus seems to require a distinction between the state of knowing a proposition, or even the state of being justified in believing a proposition, and the *ultimate* or objective *ground* or justification of the proposition. The foundational framework concerns the latter.

So the present orientation has much in common with the ontological, or metaphysical perspective of the previous section. Here, we are not directly querying the metaphysical nature of the *objects* of arithmetic—the natural numbers—but we are pursuing the metaphysical nature of arithmetic *propositions*.

To stay in the realm of epistemology, let us call knowledge that is based on objective grounding relations among the known propositions *proper foundational knowledge*. On Frege's perspective, when we have proper foundational knowledge, and are aware that we do, then we know *why* the boulder cannot be moved. In the case of arithmetic, according to Frege, this would be because the propositions are analytic.

This perspective does not make sense of the set-theoretic foundation. We can, of course, derive the "translations" of the Dedekind-Peano axioms from the axioms of set theory. I suppose one might hold that this gives us the proper foundational knowledge of arithmetic. But we cannot go from there to the proper metaphysical-cum-epistemological status of the propositions of arithmetic until we know the status of set theory.

The present metaphysical-epistemological perspective is also not pursued by advocates of categorical foundations. Their interests do not lie in proper foundational knowledge, nor in the epistemic pedigree of mathematical propositions. The goals of advocates of category theory are more directly mathematical.

In contrast, Scottish neo-logicism is an epistemological program, and its stated goal is something in the neighborhood of the present epistemological foundations. According to its advocates, the derivation of the Dedekind-Peano axioms from Hume's principle shows arithmetic to be knowable a priori. It shows that the basic theorems of arithmetic are "analytic of" the concept of *natural number*. As Crispin Wright (1997, pp. 210–211) puts it:

> Frege's theorem will ... ensure ... that the fundamental laws of arithmetic can be derived within a system of second-order logic augmented by a principle whose role is to *explain*, if not exactly to define, the general notion of identity of cardinal number, and that this explanation proceeds in terms of a notion which can be defined in terms of second-order logic. If such an explanatory principle ... can be regarded as *analytic*, then that should suffice ... to demonstrate the analyticity of arithmetic. Even if that term is found troubling, ... it will remain that Hume's Principle—like any principle serving implicitly to define a certain concept—will be available without significant epistemological presupposition ... So one clear a priori route into a recognition of the truth of ... the fundamental laws of arithmetic ... will have been made out. And if in addition [Hume's Principle] may be viewed as a *complete* explanation—as showing how the concept of cardinal number may be fully understood on a purely logical basis—then arithmetic will have been shown up by Hume's Principle ... as transcending logic only to the extent that it makes use of a *logical* abstraction principle—one [that] deploys only logical notions. So, ... there will be an a priori route from a mastery of second-order logic to a full understanding and grasp of the truth of the fundamental laws of arithmetic. Such an epistemological route ... would be an outcome still worth describing as logicism ...

There is an intense literature over how successful the program is, on its own goals. I will refrain from recapitulating it here.

3 Organizing Things

Awodey (2004) sketches an account of mathematics that eschews "foundations", at least in any of the senses articulated above. The central thought is that every mathematical theorem is of the form

(*) if such-and-such is the case, then so-and-so holds (Awodey, 2004, p. 58).

Consider, for example, the statement, "in any ring, if $x^2+x+1=0$ then $x^5=x$". In distinguishing his position from old-fashioned "if-thenism", Awodey insists that in the relevant instances of (*), such as this one, the parameters ("any ring", "x") are not to be understood as implicitly bound variables, which would presuppose a domain or realm of structures, sets, or the like, to serve as the range of the quantified variable. In our example, the rejected thesis is that the theorem presupposes a universe of all rings. The idea, rather, is that instances of (*) are "schematic", something akin to Russell's "typical ambiguity". We can say that the theorem holds of any ring, without invoking a domain of rings, even implicitly.

Thus, Awodey demurs from the perspective of ontological foundations of §1 above. He holds that mathematics has no distinctive subject matter, at least in that sense of "subject matter". There is no once-and-for-all realm of sets, structures, toposes, general categories, or whatever, from which all mathematical objects are to be constructed. Instead, the mathematician specifies (or presupposes) some conditions on properties and relations, and investigates what holds in systems that meet the conditions. As noted above, Awodey calls this a "top down" approach to mathematics, in opposition to the ontological foundationalist's "bottom up" perspective.

Awodey also rejects the epistemological foundational perspectives outlined in §2. To be sure, mathematics is an objective enterprise, with standards of reasoning that can be and are evaluated on generally agreed, public standards. There are correct and incorrect ways of going from the "such-and-such" to the "so-and-so" of each theorem. However, Awodey insists that the "methods of reasoning involved in different parts of mathematics are not 'global' and uniform across fields or even between different theorems, but are themselves 'local' or relative" (p. 56). This relativity explicitly applies to the logic of mathematics. There is no once-and-for-all system of correct inference:

> Of course, establishing any 'if ... then ... ' implication requires some tacitly assumed methods of reasoning, from simple chains of equations, to, say, ZFC. And where these methods are not conventionally assumed or obviously inferred, the statement of the theorem will generally include them: 'assuming the axiom of choice, ... ' or 'given a measurable cardinal ... ' or 'intuitionistically, ... ' (p. 60).

So the logic for a given piece of mathematics—a given theorem or group of theorems, say—is either conventional, or it is obvious what it is, or else the logic will be (or at least should be) explicitly stated as part of the theorem. And, I presume, that once the logic is explicitly stated, there is no further issue concerning how it is to be applied in the case at hand. The mathematician is not expected to resolve Wittgensteinian or Lewis Carroll-type issues concerning rule-following.

What of category theory? Notice, first, that category theory is a branch of mathematics, and so it, too, is to be understood schematically. Second, Awodey does not tout category theory as *the*, or even, *a* foundation of mathematics: "No one doing category theory thinks that we are someday going to find the one 'true topos', in which all mathematics happens" (p. 55); "No one claims that category theory is the *only* way to talk about structures of structures of ... . Or even that is the *best* way (although I know of no better one)."

As indicated by the parenthetical comment at the end of this passage, there is a special role for category theory, at least for the time being. The claim is that category theory is a good way in which to formulate the various "such-and-such" antecedents and "so-and-so" consequents of the schematic theorems of mathematics—and perhaps also any substantial logical theses about how we are to go from one to the other. That is, category theory is a good way to articulate the relevant statements in the form (*), and that, at present, no better way is known to articulate these statements, at least with a sufficient range of generality. In a sense, the idea seems to be that category theory provides a good *language* in which to formulate mathematics. It seems to me that this is at least one sense in which category theory is a "foundation", but there is no need to quibble over terminology. It surely is a very different sense of "foundation" than those delimited in previous sections.

In sum, then, there are two theses on offer here. One is that theorems of mathematics have the schematic character articulated above (and in Awodey (2004)), and the other is that category theory is a very good way—the best way known at present—in which to formulate mathematics, so construed. Advocates of the program take the second claim to be obvious and uncontentious, as do many working mathematicians. That thesis has been challenged, however, but I will say no more about it here. I'll close with some remarks on the first thesis, that mathematical theorems are all schematic (see Shapiro (2005)).

First, I take the claim that mathematical theorems are schematic to be descriptive, not normative. Advocates of Awodey's approach do not claim that mathematics ought to be pursued in the schematic manner, nor that it always was, nor that it always will be. Such disputes can degenerate into arguments over when to apply the honorific label "mathematics" to some activity or other. The claim in question is that contemporary mathematics can be accurately described as schematic, in the way indicated. It is a perspicuous way to *interpret* the enterprise, as it is now practiced.

We will need a term to contrast with "schematic", at least for purposes of discussion. Following Hellman, let us say that a sentence or utterance is *assertory* if it expresses a proposition about a fixed subject matter, with a fixed truth value. Consider the above example, "in any ring, if $x^2+x+1=0$ then $x^5=x$". As Awodey notes, an ontological, set-theoretic foundationalist would interpret this sentence as assertory. On that view, the metaphysical universe of mathematics contains a proper class of structures that satisfy the ring axioms. Each such structure is, or can be thought of as, a set together with some functions, with the functions themselves construed as sets of ordered pairs. Our proposition is a (true) statement about *all* such sets. Awodey finds this interpretation preposterous, or at least extravagant. He argues that the sentence in question is better understood as schematic, and thus not assertory. We prove that in any ring, if $x^2+x+1=0$ then $x^5=x$, and that is the end of it (for now). We do not presuppose a realm of "rings" or "sets" or anything else. As Awodey puts it, "One could say, rather speculatively, that the difference ... seems to be related to that between what is *true* for all of a fixed range of values, as opposed to what can be *proved* for an indeterminate value".

I take it as uncontroversial that much of mathematics is well-interpreted as having this schematic character. At the very least, it surely is the straightforward reading of the branches of abstract algebra—ontological, set-theoretic foundationalism aside. The question at hand is whether everything recognizable as mathematics can be interpreted in this schematic way. Let us consider meta-mathematical matters. As noted above, on occasion questions concerning the existence, or coherence, of various mathematical theories naturally arise. Suppose we are given a schematic sentence, in the form (*). One can wonder whether there are, or could be, any such-and-such's (in the relevant sense of "could be"). Admittedly, such questions do not arise all the time, but they do arise, as noted above. Awodey himself, of course, is aware of this. He writes:

> There is usually no question about *whether* such conditions are ever satisfied; rather, like axiomatic definitions, they serve to specify the range of application of the subsequent statement ... In cases where we are not sure whether the conditions are ever satisfied, i.e., whether they are consistent, we have no recourse but to investigate their consequences in order to gain more information. (p. 60)

Suppose, then, that we do wonder whether a certain specification S is coherent—whether there is (or could be) a structure of the specified kind. If we investigate the consequences of S and find them inconsistent, then, of course, we are satisfied that there is no such structure (dialetheism notwithstanding). But how do we assure ourselves that there is such a structure? If S is, in fact, consistent, then no amount of exploring the consequences of it will convince us of its consistency. For one thing, if the specification is sufficiently rich, the consistency of S will not be a deductive consequence of it, thanks to Gödel's second incompleteness theorem. Even putting that aside, how could the deduction of consistency assure us that the specification is, in fact consistent? If the specification is not consistent, then everything will be a consequence of it (assuming classical or intuitionistic logic).

As indicated above, the more usual way to establish coherence is to find an instance of the specification in an established mathematical theory. For example, we assure ourselves that complex analysis is coherent once we see that it can be modeled in R^2. Typically, the iterative hierarchy is the breeding ground for satisfiability, at least for structures with classical logic. Other structures are sanctioned by finding various category-theoretic models of them.

Consider a statement Π that a given specification S is coherent (i.e., consistent or satisfiable). The sentence Π at least looks like a piece of mathematics, and it is often settled by mathematicians doing what looks like mathematical work. Can we put Π in the form (*)? What would the "such-and-such" antecedent be? Suppose that some clever mathematician manages to find a model of S in the iterative hierarchy, using some fancy techniques. Then we would indeed have a statement in the form (*), namely

(i) If ZFC then Π.

Intuitively, I would think that this settles the matter Π, of consistency. But to think that, we would have to invoke modus ponens on (i), with a belief that ZFC is itself true, or at least coherent. The schematic perspective prevents us from doing this. To apply modus ponens to (i), we would not think of *it* as schematic. Instead, we think of (i) as itself assertory. And we assert its antecedent. But, admittedly, this is a matter of interpretation. The advocate of the present schematic perspective is free to insist that even statements of existence or consistency, or at least resolutions of questions of existence or consistency, have only the schematic form indicated, and are not assertory.

For a closely related matter, it often happens that we learn about a given structure by embedding it in a richer structure, and using information from the richer context to shed light on the original structure. A spectacular instance of this is the Wiles proof of Fermat's last theorem. Wiles did not show that the statement of Fermat's last theorem can be deduced from the Dedekind-Peano axioms. As I understand things, that question remains open. Let PA be a statement of the conditions for the natural numbers, such as the Dedekind-Peano axioms, plus any conditions on the logic. Then, from the schematic perspective, Fermat's last theorem is:

(F) If PA then $\forall x \forall y \forall z \forall n$(if $n>2$ then $x^n+y^n \neq z^n$).

As noted, Wiles did not establish (F). Let Z be a specification of the entire structure that was invoked in the proof. Then what Wiles showed was something like this:

(W) If Z then (If PA then $\forall x \forall y \forall z \forall n$(if $n>2$ then $x^n+y^n \neq z^n$)).

Of course, this is also in the form (*). But it is not the original formulation of Fermat's last theorem. So, on the schematic view, Wiles did not settle Fermat's last theorem. He showed something else. Awodey (2004) notes that there are many instances where we take one category and see it as part of a larger category, or perhaps an object of a larger category. But he insists that cross-category identifications make no sense. So, it would seem, Fermat's last theorem, in its original form, is not settled.

The accepted wisdom is that Wiles *did* settle Fermat's last theorem. I think that part of the reason for this is the belief that isomorphic structures are equivalent—wherever the instances may be found. So any instance of the Dedekind-Peano axioms can be used to study the natural numbers. The richer context allows us to shed more light on the original structure. As noted, the situation is typical in mathematics.

Consider, then, the statements that isomorphic structures are equivalent, and that if Φ is a statement in the language of second-order arithmetic, then either Φ is a model-theoretic consequence of the (classical) second-order Dedekind Peano axioms or $\neg\Phi$ is a model-theoretic consequence of those axioms. I take it that these statements are assertory. That is why they can be invoked in cases like this, to adjudicate arithmetic questions via the richer contexts. The question here is how the present, schematic account of mathematics can account for this feature of contemporary mathematics, or, failing that, how it can account for why the wisdom is mistaken.

The closing section of Awodey (2004) concerns "universal structures", like that of the natural numbers. These structures "reappear in many different categories". Statements about them are "absolute". The issue is to square this with the claim that all of mathematics is schematic. Awodey may have had something like this in mind with his closing remark:

> The study of ... absolute and structural properties of universal structures has not yet been developed. In fact, it is not even known whether, e.g., the real numbers or larger categories of sets than just the finite ones are universal, in the sense of being determined by universal properties. Only then could one sensibly ask which of their properties are not just typical but absolute. This seems to me the sort of question philosophers of mathematics might fruitfully pursue. (p. 64)

Amen to that.

Consider, finally, some other stock meta-mathematical results, such as the completeness of classical, first-order logic, and the incompleteness of arithmetic and higher-order logic. These results can be, and are, invoked throughout mathematics.

I invoked one of them, just above, in a general context. Prima facie, at least, these theorems are assertory, and not merely schematic.

One can, of course, challenge the claim that the indicated meta-mathematical statements, relating to consistency, isomorphism, completeness, and the like, are merely schematic and not assertory. I do not intend to take sides on this, or any other issue here. I will rest content if I have sufficiently illuminated the terrain.

Acknowledgments This note is a spinoff and extension of parts of Shapiro (2004) and Shapiro (2005). I am indebted to Colin McLarty, Steve Awodey, and Geoffrey Hellman for discussion and insight.

References

Awodey, S. (2004) "An answer to Hellman's question: 'Does category theory provide a framework for mathematical structuralism?' ", *Philosophia Mathematica* **12**, 54–64.

Frege, G. (1884) *Die Grundlagen der Arithmetik*, Breslau: Koebner; *The Foundations of Arithmetic*, translated by J. Austin, second edition, New York: Harper, 1960.

Hale, B. and Wright, C. (2001) *The Reason's Proper Study*, Oxford: Clarendon Press.

Jeshion, R. (2001) "Frege's notions of self-evidence", *Mind* **110**, 937–976.

Maddy, P. (1997) *Naturalism in Mathematics*, Oxford: Oxford University Press.

McLarty, C. (2004) "Exploring categorical structuralism", *Philosophia Mathematica* **12**, 37–53.

Moschovakis, Y. (1994) *Notes on Set Theory*, New York: Springer.

Shapiro, S. (2004) "Foundations of mathematics: metaphysics, epistemology, structure", *Philosophical Quarterly* **54**, 16–37.

Shapiro, S. (2005) "Categories, structures, and the Frege-Hilbert controversy: the status of metamathematics", *Philosophia Mathematica* **13**, 61–77.

Wilson, M. (1981) "The double standard in ontology", *Philosophical Studies* **39**, 409–427.

Wilson, M. (1993) "There's a hole and a bucket, dear Leibniz", *Midwest Studies in Philosophy* **18**, 202–241.

Wright, C. (1997) "On the philosophical significance of Frege's theorem", *Language, Thought, and Logic*, edited by Richard Heck, Jr., Oxford: Oxford University Press, 201–244; reprinted in Hale and Wright (2001), 272–306.

Part II
Foundations of Classical Mathematics

From Sets to Types, to Categories, to Sets

Steve Awodey

Three different styles of foundations of mathematics are now commonplace: set theory, type theory, and category theory. How do they relate, and how do they differ? What advantages and disadvantages does each one have over the others? We pursue these questions by considering interpretations of each system into the others and examining the preservation and loss of mathematical content thereby.

In order to stay focused on the "big picture", we merely sketch the overall form of each construction, referring to the literature for details. Each of the three steps considered below is based on more recent logical research than the preceding one. The first step from sets to types is essentially the familiar idea of set theoretic semantics for a syntactic system, i.e. giving a model; we take a brief glance at this step from the current point of view, mainly just to fix ideas and notation. The second step from types to categories is known to categorical logicians as the construction of a "syntactic category"; we give some specifics for the benefit of the reader who is not familiar with it. The third step from categories to sets is based on quite recent work, but captures in a precise way an intuition from the early days of foundational studies.

With these pieces in place, we can then draw some conclusions regarding the differences between the three schemes, and their relative merits. In particular, it is possible to state more precisely why the methods of category theory are more appropriate to philosophical structuralism.

1 Sets to Types

We begin by assuming a system of elementary set theory as given. The details of the set theory need not concern us for the moment, but will be specified later. We want to show how to construct a type theory from the sets, and it is the details of the type theory that are important for this step.

G. Sommaruga (ed.), *Foundational Theories of Classical and Constructive Mathematics*, The Western Ontario Series in Philosophy of Science 76,
DOI 10.1007/978-94-007-0431-2_5,

There are various different type theories that could be considered at this point: many-sorted first-order logic, simply-typed lambda calculus, dependent type theory à la Martin-L'of, the calculus of constructions, etc. We shall consider the traditional system of higher-order logic with "powertypes", i.e. higher types of "properties" or collections of objects of lower types. Since we also assume pairing and first-order logic, there are also higher types of relations and functions. It will be convenient to consider intuitionistic rather than classical logic, dropping the law of excluded middle, for reasons that will be clear later. This is of course no restriction but rather a generalization, since classical logic results simply from adding that law. Our entire discussion here could be adjusted to a different choice of type theory, however, and analogous conclusions to those arrived here would hold, *mutatis mutandis*, for that theory and suitably adjusted set theories and categories.[1]

1.1 IHOL

To be specific, let us consider the following system IHOL of (intuitionistic) higher-order logic.[2] This type theory consists of the following data.

Basic type symbols: $\mathsf{B}_1, \mathsf{B}_2, \ldots$
Type constructors: $\mathsf{A} \times \mathsf{B}$, $\mathsf{P}(\mathsf{A})$ for given type symbols A and B.
Variables: $\mathsf{x}_1, \mathsf{x}_2, \cdots : \mathsf{A}$ for each type A.
Basic terms: $\mathsf{b}_1 : \mathsf{A}_1$, $\mathsf{b}_2 : \mathsf{A}_2, \ldots$ where the types A_i are constructed from the basic ones.
Term constructors: given terms $\mathsf{a} : \mathsf{A}$, $\mathsf{b} : \mathsf{B}$, $\mathsf{c} : \mathsf{A} \times \mathsf{B}$ there are terms:

$$\langle \mathsf{a}, \mathsf{b} \rangle : \mathsf{A} \times \mathsf{B}, \; \pi_1(\mathsf{c}) : \mathsf{A}, \; \pi_2(\mathsf{c}) : \mathsf{B}.$$

Also, if φ is any formula, then $\lambda \mathsf{x}{:}\mathsf{A}.\, \varphi$ is a term of type $\mathsf{P}(\mathsf{A})$.
Formulas: include the following, where $\mathsf{a}, \mathsf{b} : \mathsf{A}$ and $\mathsf{p} : \mathsf{P}(\mathsf{A})$ are terms, and φ, ψ are formulas:

$$\mathsf{a} = \mathsf{b}, \; \mathsf{p}(\mathsf{a}), \; \neg\varphi, \; \varphi \wedge \psi, \; \varphi \vee \psi, \; \varphi \Rightarrow \psi, \; \forall \mathsf{x}{:}\mathsf{A}.\, \varphi, \; \exists \mathsf{x}{:}\mathsf{A}.\, \varphi$$

Theorems: Some formulas $\vartheta, \ldots$ are distinguished as theorems, written $\vdash \vartheta$.

We shall assume that the theorems always include the general laws of intuitionistic higher-order logic.

[1] See e.g. Awodey and Warren (2005) for the details of such an adjustment of step 3 below, which is the most novel of the three.

[2] This informal sketch is not intended as a precise specification of a system of type theory; for that, see e.g. (Johnstone, 2003; Lambek & Scott, 1986).

1.2 Semantics

Given a type theory $\mathbb{T}$, there is a familiar way of interpreting it in set theory; this consists essentially in giving a *model* of the theory, i.e. an interpretation that satisfies the theorems. We start with some sets $B_1, B_2, \ldots$ interpreting the basic types $\mathsf{B}_1, \mathsf{B}_2, \ldots$. Let us use the "Scott-brackets" notation $[\![X]\!]$ to indicate semantic interpretation of a bit of syntax X:

$$\begin{aligned} [\![\mathsf{B}_1]\!] &= B_1 \\ [\![\mathsf{B}_2]\!] &= B_2 \\ &\vdots \end{aligned}$$

We extend the interpretation to all types using the set-theoretic cartesian product and powerset operations:

$$\begin{aligned} [\![\mathsf{A} \times \mathsf{B}]\!] &= [\![\mathsf{A}]\!] \times [\![\mathsf{B}]\!] \\ [\![\mathsf{P}(\mathsf{A})]\!] &= \mathcal{P}([\![\mathsf{A}]\!]) \end{aligned}$$

Fixing interpretations for the basic terms $[\![\mathsf{b}_i]\!] \in [\![\mathsf{A}_i]\!]$, the constructed terms have a natural interpretation using corresponding set-theoretic operations. For instance,

$$[\![\langle \mathsf{a}, \mathsf{b} \rangle]\!] = ([\![\mathsf{a}]\!], [\![\mathsf{b}]\!])$$

using set theoretic pairing. Given a formula φ, we set:

$$[\![\lambda \mathsf{x} : \mathsf{A}.\, \varphi]\!] = \{x \in [\![\mathsf{A}]\!] \mid [\![\varphi]\!]\}$$

Finally, formulas are interpreted by set theoretic formulas, e.g. given terms $\mathsf{a} : \mathsf{A}$ and $\mathsf{p} : \mathsf{P}(\mathsf{A})$ and formulas φ, ψ, we let

$$\begin{aligned} [\![\mathsf{p}(\mathsf{a})]\!] &= [\![\mathsf{a}]\!] \in [\![\mathsf{p}]\!] \\ [\![\varphi \wedge \psi]\!] &= [\![\varphi]\!] \wedge [\![\psi]\!] \\ [\![\forall \mathsf{x} : \mathsf{A}.\, \varphi]\!] &= \forall \mathsf{x} \in [\![\mathsf{A}]\!].\, [\![\varphi]\!] \end{aligned}$$

and so on. Note for later reference that the set theoretic formulas $[\![\varphi]\!]$ coming from type theory are always Δ_0, i.e. all quantifiers are bounded by sets.

Every interpretation determines a theory, the theorems of which are all the formulas ϑ that come out true under the interpretation, i.e. such that $[\![\vartheta]\!]$ holds in the set theory. As an example, consider the theory PA of Peano arithmetic, with one basic type N for the natural numbers and basic constants $\mathsf{o} : \mathsf{N}$ for zero and $\mathsf{s} : \mathsf{N} \to \mathsf{N}$ for successor (as usual, the function type $\mathsf{N} \to \mathsf{N}$ can be constructed from the type of relations $\mathsf{P}(\mathsf{N} \times \mathsf{N})$). The interpretation is the evident one assigning these symbols to the set of natural numbers, the zero element, and the successor function, respectively. The theorems are all the formulas in this language that are true under this interpretation.

A system of type theory $\mathbb{T}$ can be modeled in set theory in various different ways, each determined by the interpretation of the basic types and terms $[\![\mathsf{B}]\!], \ldots, [\![\mathsf{b}]\!], \ldots$. The formulas that *always* come out true, under every interpretation, will of course include the general laws of intuitionistic higher-order logic usually specified by a deductive system. This is just the "soundness" of the system of deduction.

Now, given a system of set theory **S** (more precisely, a model of our assumed elementary set theory), is there a distinguished type theory $\mathbb{T}(\mathbf{S})$ with a distinguished interpretation in **S**? As basic types, we take symbols $\ulcorner A \urcorner, \ulcorner B \urcorner, \ulcorner C \urcorner, \ldots$ for all the sets $A, B, C, \ldots$ of **S**; as basic terms, we take symbols $\ulcorner a \urcorner, \ulcorner b \urcorner, \ulcorner c \urcorner, \ldots$ for all the elements $a, b, c, \ldots$ of the sets, whereby we of course set $\ulcorner a \urcorner : \ulcorner A \urcorner$ just if $a \in A$.

This type theoretic language has an obvious interpretation back into **S** by setting $[\![\ulcorner A \urcorner]\!] = A$ and $[\![\ulcorner a \urcorner]\!] = a$, etc. As theorems, we take all the formulas of $\mathbb{T}(\mathbf{S})$ that hold under this interpretation,

$$\mathbb{T}(\mathbf{S}) \vdash \varphi \quad \text{iff} \quad \mathbf{S} \models [\![\varphi]\!].$$

Note that for each set A there will be both a powertype $\mathsf{P}(\ulcorner A \urcorner)$ and a basic type $\ulcorner \mathcal{P}(A) \urcorner$ for the powerset. However, since clearly

$$[\![\mathsf{P}(\ulcorner A \urcorner)]\!] = \mathcal{P}(A) = [\![\ulcorner \mathcal{P}(A) \urcorner]\!],$$

we will have theorems of the form $\vdash \mathsf{P}(\ulcorner A \urcorner) \cong \ulcorner \mathcal{P}(A) \urcorner$ for each set A. Thus the types $\mathsf{P}(\ulcorner A \urcorner)$ and $\ulcorner \mathcal{P}(A) \urcorner$ are syntactically isomorphic, since their interpretations are equal, and thus isomorphic. The same is true for product types $\ulcorner A \urcorner \times \ulcorner B \urcorner \cong \ulcorner A \times B \urcorner$, and for all other type theoretic constructions that are definable in set theory. So although there is a great duplication of data, the type theory holds there to be isos relating old to new. Of course, it also holds that everything true in the original set theory is true in the type theory, to the extent it can be stated there. Indeed, this type theory captures all of the *type theoretic* information of **S**; what it omits cannot be expressed in type theory.

2 Types to Categories

Next, given a system of type theory $\mathbb{T}$, we shall construct from it a category $\mathcal{E}(\mathbb{T})$ by identifying certain terms as the objects and arrows. This category, it turns out, is of a very special kind known as a "topos". This means that it has a certain categorical structure typical of the categories of sheaves that arise in geometry. These categories were first identified and studied by the Grothendieck school of algebraic geometry, and have been axiomatized (by F.W. Lawvere and M. Tierney) and investigated for their fascinating logical properties (see Mac Lane & Moerdijk, 1992). We follow roughly the exposition of (Lambek & Scott, 1986) for a sketch of the construction of the "syntactic topos" $\mathcal{E}(\mathbb{T})$. First, let us recall the basic definition.

2.1 Topoi

A *topos* is a category $\mathcal{E}$ such that:

- $\mathcal{E}$ has all finite limits: in particular, it has a terminal object 1 and all binary products $A \times B$, as well as all pullbacks,

$$\begin{array}{ccc} A \times_C B & \longrightarrow & B \\ \downarrow & & \downarrow \\ A & \longrightarrow & C, \end{array}$$

  which are products in the slice categories $\mathcal{E}/C$.
- $\mathcal{E}$ has exponentials: for every pair of objects A, B, there is an object B^A, and an isomorphism between arrows of the forms

$$\frac{X \to B^A}{X \times A \to B}$$

  Moreover, this correspondence is natural in X, making $(-)^A : \mathcal{E} \to \mathcal{E}$ into a functor right adjoint to $(-) \times A : \mathcal{E} \to \mathcal{E}$.
- $\mathcal{E}$ has a subobject classifier: there is an object Ω with an arrow $t : 1 \to \Omega$ such that every subobject $S \rightarrowtail A$ fits into a pullback diagram

$$\begin{array}{ccc} S & \longrightarrow & 1 \\ \downarrow & & \downarrow t \\ A & \xrightarrow[\phi_S]{} & \Omega \end{array}$$

  for a unique "classifying arrow" $\phi_S : A \to \Omega$.

The subobject classifier axiom can be thought of as saying that every subset has a unique characteristic function. The concept of a topos has proven to be extremely rich and versatile. The combination of exponentials and a subobject classifier provides for powerobjects in the form $P(A) = \Omega^A$, and one can show that a topos must also have all finite colimits, as well as the structure required to interpret first-order logic. There are topoi such as the category **Sets** of all sets and functions, and the functor categories $\mathbf{Sets}^{\mathbb{C}}$ for any small category $\mathbb{C}$, as well as the "geometric" categories of sheaves mentioned earlier; but there are also topoi arising naturally in logic from forcing, permutation, and Kripke models, and realizability, as well as from systems of type theory, as we now indicate.

2.2 Syntactic Topos

Given the type theory $\mathbb{T}$, we shall construct from it a topos $\mathcal{E}(\mathbb{T})$ comprised of syntactic material from $\mathbb{T}$.

First, it is convenient to add a unit type $\mathsf{1}$ with a basic term $* : \mathsf{1}$ and an axiom

$$\vdash \forall \mathsf{x} : \mathsf{1}.\, * = \mathsf{x}.$$

The type $\mathsf{P(1)}$ now acts as a type of formulas, in that for every formula φ there is an associated term $\overline{\varphi} : \mathsf{P(1)}$,

$$\overline{\varphi} = \lambda \mathsf{x} : \mathsf{1}.\, \varphi$$

such that

$$\vdash \varphi \Leftrightarrow (\overline{\varphi} = \top)$$

where $\top = \overline{(* = *)}$. The term $\overline{\varphi}$ can be thought of as the characteristic function of the "extension" of φ.

The topos $\mathcal{E}(\mathbb{T})$ is now defined as follows.

objects: are equivalence classes under provable equality of closed terms $\lambda \mathsf{x} : \mathsf{A}.\, \varphi$ of various powertypes $\mathsf{P(A)}$, which we write as $[\, \mathsf{x} : \mathsf{A} \mid \varphi \,]$ (or simply $[\, \mathsf{x} \mid \varphi \,]$ when the type A can be inferred).

arrows: of the form $[\, \mathsf{x} : \mathsf{A} \mid \varphi] \rightarrow [\, \mathsf{y} : \mathsf{B} \mid \psi]$ are (provable-equality equivalence classes of) provably functional relations $[\, (\mathsf{x}, \mathsf{y}) : \mathsf{A} \times \mathsf{B} \mid \rho \,]$ from φ to ψ,

$$\vdash (\forall \mathsf{x}\, \exists ! \mathsf{y}.\, \rho) \wedge (\forall \mathsf{x}, \mathsf{y}.\, (\rho \Rightarrow \varphi \wedge \psi))$$

units:

$$1_A = [\, \mathsf{x}, \mathsf{y} \mid \mathsf{x} = \mathsf{y} \,] : \mathsf{A} \rightarrow \mathsf{A}$$

composition:

$$[\, \mathsf{y}, \mathsf{z} \mid \sigma \,] \circ [\, \mathsf{x}, \mathsf{y} \mid \rho \,] = [\, \mathsf{x}, \mathsf{z} \mid \exists \mathsf{y}.\, \sigma \wedge \rho \,]$$

products:

$$\begin{aligned} 1 &= [\, \mathsf{u} : \mathsf{1} \mid \mathsf{u} = \mathsf{u} \,] \\ [\, \mathsf{x} : \mathsf{A} \mid \varphi \,] \times [\, \mathsf{y} : \mathsf{B} \mid \psi \,] &= [\, (\mathsf{x}, \mathsf{y}) : \mathsf{A} \times \mathsf{B} \mid \varphi \wedge \psi \,] \end{aligned}$$

exponentials:

$$[\, \mathsf{y} : \mathsf{B} \mid \psi \,]^{[\mathsf{x} : \mathsf{A} \mid \varphi]} = \\ [\, \mathsf{r} : \mathsf{P}(\mathsf{A} \times \mathsf{B}) \mid (\forall \mathsf{x}\, \exists ! \mathsf{y}.\, \mathsf{r}(\mathsf{x}, \mathsf{y})) \wedge (\forall \mathsf{x}, \mathsf{y}.\, (\mathsf{r}(\mathsf{x}, \mathsf{y}) \Rightarrow \varphi \wedge \psi)) \,]$$

subobject classifier:

$$\Omega = [\, \mathsf{p} : \mathsf{P(1)} \mid \mathsf{p} = \mathsf{p} \,]$$

It is straightforward to verify that this actually *is* a category, and that the indicated constructions have the required universal properties making it a topos. The syntactic topos $\mathcal{E}(\mathbb{T})$ itself also has a universal mapping property, somewhat analogous to that of a polynomial ring, characterizing it as the free topos with a model of the theory $\mathbb{T}$. If we take as a theory, for instance, the empty theory $\mathbb{T}_0$ without any basic types or terms, and as theorems just the deductive consequences of the conventional axioms and rules of classical higher-order logic, then the syntactic topos is the category $\mathbf{Sets}_{\mathrm{fin}}$ of finite sets:

$$\mathcal{E}(\mathbb{T}_0) = \mathbf{Sets}_{\mathrm{fin}}$$

This follows from a classical result of L. Henkin, the completeness of the theory of propositional types (Henkin, 1963). (Note that here we really needed to add the unit type 1 to get things going!) The general construction of a topos out of a type theory demonstrates a completeness theorem for general deductive higher-order logic with respect to topos models (see Johnstone, 2003; Lambek & Scott, 1986).

3 Categories to Sets

For the final step, we indicate how to extract an elementary set theory from a topos. The resulting set theory will have the property that its sets and functions are essentially the objects and arrows of the topos we started with, and its theorems all hold in the topos. This construction, which was only recently given in (Awodey et al., 2007a, 2007b; Awodey & Forssell, 2005), involves some technical methods from category and sheaf theory, and so we cannot give the details here; but it is similar in spirit to an old idea from type theory, which we can use as motivation for our sketch.

The motivating idea, which can be found e.g. in (Quine, 1963) and elsewhere, is that one can "sum the types" of a type theory $\mathbb{T}$ to obtain a universal type

$$\mathsf{U} = \bigcup_{\mathsf{A} \in \mathbb{T}} \mathsf{A},$$

into which all of the original types then embed $\mathsf{A} \subseteq \mathsf{U}$. Moreover, if the "sum" is taken in the right way, there will also be a powertype

$$\mathsf{P}(\mathsf{U}) = \bigcup_{\mathsf{A} \in \mathbb{T}} \mathsf{P}(\mathsf{A}),$$

which in turn will also embed $\mathsf{P}(\mathsf{U}) \subseteq \mathsf{U}$. The universal type U thus admits an *untyped* membership relation

$$\in_{\mathsf{U}} \subseteq \mathsf{U} \times \mathsf{P}(\mathsf{U}) \subseteq \mathsf{U} \times \mathsf{U}.$$

This binary relation then models an elementary set theory, the theorems of which depend on the type theory $\mathbb{T}$ with which we began.

3.1 Category of Ideals

In the type theoretic setting, the scheme of "summing the types" is more of a figurative, guiding idea than an actual construction. But if we start from a topos $\mathcal{E}$ instead of a system of type theory, we can apply certain constructions from sheaf theory which capture that intuition in a rigorous way and allow us in the end to actually read off an elementary set theory describing $\mathcal{E}$. The main new concept is that of an *ideal* in the category $\mathcal{E}$, which is essentially a order ideal in the partial ordering of monomorphisms of $\mathcal{E}$, i.e. a non-empty subcategory $\mathbf{C} \hookrightarrow \mathcal{E}$ of objects and monomorphisms such that $A \in \mathbf{C}$ and $A' \rightarrowtail A$ implies $A' \in \mathbf{C}$, and $A, B \in \mathbf{C}$ implies $C \in \mathbf{C}$ for some $A \rightarrowtail C \leftarrowtail B$. The actual definition requires either some care in specifying choices of monomorphisms, as is done in (Awodey et al., 2007b), or a sheaf-theoretic approach as in (Awodey & Forssell, 2005). In either case, the category $\mathrm{Idl}(\mathcal{E})$ of all ideals, called the *ideal completion* of $\mathcal{E}$, is characterized by a universal property: it is the completion of $\mathcal{E}$ under filtered colimits of monomorphisms. It is a generalization to categories of the ideal completion of a poset, and like that construction it has some very good logical properties.

An ideal in $\mathcal{E}$ can be thought of as being "patched together" out of pieces consisting of objects of $\mathcal{E}$; indeed, the notion of a scheme in algebraic geometry is closely related to that of an ideal. In the logical case of a topos, we can think of the ideals as (abstract) classes, with the principle ideals $\downarrow(A)$ for $A \in \mathcal{E}$ as the "sets". The category $\mathrm{Idl}(\mathcal{E})$ of all ideals then has a somewhat weaker logical structure than the original topos $\mathcal{E}$ (e.g. it does not have all exponentials) but it does still support an interpretation of first-order logic. It also has something that the topos $\mathcal{E}$ cannot have: an object $\mathbf{U}$ with a monomorphism $P(\mathbf{U}) \rightarrowtail \mathbf{U}$. This is made possible by the fact that the powerobject $P(\mathbf{C})$ in ideals is in effect the class of all sub*sets* of the ideal $\mathbf{C}$ (rather than all sub*classes*). The universal object $\mathbf{U}$ is just the total ideal, and thus also embeds all the "sets" $\downarrow(A) \rightarrowtail \mathbf{U}$, as desired. (This is the "right" way of "summing the types" mentioned above.)

In particular, there is then a membership relation

$$\in_{\mathbf{U}} \rightarrowtail \mathbf{U} \times P(\mathbf{U}) \rightarrowtail \mathbf{U} \times \mathbf{U}$$

on $\mathbf{U}$ as desired. In this way, we construct from the topos $\mathcal{E}$ a category $\mathrm{Idl}(\mathcal{E})$ containing a model $(\mathbf{U}, \in_{\mathbf{U}})$ of an elementary set theory, the sets and functions of which form a category equivalent to $\mathcal{E}$. It is quite surprising that this set theory can be axiomatized in a simple and familiar way.

3.2 Basic Intuitionistic Set Theory

The elementary set theory that is modeled by every topos is a variant of conventional Zermelo-Fraenkel set theory **ZF**, which we call **BIST** for Basic Intuitionistic Set Theory. It differs from **ZF** in the following three respects:

1. it is formulated in intuitionistic rather than classical logic,
2. it allows for "urelements", or atoms,
3. the axiom scheme of separation is restricted to formulas with bounded quantifiers, the so-called Δ_0 formulas.

Apart from these changes, it agrees with **ZF** in having axioms of extensionality, emptyset, singletons, pairs, unions, powersets, foundation, and replacement. An axiom of infinity holds for topoi with an infinite object, but otherwise not, so we do not include it in the definition of **BIST**.

The use of intuitionistic logic (1) is required by the fact that the logic of topoi is generally intuitionistic; this is not a philosophical decision, but a fact of nature. The topoi that arise from notions of variation and continuity in geometry, for instance, just naturally satisfy intuitionistic rather than classical logic. The possible presence of atoms (2) is required to accommodate topoi based on some given objects and arrows, such as the representable functors in a functor category $\mathbf{Sets}^{\mathbb{C}}$, or the basic types and terms in a syntactic topos $\mathcal{E}(\mathbb{T})$ coming from a type theory. The restriction in the separation scheme (3) arises algebraically from the fact that a subideal of a principle ideal need not itself be principle, and so not every subclass of a set need be a set. It is interesting to note that bounded separation and full (unbounded) replacement are compatible under intuitionistic logic; by contrast (full) separation follows classically from replacement. Indeed, replacement itself can even be given a stronger (intuitionistically) formulation, called collection. The specific formulations of some of the other axioms are also adjusted to account for intuitionistic logic and the possibility of atoms (see (Awodey, 2008; Awodey, 2007a, 2007b) for details).

For our purposes, the remarkable fact about **BIST** is that it is not only sound but also deductively *complete* with respect to topoi, modeled in their ideal completions as indicated above:

$$\mathbf{BIST} \vdash \varphi \quad \text{iff} \quad (\mathbf{U}, \in_{\mathbf{U}}) \models_{\mathrm{Idl}(\mathcal{E})} \varphi \quad \text{for all } \mathcal{E}.$$

Of course, for a particular topos $\mathcal{E}$, the set theory $\mathbf{S}(\mathcal{E}) = (\mathbf{U}, \in_{\mathbf{U}})$ in the ideal completion $\mathrm{Idl}(\mathcal{E})$ will also model more set theoretic formulas than just the deductive consequences of **BIST**. If $\mathcal{E}$ is boolean, for instance (like the classical category of sets), then the set theory $\mathbf{S}(\mathcal{E})$ will satisfy excluded middle for sets, and if $\mathcal{E}$ satisfies the axiom of choice, then $\mathbf{S}(\mathcal{E})$ will also satisfy that axiom for sets. In fact, given any property of objects and arrows in a topos that is expressible by a formula of set theory, the property holds in $\mathcal{E}$ if and only if the corresponding formula holds in the set theory $\mathbf{S}(\mathcal{E})$. In that sense, $\mathcal{E}$ can be regarded as a category of sets and functions of the set theory $\mathbf{S}(\mathcal{E})$. For example, if $\mathcal{E} = \mathcal{E}(\mathsf{PA})$ is the syntactic topos of (intuitionistic) Peano arithemic, then the set theory $\mathbf{S}(\mathcal{E}(\mathsf{PA}))$ is intuitionistic **ZF** with bounded separation, sometimes called $\mathbf{IZF}_0$, in which the arithmetic of the natural numbers agrees with that provable in PA.

Finally, let us tie up a loose end. In section 1 we said that the details of the set theory assumed were to be specified later. Now we can do so succinctly: it should be (at least) **BIST**.

4 Composites

The three constructions that we have just sketched,

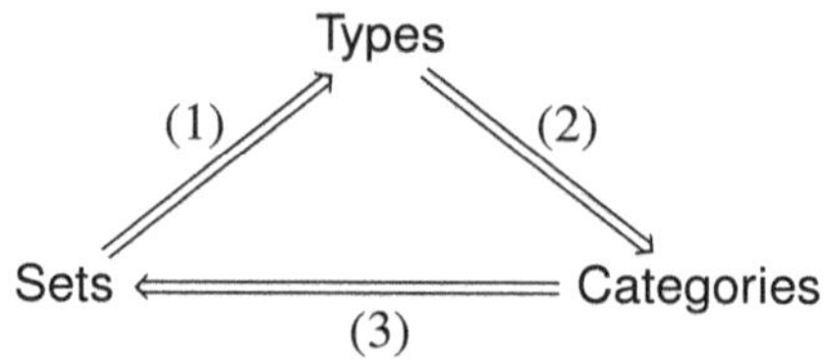

(4)

can now be composed to yield three more interpretations,

$$
\begin{aligned}
(2)\circ(1) &: \mathsf{Sets} \Rightarrow \mathsf{Categories}\\
(3)\circ(2) &: \mathsf{Types} \Rightarrow \mathsf{Sets}\\
(1)\circ(3) &: \mathsf{Categories} \Rightarrow \mathsf{Types}
\end{aligned}
$$

Let us briefly consider each of these in turn.

4.1 Sets to Categories

Starting from a set theory $\mathbf{S}$ we compose the construction (1) of a type theory $\mathbb{T}(\mathbf{S})$ with (2) from type theory to categories, i.e. the syntactic topos construction. What is the syntactic topos $\mathcal{E}(\mathbb{T}(\mathbf{S}))$? It is not hard to see that, up to equivalence of categories, it is just the category of sets and functions of $\mathbf{S}$. The objects of the syntactic topos are the definable sets $[\,\mathsf{x} : \mathsf{A} \,|\, \varphi\,]$ in $\mathbb{T}(\mathbf{S})$, all of which are isomorphic to sets coming from $\mathbf{S}$; and the arrows between these are all given by functional relations, which are all uniquely determined by functions in $\mathbf{S}$. Thus the composite of these two constructions is a familiar construction from sets to categories, namely, taking the category of sets and functions of a set theory.

4.2 Types to Sets

Here we start with a type theory $\mathbb{T}$, make the syntactic topos $\mathcal{E}(\mathbb{T})$, take its ideal completion $\mathrm{Idl}(\mathcal{E}(\mathbb{T}))$, and find there the universal object model $(\mathbf{U}, \in_{\mathbf{U}})$ determining the set theory $\mathbf{S}(\mathcal{E}(\mathbb{T}))$. But the guiding idea of the ideals construction was that it gives a rigorous treatment of the informal scheme of "summing the types" of a type theory to get a set theory, with the definable collections of all types as the sets.

So this composite $(3)\circ(2)$ is just a precise formulation of that informal idea of turning a type theory into a set theory by "summing the types".

4.3 Categories to Types

If we start with a topos $\mathcal{E}$ and apply the constructions (3) and (1) in turn, what results is essentially what the categorical logician calls the "internal logic" of the topos (see Lambek & Scott, 1986). It is a type theory in which the basic types are the objects of the topos, the basic terms are the arrows, along with some coordinating terms as in (1), and the axioms are all the formulas that hold under the evident standard interpretation back into the topos. This well-known construction of a type theory out of a topos is a sort of "inverse" to the syntactic topos construction, lacking only a suitable notion of equivalence of type theories in order to be an actual inverse.

5 Conclusions

We are now in a position to make a more informed comparison between these three different approaches to foundations.

First, let us note that there are of course further composites of the translations (1), (2), and (3), namely, going "once around" to the starting point. In each case, the result is a system extending the original one by further data, the behavior of which is determined by isomorphic data in the original system. For categories, this familiar situation is just what the notion of *equivalence* was invented for. Starting from a topos $\mathcal{E}$ and going once around the diagram (4) of translations results in a category equivalent to $\mathcal{E}$. For the other three-fold composites, the situation is not as succinctly expressed. The resulting systems of type or set theory are "equivalent" to the original ones, in some sense that needs to be made precise. This involves additional basic data such as basic types and terms or atoms, which are copies of preexisting objects determining them; something like a notion of a "definitional expansion" of the original system is about right. Perhaps the clearest thing that one can say is that these composite constructions result in systems which, under the further translation to categories, are categorically equivalent to the original ones.[3] (One fine point: if we start with the *empty* theory $\mathbb{T}_0$ and go once around, the result is not equivalent, since we have added the unit type $\mathsf{1}$. To smooth things out, every type theory should really have a unit type—which has other benefits as well.)

The first and most obvious conclusion to be drawn from this is that the three systems of foundations are therefore mathematically equivalent. Elementary set theory at least as strong as our basic theory **BIST**, type theory in the form of higher-order logic, and category theory as represented by the notion of a topos, all permit the same mathematical definitions, constructions, and theorems—to the extent that these do not depend on the specifics of any one system. This is perhaps the definition of the "mathematical content" of a system of foundations, i.e. those definitions,

[3] A more careful analysis shows that deductive IHOL is not only sound with respect to models in **BIST** and complete with respect to models in topoi, but both sound and complete with respect to both.

theorems, etc. that are independent of the specific technical machinery, that are invariant under a change of foundational schemes. The very constructions that we have been discussing, for instance, in order to be carried out precisely, would have to be formulated in some background theory; but what should *that* be? Any of the three systems themselves would do for this purpose, and the results we have mentioned would not depend on the choice.

Another conclusion to be drawn is this: the objects of type theory and set theory are structured by the operations of their respective systems in certain ways that are not mathematically salient. That additional information is essentially what is lost by our comparisons, e.g. distinctions between basic data and derived objects, between types of different complexity, ordinal rank of a set, membership chains within a set, etc. Categorical structure is closer to the mathematical content, and it is not lost in translation. Equivalence of categories preserves categorical properties and structures, because these are determined only up to isomorphism in the first place.

The structural approach implemented by category theory is thus more stable, more robust, more invariant than type or set theoretic constructions. On the other hand, type and set theory have certain distinctive advantages as well. Type theory has something of a concrete, "nominalistic" character, owing to the fact that one actually constructs its objects syntactically—although in impredicative systems, it is of course not really the case that everything the theory posits can be written down. Nonetheless, there is the idea that the objects are systematically generated from some basic data by repeated iteration of the operations, making them more managable. Set theory sacrifices the nominalistic pretense in favor of greater flexibility and range of set formation, while retaining the conception of a systematic generation of its objects "from below", i.e. iteratively, from basic data. This still allows for some degree of control over the objects in the form of ordinal ranks, ϵ-induction, and the like. Although these additional logical structures do not have a stable mathematical content—no topologist or algebraist is concerned with the logical type or ordinal rank of a manifold or module—they can serve a useful purpose in foundational work by providing the concrete data for specifications and calculations, facilitating constructions and proofs.

By contrast, the purely structural approach of category theory sometimes offers comparatively little such "extra" structure to hold on to. Practically speaking, it can be harder to give an invariant proof. That is why it's good to know that such logical structure can always be introduced into a category when needed; the devices of introducing an internal logic or a set theoretic structure into a category, as sketched in the foregoing sections, were originally developed in order to benefit from their advantages, much like introducing local coordinates on a manifold for the sake of calculation. The analogy is quite a good one: no one today regards a manifold as involving specific coordinate charts, and one generally works with coordinate free methods so that the results obtained will apply directly—this is the modern, structural approach. But at times it can still be useful to introduce coordinates for some purpose, and this is unobjectionable, as long as the results are invariant. So it is with categorical versus logical foundations: category theory implements the structural approach directly. It admits interpretations of the conventional logical systems, without being tied to them. Category theory presents the invariant content of logical foundations.

References

Awodey, S. (2008) A brief introduction to algebraic set theory, *Bulletin of Symbolic Logic* **14**(3), 281–298.

Awodey, S., Butz, C., Simpson, A. and Streicher, T. (2007a) Relating first-order set theories and elementary toposes, *Bulletin of Symbolic Logic* **13**(3), 340–358.

Awodey, S., Butz, C., Simpson, A. and Streicher, T. (2007b) Relating first-order set theories, toposes and categories of classes, in preparation, preliminary version available at Algebraic Set Theory. Web site: `www.phil.cmu.edu/projects/ast.`

Awodey, S. and Forssell, H. (2005) Algebraic models of intuitionistic theories of sets and classes, *Theory and Applications of Categories* **15**(1), 147–163.

Awodey, S. and Warren, M. A. (2005) Predicative algebraic set theory, *Theory and Applications of Categories* **15**(1), 1–39.

Henkin, L. (1963) A theory of propositional types, *Fundamenta Mathematicae* **52**, 323–344.

Johnstone, P. T. (2003) *Sketches of an Elephant*, Oxford: Oxford University Press.

Lambek, J. and Scott, P. (1986) *Introduction to Higher-Order Categorical Logic*, Cambridge: Cambridge University Press.

Mac Lane, S. and Moerdijk, I. (1992) *Sheaves in Geometry and Logic*, Berlin: Springer.

Quine, W. V. O. (1963) *Set Theory and its Logic*, Cambridge, MA: Harvard University Press.

Enriched Stratified Systems for the Foundations of Category Theory

Solomon Feferman

1 Introduction

This is the fourth in a series of intermittent papers on the foundations of category theory stretching back over more than 35 years. The first three were "Set-theoretical foundations of category theory" (1969), "Categorical foundations and foundations of category theory" (1977), and much more recently, "Typical ambiguity: Trying to have your cake and eat it too" (2004). The present paper summarizes the results from a long (in two senses) unpublished manuscript,"Some formal systems for the unlimited theory of structures and categories" (1974), referred to below simply as "Unlimited". That MS can be found in full on my home page at `http://math.stanford.edu/~feferman/papers/Unlimited.pdf`; the lengthy proof of its main consistency result is omitted here but the methods involved are outlined briefly in the Appendix below.

I have been concerned in these papers with set-theoretical foundations of category theory not because I am a proponent of set theory—on the contrary I am opposed to it on fundamental philosophical grounds[1]—but rather because it is currently widely accepted, its ins and outs are well understood, and it has dealt successfully with the problems surrounding objects that are somehow "too large". It is just such problems in what is sometimes called "naïve" category theory that require foundational attention. Namely, objects like the category of all groups, the category of all topological spaces, etc., seem natural enough mathematically, but what about the category of all categories? And, further, what about the category of all functors between any two categories? Several proposals have been made for dealing with these within the general framework of axiomatic set theory, most notably the familiar ones of Mac Lane (1961, 1971) and Grothendieck (in Gabriel, 1962). This is one reason that alternatives, such as my (1969, 2004) and the present one, are best explored within the same framework for purposes of comparison. I do think that the foundations of category theory ought also to be explored within other frameworks

[1] See, for example, my collection of essays, *In the Light of Logic* (Feferman, 1998).

G. Sommaruga (ed.), *Foundational Theories of Classical and Constructive Mathematics*, The Western Ontario Series in Philosophy of Science 76,
DOI 10.1007/978-94-007-0431-2_6,

such as those of constructive or semi-constructive mathematics of various stripes, but the directions those might take is not touched on here.

There are some workers in the field who think that category theory does not need foundations and in fact that it is category theory itself that provides the proper foundations for all of mathematics, including itself; see, for example, Lawvere (1966), Bénabou (1985) and, more recently, McLarty (2004). In my 1977 paper cited above, I have argued against that position in a way that I think is no less compelling now than then. Those arguments are not repeated here; in addition to my (1977), the interested reader should also see their extension by Hellman (2003). But there are further objections to be made. It is not clear what exactly is meant by categorical foundations for category theory and how it proposes to handle the problem of the category of all categories and that of arbitrary functor categories. There is also a specific mathematical objection that has been raised by Rao (2006) concerning the construction of localizations in homotopical algebra that make use of transfinite induction and recursion. As he says, "[i]t is not clear how to formulate these in categorical terms....Solving these problems [by such means] looks remote at the moment."

2 What the Various Proposals Do and Don't Do

In my 1977 paper cited above, on pp. 154–156 I suggested three requirements on a system S for the foundations of category theory.[2] Rephrased from there, S should:

(R1) Allow us to construct the category of all structures of a given kind, e.g. the category **Grp** of all groups, **Top** of all topological spaces, and **Cat** of all categories.

(R2) Allow us to construct the category B^A of all functors from A to B, where A and B are any categories.

(R3) Allow us to establish the existence of the usual basic mathematical structures and carry out the usual set-theoretical operations.

A further requirement was not stated there, but is implicit in the above:

(R4) S should be established to be consistent relative to currently accepted systems of set theory.

Let us see how the existing proposals stack up against these requirements. Mac Lane's proposal was to work in the Bernays-Gödel (BG) system of sets and classes, using the distinction between "small categories" and "large categories" according to whether the categories are sets or proper classes. This meets (R1) in a rather modified form: one can only talk about the large categories $\mathbf{Grp}_{\mathrm{sm}}$, $\mathbf{Top}_{\mathrm{sm}}$, and $\mathbf{Cat}_{\mathrm{sm}}$ of all small groups, small topological spaces and small categories, respectively. (R2) can be met only for A small, since the class of all functions from one proper class

[2] Bénabou (1985) proposes more specific requirements which need to be considered for a full scale foundation of naïve category theory.

into another does not exist in BG. (R3) and (R4) are of course met as is: BG is a conservative extension of Zermelo-Fraenkel set theory ZF, and the same holds when the Axiom of Choice AC is added to both systems.

Grothendieck's proposal was to work in ZFC (= ZF + AC) with the addition of a strong axiom of "universes". Roughly speaking, a universe U is a transitive set that contains the set ω of natural numbers, is closed under subsets, satisfies the ZFC axioms, and in addition satisfies the inaccessibility condition that whenever $a \in U$ and $f : a \to U$ then the range of f is in U. These conditions imply that the cardinal number of U is a strongly inaccessible cardinal. The Grothendieck axiom is that there are arbitrarily large universes, i.e. for every set a there is a universe U with $a \in U$. Again, requirements (R1) and (R2) are met only in a restricted form. Namely, for any universe U, we may speak only of the category of all categories that lie in U; it belongs to any larger universe U'. Also if A and B are two categories whose objects and morphisms all lie in U, then B^A lies in U'. (R3) is met by assumption; (R4) is met by the reduction to ZFC + "there exist arbitrarily large strongly inaccessible cardinals."[3]

In my papers (1969, 2004), I worked in ZFC with one or more additional constant symbols for universes U that are transitive, closed under subsets and satisfy the reflection principle, i.e. the scheme for each formula $\varphi(x_1 \ldots x_k)$ of the language of ZFC (without the additional symbols):

$$\forall x_1 \ldots \forall x_k[x_1 \ldots x_k \in U \to (\varphi^{(U)}(x_1 \ldots x_k) \leftrightarrow \varphi(x_1 \ldots x_k))]$$

This scheme insures that U satisfies the ZFC axioms. But it is *not* assumed that U is closed under the inaccessibility condition. Requirements (R1) and (R2) are met as in the Grothendieck proposal. (R3) is met by assumption. For (R4) it is shown in both the cited papers that the system is conservative over ZFC. Thus one need not assume the existence of inaccessible cardinals, though a few applications (such as the Kan Extension theorem) apparently need U to satisfy the inaccessibility condition. The principal advantage of my proposal over Grothendieck's is a conceptual one: any U satisfying the above reflection condition looks, from the point of view of the set-theoretical language, exactly like the universe V of all sets, and thus serves as a stand-in for it. Thus anything we can contemplate doing over V can already be done over U and in that way be fully expressed in ZFC.[4] This approach is also taken by Rao (2006).

Though each of these solutions is adequate for normal applications of category theory as in Mac Lane (1971), none of them satisfies (R1) and (R2) without modification. The purpose of this paper is to show how those two requirements can be met in full by working in certain systems of set theory extending Quine's idea of strati-

[3] Though inaccessible cardinals are not met in ordinary mathematical practice, working set-theorists accept their existence without hesitation as constituting a natural extension of the ZFC axioms, and indeed as only the first in a series of progressively stronger extensions. Gödel (1947) was an early proponent of this idea.

[4] Just one universe of this kind is assumed in my 1969 paper; that is all one needs for the applications. In the 2004 paper, I assumed a sequence of such universes $U_n \in U_{n+1}$ for each $n \in \omega$, in order to relate the idea more directly to Russell's idea of typical ambiguity.

fication, as explained in the next section. These systems are shown to be consistent relative to standard systems of set theory (as proved in detail in the "Unlimited" MS and outlined in the Appendix), so (R4) is also met. Finally, while (R3) is met to a considerable extent, we shall see that there are two ubiquitous set-theoretical constructions that can't be carried out in these systems without ad hoc modification: passage to equivalence classes under an equivalence relation, and formation of the Cartesian product of an indexed family of classes. In addition, certain basic results of category theory such as the Cartesian closedness of the category of all sets and Yoneda's Lemma can't be formulated unrestrictedly. These drawbacks are the price paid under the existing stratified approach in order to satisfy (R1) and (R2) in full. It may be that there can be no solution to (R1)–(R4) without such trade-offs, but nothing I know currently excludes that.

What is to be emphasized from all this work is not that naïve category theory ought further to be pursued within the framework of stratified systems (nor, equally, that it ought *not* to be pursued in that way), but rather that it serves to illustrate how one can meet at least some of the basic requirements without restriction, unlike current standard set-theoretical approaches. Thus emboldened, one should seek ways to meet *all* of the requirements without restriction.

3 The System NFU With Stratified Pairing

The system NF of "New foundations for mathematical logic" has a single sort of variable and the basic relations = and $\in$; its axioms are Extensionality and Stratified Comprehension. For reasons below, I shall use capital letters $A, B, C, \ldots, X, Y, Z$ for its variables; the objects these range over will be called *classes*.[5] A formula φ is said to be *stratified* if it results from a formula of simple type theory by suppressing type distinctions, equivalently if it is possible to assign natural number *type* superscripts to each variable in φ in such a way that (i) each variable is assigned the same type at all its occurrences, (ii) for each subformula of φ of the form $X = Y$, the variables X and Y are assigned the same type, and (iii) for each subformula of φ of the form $X \in Y$, the variable Y is assigned type $n + 1$ when X is assigned type n.

Examples: $X \in Y$ and $Y \in X$ are stratified (for X, Y distinct variables) but $X \in X$ is not and $(X \in Y \wedge Y \in X)$ is not.

The Stratified Comprehension Axiom scheme consists of (the universal closures of) all formulas of the form

$$\text{(SCA)} \quad \exists A \forall X [X \in A \leftrightarrow \varphi]$$

where φ is stratified and the variable A does not occur in φ.

Extensionality (Ext) is stated as usual. Thus NF = Ext + SCA. To this day it is not known whether NF is consistent. For an exposition of the considerable work that has been done exploring NF and some of its variants, see Forster (1995) and

[5] Lower case letters will also be used for classes in some contexts below.

Holmes (1998). The variant that occupies our attention here and that *has* been shown to be consistent by Jensen (1969) is called NFU, because it allows for the possible existence of more than one "urelement", i.e., a class which has no members. This is done by weakening Extensionality as follows to apply to non-urelements:

$$(\text{Ext}')\quad \exists X(X \in A) \wedge \forall X(X \in A \leftrightarrow X \in B) \rightarrow A = B$$

Thus NFU = Ext′ + SCA. NFU is very weak as systems go; Jensen proved its consistency relative to Peano Arithmetic, PA. One cannot prove the existence of an infinite class in NFU.[6]

By (SCA) there is at least one empty class; fix any such and denote it by Λ. For each stratified φ with free variables included in $\{X, Y_1, \ldots, Y_n\}$, we define $\{X \mid \varphi(X, Y_1, \ldots, Y_n)\}$ to be the unique A satisfying SCA for φ if $\exists X\varphi(X, Y_1, \ldots, Y_n)$, otherwise Λ.

In particular, we can define the familiar set-theoretical operations as usual in NFU: $\{Y\}$, $\{Y_1, Y_2\}$, $Y_1 \cup Y_2$ and $Y_1 \cap Y_2$; more generally we can define the union of any class of classes $\bigcup Y$ as $\{X \mid \exists Z(X \in Z \wedge Z \in Y)\}$. Writing $X \subseteq Y$ for $\forall Z(Z \in X \rightarrow Z \in Y)$, we can also define $\wp(Y) = \{X \mid X \subseteq Y\}$. Constructions that are distinctive to classes are $-Y = \{X \mid X \notin Y\}$ and $V = \{X \mid X = X\}$; we have $-\Lambda = \bigcap \Lambda = V$. Also, self-membership makes its first appearance with $V \in V$.

When dealing with relations R in a typed or stratified set-up, for example those that are binary, it is natural to consider them as classes of ordered pairs (X, Y) such that

(*) the types assigned to X, Y, and (X, Y) are all the same.

The usual way of defining pairs in set theory as $(X, Y) = \{\{X\}, \{X, Y\}\}$ is not stratified in this sense. Quine (1945) showed how to define a pairing operation in NF to satisfy (*), but his definition requires full Extensionality and an Axiom of Infinity (Rosser, 1952). Let L_p be the language L augmented by a binary operation $(.,.)$ symbol. By the Pairing Axiom in L_p is meant the following:

(P) $(X_1, X_2) = (Y_1, Y_2) \rightarrow X_1 = Y_1 \wedge X_2 = Y_2$.

The terms $s, t, \ldots$ of L_p are generated from the variables by closing under the pairing operation: whenever s, t are terms, so also is (s, t). The system NFU_P consists of Ext′ + SCA + P, where now the notion of stratification in SCA has to be expanded to accord with (*); this can be achieved by modifying the definition of a formula φ being stratified as follows:

1. Each term t occurring in φ is assigned a natural number as type
2. The type assigned to a term t of φ is the same as the type assigned to each variable occurring in t
3. Each variable of φ has the same type assigned to it at all occurrences

[6] Actually, NFU is quite weak, proof-theoretically, compared to PA (Solovay, unpublished). As shown by Enayat (2004), one can obtain an extension of NFU equivalent in strength to PA by adding "every set is finite" and "every Cantorian set is strongly Cantorian" as axioms (cf. the final section below for the notions of Cantorian and strongly Cantorian sets in the framework of NFU).

4. For each subformula of φ of the form $s = t$, the types assigned to s and t are the same
5. For each subformula of φ of the form $s \in t$ and type n assigned to s, the type assigned to t is $n + 1$.

Examples: $(X, Y) \in Z$ is L_p stratified, but not $[(X, Y) \in Z \wedge X \in Y]$.

Theorem 1 NFU$_P$ *is consistent.*

This theorem may be proved by a straightforward modification of the proof of consistency of NFU + Inf in Theorem 1 of Jensen (1969), where Inf is an Axiom of Infinity. Consistency of a much stronger system than NFU$_P$ is stated in the penultimate section below and an outline of how that is proved is given in the Appendix.[7]

4 First-Order Structures in NFU$_P$

For any classes A, B, define $A \times B$ to be the class of all (X, Y) with $X \in A \wedge Y \in B$. Define n-tuples inductively by $(X_1) = X_1$ and $(X_1, \ldots, X_n, X_{n+1}) = ((X_1, \ldots, X_n), X_{n+1})$. Then for any A, n, define A^n to be the class of all n-tuples $(X_1, \ldots, X_n)$ with $X_i \in A$. An n-ary relation R on A is a subclass of A^n. A function F on A into B, in symbols, $F : A \to B$, is a subclass of $A \times B$ such that for each $X \in A$ there is exactly one Y with $(X, Y) \in F$; we write $F(X) = Y$ in this case. Note that $B^A = \{F \mid F : A \to B\}$ exists by SCA. An n-ary function from A to B is an $F : A^n \to B$.

A single-sorted first-order structure is a tuple

$$A = (O, R_1, \ldots, R_j, F_1, \ldots, F_k, K_1, \ldots K_l)$$

where the domain O of objects of A is non-empty and each R_i is an n_i-ary relation on O for some n_i, each F_i is an m_i-ary function on O to O for some m_i, and each K_i is a singleton, $K_i = \{C_i\}$ for some $C_i \in O$. This is generalized in the obvious way to many-sorted first-order structures. Given a sentence θ in the first-order language of such structures, we write as usual, $A \models \theta$ to express that A satisfies θ, or is a model of θ. By Model(θ) we mean the class of all A such that $A \models \theta$. Associated with each such θ is an L_P stratified formula $\theta^*(X)$ such that Model(θ) $= \{X \mid \theta^*(X)\}$

Examples:

(i) Consider structures $A = (O, R)$ with R a binary relation on O. The class PO is defined to be the class of all such A that are partially ordered. Then PO $=$ Model(θ) $= \{X \mid \theta^*(X)\}$ where $\theta^*(X)$ is the following formula:

[7] Independently, Holmes (1991) showed that NFU$_P$ is interpretable in NFU + Inf, giving a more direct proof of Theorem 1 assuming Jensen's work.

$$\exists Y, Z[X = (Y, Z) \wedge \exists V(V \in Y)$$
$$\wedge \forall U(U \in Z \to \exists V, W(V \in Y \wedge W \in Y \wedge U = (V, W))$$
$$\wedge \forall V(V \in Y \to (V, V) \in Z)$$
$$\wedge \forall V, W((V, W) \in Z \wedge (W, V) \in Z \to V = W)$$
$$\wedge \forall V, W, U((V, W) \in Z \wedge (W, U) \in Z \to (V, U) \in Z)]$$

θ^* is L_P stratified by assigning type 1 to the variables X, Y, Z and type 0 to the variables U, V, W.

(ii) We treat similarly the class Equiv of all $A = (O, R)$ such that R is an equivalence relation on O. $A \in$ Equiv iff $A \models \theta$ where θ is a first-order formula, and then Equiv $=$ Model$(\theta) = \{X \mid \theta^*(X)\}$ with L_P stratified θ^*.

(iii) Consider structures $A = (O, F, G, \{E\})$ where F is a binary operation on O, G is a unary operation on O and $E \in O$. The class Grp is defined to be the class of all such A that are groups, in which F is the multiplication operation on O, G is the inverse operation on O, and E is the identity element for F. Then Grp $=$ Model$(\theta) = \{X \mid \theta^*(X)\}$ for a first-order θ as usual.

(iv) We here treat categories as two-sorted structures $A = (O, M, C, D_0, D_1)$ where O is the collection of its objects, M is the collection of its morphisms, C is the composition operation on morphisms and D_0, D_1 give the domain and codomain, resp., of a morphism, to tell when composition is defined. Thus each D_i is a function from M to O and the ternary relation $C \subseteq M^3$ is a partial function from M^2 to M, with $C(f, g)$ or fg defined for $f, g \in M$ when $D_1(f) = D_0(g)$.[8] The defining condition for A to be a category is given by a first-order formula θ and we can take Cat $=$ Model$(\theta) = \{X \mid \theta^*(X)\}$ to be the class of all categories.

As a warm-up for meeting requirement (R1) in the next section, consider the statement that the class PO is partially ordered under the substructure relation Sub, where $((O, R), (P, S)) \in$ Sub iff $O \subseteq P$ and $R \subseteq S$ and $S \cap O^2 = R$. The relation Sub is provably a class in NFU$_P$ since it is defined by an L_P stratified formula. Then the informal statement can be written as:

$$(\text{PO}, \text{Sub}) \in \text{PO}$$

Similarly, we can re-express the statement that the relation Isom of isomorphism is an equivalence relation on the class Equiv of structures as:

$$(\text{Equiv}, \text{Isom}) \in \text{Equiv}$$

[8] As in Mac Lane (1971) we use lower-case letters $f, g, h, \ldots$ for morphisms in an abstract category, but this does not signal a new kind of variable in NFU$_P$. Similarly, in the next section, where we use $a, b, \ldots$ for objects in a category and η for natural transformations.

5 Meeting Requirements (R1) and (R2) in $\mathrm{NFU_P}$

(R1)

The category of all groups has the form

$$\mathbf{Grp} = (\mathrm{Grp}, \mathrm{Hom}, C, D_0, D_1)$$

where Grp is the class of all groups as in the preceding section, Hom is the class of all $F = (F_0, A, B)$ such that $A = (O_A, \ldots)$ and $B = (O_B, \ldots)$ are groups and $F_0 : O_A \to O_B$ is a group homomorphism from A into B, $D_0(F) = A$ and $D_1(F) = B$, and the composition $C(F, G)$ of F and G in Hom is defined as usual when $D_1(F) = D_0(G)$. Since the classes Grp and Hom and the functions C, D_0, and D_1 exist by SCA, and the structure **Grp** satisfies the conditions to be a category, we may state:

$$\mathbf{Grp} \in \mathrm{Cat}$$

Similarly we can define the category **Top** of all topological spaces and verify that

$$\mathbf{Top} \in \mathrm{Cat}$$

The category of all categories has the form

$$\mathbf{Cat} = (\mathrm{Cat}, \mathrm{Funct}, C, D_0, D_1)$$

where Cat is the class of all categories as in the preceding section, Funct is the class of all $F = (F_0, F_1, A, B)$ such that $A = (O_A, M_A, \ldots)$ and $B = (O_B, M_B, \ldots)$ are categories and the pair $F_0 : O_A \to O_B$, $F_1 : M_A \to M_B$ determines a functor from A into B, $D_0(F) = A$ and $D_1(F) = B$, and the composition $C(F, G)$ of F and G in Funct is defined as usual when $D_1(F) = D_0(G)$. Since **Cat** satisfies the conditions to be a category, we have:

$$\mathbf{Cat} \in \mathrm{Cat}$$

In this way, requirement (R1) is satisfied in $\mathrm{NFU_P}$.

(R2)

Given any two categories $A = (O_A, M_A, \ldots)$ and $B = (O_B, M_B, \ldots)$, the category of all functors from A to B has the form

$$B^A = (\mathrm{Funct}(A, B), \mathrm{Nat}, C, D_0, D_1)$$

where Funct(A, B) is the class consisting of all functors F from A to B, Nat is the class of all natural transformations from one such functor F to another, and C, D_0,

D_1 are explained below. As usual we write $f : a \to b$ for $f \in M_A$ and $a, b \in O_A$ when $D_0(f) = a$ and $D_1(f) = b$, and similarly for morphisms in B. The class Funct(A, B) is the subclass of the class of all pairs (F_0, F_1) for which $F_0 : O_A \to O_B$ is such that for each $a, b \in O_A$ and $f \in M_A$ with $f : a \to b$, we have $F_1(f) \in M_B$ with $F_1(f) : F_0(a) \to F_0(b)$ and the usual conditions on preservation of composition and identity morphisms are satisfied. Natural transformations are taken to be triples (η, F, G) where F, G are two such functors, and $\eta : O_A \to M_B$ in such a way that for each $a \in O_A$, $\eta(a) : F_0(a) \to G_0(a)$ in B and we have the usual commutative square associated with any $f : a \to b$ in A; $D_0(\eta) = F$ and $D_1(\eta) = G$. Composition C of natural transformations is defined in the natural way. Once again we can check that Nat, C, D_0 and D_1 all exist and that B^A is indeed a structure in NFU$_P$. Moreover it satisfies the conditions to be a category so, finally, we can state

$$B^A \in \text{Cat}$$

as a theorem in NFU$_P$, just as required by (R2).

6 The Requirement (R3); Type-Shifting Problems in NFU$_P$

One can establish the existence of the class N in NFU$_P$ and thus the finite type-theoretic hierarchy over N obtained by iterating the power class operation $\wp$ and the construction of function types. More is needed to go to transfinite types as in ZFC; how that is done is dealt with in the next section. Otherwise, for (R3), we have seen in sections 3 and 4 that many of the standard set-theoretic constructions can be carried out without any obstacle in NFU$_P$. The fact that Extensionality is weakened to Ext′ does not hinder usual arguments either. Here we concentrate on operations that *can't* be carried out without ad hoc adjustments.

i) *Equivalence classes.* Suppose $(O, E) \in$ Equiv, i.e., E is an equivalence relation on the class O. For each $X \in O$, define $X/E = \{Y \mid (X, Y) \in E\}$ and $O/E = \{X/E \mid X \in O\} = \{Z \mid \exists X(X \in O \wedge \forall Y(Y \in Z \leftrightarrow (X, Y) \in E))\}$. This exists by SCA, assigning type level 1 to O, E, and Z and type level 0 to X and Y. However, the usual function F from O to O/E cannot be shown to exist since it consists of pairs (X, Z) such that $Z = X/E$ is of type level higher than that of X. The ad-hoc modification in this case is to introduce a new kind of function, from the class of singletons associated with O, Sing(O) $= \{W \mid \exists X(X \in O \wedge W = \{X\}\}$, to O/E. Alternatively, in the presence of a suitably universal choice function (see the penultimate section below for the consistency of that), we can deal in a stratified way with a function to representatives of equivalence classes.
ii) *Cartesian products.* A sequence of classes O_X indexed by X ranging over a class I is given by a function $F : I \to V$ with $F(X) = O_X$ for each $X \in I$. The Cartesian product of this sequence is supposed to be a class $\prod O_X(X \in I)$ whose members are all $G : I \to V$ such that for all $X \in I$, $G(X) \in O_X$. Thus each such G consists of pairs (X, Y) such that $Y \in Z$ where $(X, Z) \in F$; this cannot be arranged in

a stratified way in NFU_P. Again, an ad hoc solution is to modify the notion of function, say by taking $F : \text{Sing}(I) \to V$ for the initial sequence of classes.

iii) *Cartesian closedness of* **Class**. In the context of NFU_P, one deals with the category **Class** of all classes in place of the category of all sets in ordinary set-theoretical foundations. The latter is Cartesian closed, one of whose conditions is that we have an adjoint to Cartesian product (cf. Mac Lane, 1971, p. 95). This yields the exponentiation operation with the evaluation morphisms $\text{ev} : b^a \times a \to b$ given by $\text{ev}(f, x) = f(x)$ for each $f : a \to b$. But that can't be done for **Class** in a stratified way in NFU_P. More definitively and more generally, McLarty (1992) showed that Cartesian closedness of **Class** and **Cat** provably fail in NF; his argument works equally well in NFU_P.

iv) *Yoneda lemma.* Given an abstract category $A = (O_A, M_A, \ldots)$, the Hom classes associated with A are the classes

$$\text{Hom}_A(a, b) = \{f \mid f \in M_A \wedge D_0(f) = a \wedge D_1(f) = b\}$$

As in iii), **Class** is the category of all classes, with the usual mappings from one class into another constituting its morphisms. For each $a \in O_A$ we have a functor H^a from A into **Class** given by

$$H^a(b) = \text{Hom}_A(a, b)$$

for each $b \in O_A$, with the obvious choice of $H^a(f) : H^a(b) \to H^a(c)$ whenever $f : b \to c$ in A. What the Yoneda Lemma does is set up a natural isomorphism between $F(a)$ and the natural transformations from H^a into F for each functor F from A into **Class** (cf. Mac Lane, 1971, p. 61). Closer inspection shows that there is a lot of mixing of types here that can't be represented in NFU_P without ad hoc modifications, to begin with of H^a as a function, since the type level of $H^a(b)$ is one higher than that of its elements $f : a \to b$, which are of the same type level as those of a and b. Like iii), this is a serious drawback to the use of NFU_P as it stands as a foundation for category theory.

There is no obvious modification of the notion of stratification for systems with pairing, that allows pairs (s, t) of mixed type and is consistent. Type-theoretically, the natural thing to try is to assign to (s, t) the type level $\max(n, m)$ when s is assigned n and t is assigned m. The problems i)-iv) all concern situations involving pairs (s, t) where s is assigned type n and t is assigned type $n + 1$. However, if SCA were expanded to allow stratification in this sense we would derive a contradiction from

$$\exists A \forall X[X \in A \leftrightarrow \exists Y, Z(X = (Y, Z) \wedge Y \in Z)]$$

and

$$\forall A \exists B \forall Y[Y \in B \leftrightarrow (Y, Y) \notin A]$$

B is just the Russell class. A possible way out is to restrict oneself to stratified proofs in a suitable sense, so that the types assigned to a class of pairs don't change in the course of the proof. In the example, the elements of A change from pairs of type

(0, 1) to pairs of type (0, 0). Even if that idea were to lead to a consistent system, it might require keeping track of things in a cumbersome way.

7 The Requirement (R3), Continued; Building in ZFC

In this section we boost NFU_P to incorporate ZFC in a certain way; the resulting system was denoted S^* in "Unlimited", and for simplicity of notation, we shall follow that here, too. The language L^* of S^* extends the language L_p of NFU_P by the adjunction of a constant symbol V_0 and *set variables* $a, b, c, \ldots, x, y, z$.[9] The terms of L^* are generated from the constant V_0 and both kinds of variables by closure under the pairing operation. The atomic formulas are $s = t$ and $s \in t$, where s, t are arbitrary terms. The notion of a formula φ being L^* stratified is modified as follows from the section that introduced NFU_P.

1. Each term t occurring in φ is assigned a natural number as type;
2. the type assigned to a term t in φ is the same as the type assigned to each class variable occurring in t;
3. each class variable of φ has the same type assigned to it at all occurrences;
4. for each subformula of φ of the form $s = t$, the types assigned to s and t are the same; and
5. for each subformula of φ of the form $s \in t$ and type n assigned to s, the type assigned to t is $n + 1$.

N.B. No restrictions are made on the types that are assigned to V_0 or set variables; these may be assigned *any* type and may be assigned different types at different occurrences in the same formula.

Examples: The formulas $(V_0 \in X \wedge X \notin V_0)$, $X \in x$, and $(x, X) \in x$ are all L^* stratified, but not the formula $(x, X) \in X$.

Axioms of S^*:

1. *Stratified comprehension:* $\exists A \forall X[X \in A \leftrightarrow \varphi]$ for each L^* stratified φ that does not contain the variable A.
2. *Weak extensionality:* $\exists X(X \in A) \wedge \forall X(X \in A \leftrightarrow X \in B) \rightarrow A = B$.
3. *Pairing.* $(X_1, X_2) = (Y_1, Y_2) \rightarrow X_1 = Y_1 \wedge X_2 = Y_2$.
4. *Sets and classes.*

    a. $\forall x \exists X(x = X)$
    b. $X \in V_0 \leftrightarrow \exists x(x = X)$
    c. $X \in x \rightarrow X \in V_0$

5. *Empty set:* $\exists! z \forall y(y \notin z)$.
6. *Operations on sets:*

[9] In the syntax of S^* lower case letters are now used only in this way.

a. $\{x, y\} \in V_0$
b. $\bigcup x \in V_0$
c. $\wp(x) \in V_0$
d. $(x, y) = \{\{x\}, \{x, y\}\}$

7. *Infinite set:* $\exists a[\exists z(z \in a \wedge \forall y(y \notin z)) \wedge \forall x(x \in a \rightarrow x \cup \{x\} \in a)]$
8. *Replacement:* $\forall x, y_1, y_2[\psi(x, y_1) \wedge \psi(x, y_2) \rightarrow y_1 = y_2] \rightarrow \forall a \exists b \forall y[y \in b \leftrightarrow \exists x(x \in a \wedge \psi(x, y))]$
9. *Foundation:* $\exists x \psi(x) \rightarrow \exists x[\psi(x) \wedge \forall y(y \in x \rightarrow \neg\psi(y))]$, where $\psi(x, \ldots)$ is any L^* formula that does not contain the variable y.
10. *Universal choice:*

$$\exists C[\forall X, Y_1, Y_2((X, Y_1) \in C \wedge (X, Y_2) \in C \rightarrow Y_1 = Y_2)$$

$$\wedge \forall X(\exists Y(Y \in X) \rightarrow \exists Y(Y \in X \wedge (X, \{Y\}) \in C))]$$

Remarks.

(i) The axioms of S^* as presented here are a slight variant of those in "Unlimited"; they are interderivable.
(ii) Axioms 1-3 make S^* an extension of NFU_P.
(iii) The "ontological" Axiom 4 tells us that the sets are exactly the classes that belong to V_0.
(iv) Since there is a unique empty set by Axiom 5, and Extensionality holds for non-empty classes by 2, we have full Extensionality for sets. As in ZF we use 0 to denote the empty set.
(v) Axiom 6 tells us that V_0 is closed under the operations of unordered pair, union and power as defined for classes above; it also tells us that the ordered pair operation coincides with its usual definition in set theory when restricted to sets.
(vi) From Axiom 7 and Separation (see next), we can define ω as the least set containing 0 and closed under the successor operation $x' = x \cup \{x\}$. Again by Separation (or Foundation) we can apply induction on ω to any formula $\psi(x, \ldots)$ of L^*, not just stratified formulas.
(vii) The Separation scheme for sets in L^* consists of all instances

$$\forall a \exists b \forall x[x \in b \leftrightarrow x \in a \wedge \theta(x)]$$

for any formula $\theta(x, \ldots)$ of L^* that does not contain the variables a, b. This is an immediate consequence of Replacement (8), using $\psi(x, y)$ as the formula $\theta(x) \wedge y = x$.
(viii) Foundation (9) is equivalent to transfinite induction on the $\in$-relation restricted to sets for arbitrary formulas $\psi(x, \ldots)$ of L^*:

$$\forall x[\forall y(y \in x \rightarrow \psi(x)) \rightarrow \psi(x)] \rightarrow \forall x \psi(x)$$

(ix) Universal Choice (10) is given in a form appropriate to the use of functions as treated in NFU_P in section 4. It implies the following form of Universal

Choice (UC) for sets:

$$\exists F[F : V_0 \to V_0 \wedge \forall x(x \neq 0 \to F(x) \in x)]$$

for, given C as in axiom X, take $F = \{(x, y) \mid (x, \{y\}) \in C\}$. Then UC implies AC, the usual Axiom of Choice for any set of non-empty sets.

Theorem 2 *(i)* S^* *is consistent.*
(ii) S^* *is an extension of both* NFU_P *and* ZFC.
(iii) The system of Morse-Kelley MK *with* UC *is interpretable in* S^*.

The proof of (i) has been given in "Unlimited"; an outline of the ingredients is presented in the Appendix below. (ii) is immediate from the preceding remarks. As to (iii), the system MK is what we obtain from BG by allowing *any* formula $\theta(x)$ in the language of BG to define a class of sets, not just predicative formulas as in BG. We can interpret it in S^* by simply taking the class variables to range over those classes X in S^* with $X \subseteq V_0$.

MK is stronger than BG, since we can establish a notion of truth for the language of ZF, and by its means prove that every theorem of ZF is true; hence MK proves the consistency of ZF, while BG (which is a conservative extension of ZF) does not. Similarly, MK + UC is stronger than ZFC, and a fortiori of BG + AC.

S^* makes up for the defects of NFU_P to a certain extent. Obviously we can deal with equivalence classes on sets and Cartesian products on sets as usual as in ZFC. More generally for equivalence relations between classes, the Universal Choice axiom provides the possibility of working with representatives rather than equivalence classes in a stratified way.[10] One can also make the kinds of distinctions used by Mac Lane to secure the applications of category theory by means of the notions of "small categories", "locally small categories", etc. So all of Mac Lane (1971) can be directly represented in S^*. But one should also revisit results like Yoneda's Lemma, the Kan Extension Theorem, the Adjoint Functor Theorem, etc., that so far have been formulated using such distinctions to see whether the use of NFU_P over ZFC as provided by S^* gives any additional flexibility or generality. That remains to be done. On the other hand, this type of work returns one to the kinds of distinctions that the aim for a direct foundation of naïve category theory is supposed to avoid. It may be that the use of stratified systems for that purpose cannot be advanced much beyond what has been illustrated here. But at least it shows that the program to satisfy such requirements as (R1)-(R4) is a reasonable one to pursue by some means or another.

8 Cantorian Classes and Extension of NFU in ZFC

By way of comparison with the preceding, much work has been done beginning in the 1990s on the study of extensions of NFU in which ZFC can be interpreted

[10] This has been suggested to me by Randall Holmes.

directly. This centers around the *Cantorian* and *strongly Cantorian* classes, as defined in Holmes (1998): a class A is Cantorian if it is in one-one correspondence with the class $\{Y \mid \exists X(X \in A \wedge Y = \{X\})\}$ of its singletons, and it is strongly Cantorian (or s.c.) if the one-one correspondence is given by the standard map sending X to $\{X\}$ on A. Stratification prevents one from showing that every Cantorian class is strongly Cantorian, let alone that every class has this property; Russell's paradox precludes V, among many other classes, from being strongly Cantorian.[11] The kinds of type-shifting problems met above with the development of category theory in $\mathrm{NFU_P}$ are avoided by restricting to s.c. classes. Thus it is possible that an enhanced development of naïve category theory in a stratified framework could be provided by restriction to s.c. classes and the associated categories defined in terms of them; clearly this would require that there exist "enough" s.c. classes. On the other hand, such a restriction would mean giving up requirement (R1), since the collection of s.c. classes does not form a class (let alone a s.c. class). It is known that the s.c. classes are closed under exponentiation, but this does not help with (R2) for large categories, if (R1) can't be satisfied.

By NFUA is meant the system NFU (with stratified pairing) together with the axioms of infinity and choice and the axiom "every Cantorian class is strongly Cantorian". In Holmes (1998), chapter 20, it is shown how to interpret ZFC in an extension of NFU stronger than NFUA via a certain class of isomorphism types of pointed well-founded extensional relations; this interpretation works in NFUA as well by recent results of Enayat (2004). In fact, much stronger extensions of ZFC come along with that interpretation: in unpublished work, Solovay established the equiconsistency of NFUA with ZFC + "there exist n-Mahlo cardinals" (for each $n \in \omega$); a published proof of that is to be found in Enayat (2004), Theorem 5.5. The strength of the full system in Holmes (1998) has been shown to be that of Morse-Kelley set theory MK plus measurability (in a suitable sense) of the proper class ordinal (the class of all ordinals considered as a virtual ordinal) Holmes (2001). An interesting intermediate system designated NFUB has been proved by Solovay (1997) to be of the same strength as MK + "the proper class ordinal is weakly compact."

Compared to these extensions of NFU, the system S^* of section 7 interprets ZFC only by the addition of a constant symbol for a class V_0 and axioms concerning its members. All the members and subclasses of V_0 are automatically strongly Cantorian. It is an open question whether there is a direct interpretation of ZFC in an extension of NFU without such an additional symbol V_0, in which the sets are taken to range over some collection **C** of classes and the membership relation is the restriction of the $\in$ relation to **C**. It is also open what the exact consistency strength is of S^*; in my original proof, I assumed the existence of two inaccessible cardinals. Since seeing a draft of this paper, Ali Enayat has been looking into that question, and has informed me that — in consistency strength — S^* lies strictly between ZFC + "there exists an inaccessible cardinal" and ZFC + "there exist at least two inac-

[11] An axiom stating that all sets are Cantorian was first studied by Henson (1973). A related "axiom of counting" was introduced by Rosser (1953) in order to develop a smooth theory of finite cardinals in NF. It states that the set of finite cardinals is strongly Cantorian; that set is Cantorian in NF and in NFU + Infinity. (I am indebted to Ali Enayat for this background information.)

cessible cardinals." He has also pointed out to me that my original proof (outlined below) also establishes the consistency of a strengthening S^{**} of it by an axiom scheme asserting that the extension on V_0 of any property given by an arbitrary formula $\varphi(x, \ldots)$ of the language of S^* is a class: $\exists X \forall x \in V_0[x \in X \leftrightarrow \varphi(x)]$. Furthermore, in the presence of this additional axiom, the Replacement scheme follows from the statement that no partial map from an initial segment of V_0 to V_0 can have a cofinal image, and the scheme of Foundation follows from the usual formulation of Foundation in ZFC (that every non-empty set contains an $\in$-minimal element). It may be more tractable to determine the exact consistency strength of S^{**} than that of S^* in terms of more or less standard extensions of ZFC.

Appendix

The methods used to prove Theorem 2(i), the consistency of S^*, in "Unlimited", are by an extension of those applied by Jensen (1969). They consist of three parts:

1. Specker (1962) reduced the consistency of NF to the existence of models $M_T = (\langle U_i \rangle, \langle \in_i \rangle)_{i \in Z}$ of type theory with types i ranging over the set of all integers, $Z = \{\ldots, -3, -2, -1, 0, 1, 2, 3, \ldots\}$, where $\in_i \subseteq U_i \times U_{i+1}$, for which M_T satisfies the axioms of typed comprehension and extensionality, and in addition has a type-shifting automorphism $\sigma : U_i \to U_{i+1}$ for all $i \in Z$. The model of NF constructed from M_T is defined to be $M^* = (U_0, \in^*)$ where for $a, b \in U_0$, $a \in^* b \leftrightarrow a \in_0 \sigma(b)$. Jensen observed that if M_T satisfies extensionality only for non-empty classes, then M^* is a model of NFU.
2. Ehrenfeucht and Mostowski (1956) applied the infinite Ramsey theorem to obtain models of first-order theories with indiscernibles $\{c_i\}_{i \in I}$ in given orderings $(I, <)$. When these models are generated by Skolem functions from the indiscernibles we get elementary substructures having automorphisms induced by those of $(I, <)$. Jensen applied the Ehrenfeucht-Mostowski theorem to obtain models M of Zermelo set theory plus the Skolem function axioms having indiscernibles c_i in order type $(Z, <)$ and shifting automorphism induced by $\sigma(c_i) = c_{i+1}$. A Z-typed model as required for the Specker construction of M^* is formed by taking $U_i = \{x \mid x \in_M c_i\}$. Jensen showed that one can also arrange to have M a model of the axioms of Infinity and Choice, which leads to M^* having the same properties. Thus NFU is consistent with Infinity and Choice. In order to satisfy $\mathrm{NFU_P}$ it is only necessary to ensure of the model M that if $x, y \in c_i$ then $\{x\}$ and $\{x, y\} \in c_i$, hence $(x, y) = \{\{x\}, \{x, y\}\} \in c_i$.
3. In part II of his paper, Jensen showed how, given any ordinal α, one can construct M^* satisfying these conditions which is an end-extension of α; this uses the Erdös-Rado (1956) generalization of the Ramsey theorem to certain infinite partitions. These methods were extended in "Unlimited" to construct M^* which are end-extensions of any given transitive set A. The main theorem needed for this and proved in the Appendix of "Unlimited" is in terms of models of $L_{\infty,\omega}$ with indiscernibles satisfying certain prescribed properties. The formulation of

that theorem is too technical to present here. The particular transitive set used in the application to Theorem 2(i) above is the cumulative hierarchy up to a strongly inaccessible cardinal κ. The proof also assumes the existence of a strongly inaccessible cardinal δ greater than κ.

Acknowledgments I wish to thank Ali Enayat, Thomas Forster, Randall Holmes, Robert Solovay and Sergei Tupailo for their helpful comments on a draft of this article. I am especially grateful to Shivaram Lingamneni for his work on preparing a LaTeX version of this paper.

References

Bénabou, J. (1985) Fibered Categories and the Foundations of Naive Category Theory, *Journal of Symbolic Logic* **50**, 10–37.

Ehrenfeucht, A and Mostowski, A. (1956) Models of Axiomatic Theories Admitting Automorphisms, *Fundamenta Mathematicae* **43**, 50–68.

Enayat, A. (2004) Automorphisms, Mahlo cardinals, and NFU, in Enayat, A. and Kossak, R., eds. *Nonstandard Models of Arithmetic and Set Theory, Contemporary Mathematics,* Vol. 361, Providence, RI: American Mathematical Society.

Enayat, A. (2006) From bounded arithmetic to second order arithmetic via automorphisms, in *Logic in Tehran*, Notes in Logic 26, Association for Symbolic Logic (Urbana, IL); Natick, MA: A.K. Peters, Ltd.

Erdös, P. and Rado, R. (1956) A Partition Calculus for Set Theory, *Bulletin of the American Mathematical Society* **62**, 427–488.

Feferman, S. (1969) Set-Theoretical Foundations for Category Theory (with an appendix by G. Kreisel), in Barr, M. et al., eds., *Reports of the Midwest Category Seminar III, Lecture Notes in Mathematics* **106**, 201–247.

Feferman, S. (1974) Some formal systems for the unlimited theory of structures and categories. Unpublished MS, available online at http : //math.stanford.edu/~ feferman/papers/Unlimited.pdf, Abstract in *Journal of Symbolic Logic* **39** (1974), 374–375.

Feferman, S. (1977) Categorical Foundations and Foundations of Category Theory, in Butts, R. and Hintikka, J., eds., *Logic, Foundations of Mathematics and Computability Theory*, Vol. 1, Dordrecht: Reidel, pp. 149–165.

Feferman, S. (1998) *In the Light of Logic*, Oxford: Oxford University Press.

Feferman, S. (2004) Typical Ambiguity: Trying to Have Your Cake and Eat It Too, in Link,G., ed., *One Hundred Years of Russell's Paradox*, Berlin: de Gruyter, pp. 135–151.

Forster, T.E. (1995) *Set Theory with a Universal Set*, Oxford: Clarendon Press.

Gabriel, P. (1962) Des catégories abéliennes, *Bull. Soc. Math. France* 90, 323–448.

Gödel, K. (1947) What is Cantor's continuum problem?, *The American Mathematical Monthly* **54**, 515–525; errata **55**, 151; reprinted in Gödel, K., *Collected Works*, Vol. 2 (1990), 176–187, along with its revised 1964 version, 254–270.

Hellman, G. (2003) Does category theory provide a framework for Mathematical structuralism?, *Philosophia Mathematica* **11**, 129–157.

Henson, W. (1973) Type-raising operations on cardinals and ordinals in Quine's 'New Foundations', *Journal of Symbolic Logic* **39**, 59–68.

Holmes, M.R. (1991) The axiom of anti-foundation in Jensen's 'New Foundations with Ur-elements', *Bull. de la Soc. Math. de Belgique (serie B)* **43**, 167–191.

Holmes, M.R. (1998) *Elementary Set Theory with a Universal Set*, Vol. 10 of the Cahiers du Centre de logique, Louvain-la-Neuve (Belgium): Academia. Corrected version available at http: //math.boisestate.edu/~ holmes/holmes/head2.ps.

Holmes, M.R. (2001) Strong axioms of infinity in NFU, *Journal of Symbolic Logic* **66**, 87–116.

Jensen, R. (1969) On the consistency of a slight (?) modification of Quine's New Foundations, in Davidson, D. and Hintikka, J., eds., *Words and Objections. Essays on the work of W.V.O. Quine*, Dordrecht: Reidel, pp. 278–291.

Lawvere, F.W. (1966) The category of all categories as a foundation for mathematics, *Proceedings of the La Jolla Conference on Categorical Algebra*, Berlin: Springer, pp. 1–20.

Mac Lane, S. (1961) Locally small categories and the foundations of mathematics, in *Infinitistic Methods*, Oxford: Pergamon Press, pp. 25–43.

Mac Lane, S. (1971) *Categories for the Working Mathematician*, Berlin: Springer.

McLarty, C. (1992) Failure of Cartesian Closedness in NF, *Journal of Symbolic Logic* **57**, 555–556.

McLarty, C. (2004) Exploring categorical structuralism, *Philosophia Mathematica* **12**, 37–53.

Quine, W.V. (1937) New foundations for mathematical logic, *The American Mathematical Monthly* **44**, 70–80.

Quine, W.V. (1945) On Ordered Pairs, *Journal of Symbolic Logic* **10**, 95–96.

Rao, V.K. (2006) On Doing Category Theory Within Set-Theoretic Foundations, in Sica, G., ed., *What is Category Theory?* Monza: Polimetrica, pp. 275–290.

Rosser, J.B. (1952) The axiom of infinity in Quine's New Foundations, *Journal of Symbolic Logic* **9**, 238–242.

Rosser, J.B. (1953) *Logic for Mathematicians*, New York: McGraw-Hill. Second edition, New York: Chelsea Publishing Co., 1978.

Solovay, R. (1997) The consistency strength of NFUB, http://front.math.ucdavis.edu/author/Solovay-R*&c=LO

Specker, E. (1962) Typical Ambiguity in Logic, in Nagel, E., et al., eds., *Methodology and Philosophy of Science. Proceedings of the 1960 International Congress*, Stanford, CA: Stanford University Press, pp. 116–123.

Tupailo, S. (2005) Monotone inductive definitions and consistency of New Foundations, Preprint.

Recent Debate over Categorical Foundations

Colin McLarty

The proposal by Shapiro (2009, p. 76) "to sharpen the battle lines a little" around categorical philosophy and foundations for mathematics suggests also extending the lines to include the original publications by mathematicians William Lawvere (1963, 1964, 1966) and Saunders Mac Lane (1986, 1998).

As a central issue, Hellman (2005) asks whether categorical foundations are asserted as true or merely posed as abstract axioms. Shapiro (2009) surveys possible answers. But neither one cites the original papers on this question. Those papers give the only truly possible answer—the same as for the Zermelo Fraenkel (ZFC) axioms—which is "both." Just like ZFC, category theoretic axioms can be asserted of the basic objects of mathematics, and they can be taken as abstract axioms with multiple interpretations to be studied for various purposes.

The difference is that ZFC is only used in foundations while categorical methods are used all over mathematics. So multiple interpretations of ZFC essentially never occur outside of foundations and they were created only after ZFC foundations.[1] Multiple different categories were used all over mathematics before anyone conceived of categorical foundations. By far the greatest part of category theory is not foundational. It requires some foundation as remarked at length in Eilenberg and Mac Lane (1945, 246–248) where they discuss many approaches available at that time including ZFC and von Neumann-Gödel-Bernays set theory.

The point of Lawvere creating categorical foundations in the early 1960s was that the abstract category axioms are *not* foundations! Logicians who prefer ZFC, or modal structuralism (Hellman, 1989, 2005), or predicativism (Feferman, 1977, 2005) all distinguish between foundational and non-foundational axioms in principle. But their favored concepts rarely occur outside foundations so they rarely need to make the distinction in fact. We will see they sometimes apply it less carefully than category theorists do.

Lawvere and Mac Lane address several goals now associated with "structuralism" in philosophy of mathematics which we discuss below. Their work also re-

[1] E.g. forcing models show the continuum hypothesis is not provable in ZFC.

G. Sommaruga (ed.), *Foundational Theories of Classical and Constructive Mathematics*, The Western Ontario Series in Philosophy of Science 76,
DOI 10.1007/978-94-007-0431-2_7,

minds us there can be no opposition of *categorical* versus *set theoretic* foundations. The first categorical foundation in print was a set theory and specifically the "Elementary Theory of the Category of Sets" (ETCS) given by Lawvere (1964). Mac Lane urged this set theory as a foundation. The contrast is between categorical foundations including categorical set theory on one hand, and on the other hand *membership-based* set theories notably including ZFC. On this matter, Feferman (2006) extends the objections to categorical set theory from his (1977) by adding a mathematical objection from Rao (2006) which in fact was answered by Mitchell (1972) decades ago. Clarity on these issues will allow the battle lines to move on to the actual differences between categorical foundations and ZFC.

1 The Founding Ideas

Graduate student William Lawvere, having prepared with a course on logic with Mendelson and logic seminars by Tarski, Rabin, and Scott, laid out first order axioms for the category of sets and a clear conception of two perspectives on them:

> (a) There is essentially only one category which satisfies these eight axioms together with the additional (nonelementary) axiom of completeness, namely, the category $\mathcal{S}$ of sets and mappings. Thus our theory distinguishes $\mathcal{S}$ structurally from other complete categories, such as those of topological spaces, groups, rings, partially ordered sets, etc. (b) The theory provides a foundation for number theory, analysis, and much of algebra and topology even though no relation $\in$ with the traditional properties can be defined. (Lawvere, 1964, p. 1506)

Lawvere's perspective (a) takes it that sets, groups, topological spaces, and so on all exist prior to any foundation. On this perspective the first order, or elementary, ETCS axioms are abstract. They admit multiple interpretations and Lawvere gives one precise measure of how they would have to be extended beyond first order terms to uniquely characterize the category of sets. Of course no first order axioms can uniquely characterize sets, for Löwenheim-Skolem reasons even before we invoke Gödel's incompleteness theorem.

At the same time Lawvere offers the second perspective (b) which "provides a foundation." On this perspective, the eight first order ETCS axioms are asserted as an explicit, self-contained account of sets. This account suffices to state and prove the standard results of mathematics as first order theorems without using the nonelementary completeness axiom. This is not just formally independent of ZFC foundations but entirely omits the basic relation $\in$ of ZFC.

He gives an equally clear account of the scope and nature of a stronger, more comprehensive foundation in the category of categories. Again he takes it that mathematical objects exist prior to foundations, while foundations aim to give formally adequate first order descriptions of our mathematics:

> Here by "foundation" we mean a single system of first order axioms in which all usual mathematical objects can be defined and all their usual properties proved.... The author believes, in fact, that the most reasonable way to arrive at a foundation [which can reckon with the Eilenberg-Mac Lane theory of categories and functors] is simply to write down

axioms descriptive of properties which the intuitively-conceived category of all categories has until an intuitively-adequate list is attained; that is essentially how the theory described below was arrived at. Various metatheorems should of course then be proved to help justify the feeling of adequacy. (Lawvere, 1966, p. 1)

Hellman (2003, pp. 136–137) asked "what axioms govern the existence of categories or topoi" and found with surprise that "the question really just does not seem to be addressed!" In fact Lawvere had long before offered ETCS and CCAF as two answers. Mac Lane had endorsed ETCS in a long philosophical account (1986, especially chapter XI) and in the standard graduate text on category theory (1998, Appendix) and in a textbook on sheaves (Mac Lane & Moerdijk, 1992, pp. 331ff.). Shapiro (2005) treats Lawvere's and Mac Lane's answers as debatable, as philosophers must. But that debate would benefit from greater familiarity with the answers as Lawvere and Mac Lane give them.

When Hellman (2005, p. 549) looked at those answers he took a strong stand, claiming Lawvere's and Mac Lane's positions are "in diametric opposition to the algebraico-structuralist reading that categorists apply to their own systems." He was mistaken. Categorists are not diametrically opposing themselves. Indeed Awodey (1996, p. 213) says, and says correctly, that "a category is *anything* satisfying" the elementary category axioms given by Eilenberg and Mac Lane (1945). We do indeed apply an "algebraico-structuralist" reading to these axioms. And so no one offers them as a foundation. Various of us have offered the ETCS and CCAF axioms as foundations. The ETCS axioms taken as a foundation are not abstract nor is the specific *category of sets* just anything that satisfies them. The CCAF axioms are not abstract nor is the specific category of categories just anything satisfying them. This is the distinction mentioned above in the introduction, between categories in the abstract sense which find very many different uses in mathematics and specific categories offered as foundations. Hellman failed to make the very distinction that he claimed the categorists fail to see.

Feferman also fails to distinguish foundations from applications which need foundations. Hellman endorsed Feferman on this point:

Feferman argued that category theory presupposes and uses, informally, notions of *collection* and *operation* both in saying what a category (or topos) is, and in relating categories to to one another through *homomorphisms* or *functors*. Moreover, a foundational framework for mathematics must provide some systematic account of these notions, something set theory does but category theory does not. (Hellman, 2005, p. 549)

It is entirely true that general category theory presupposes those things. For example, many categories used in practice are assumed to have *products* for all sets of objects. So the theory must be able to talk about all sets of objects (Feferman, 1977, p. 150).

Happily, neither ETCS nor CCAF is general category theory! They are two explicit, powerful axiom systems which do not presuppose sets or any other foundation. Each makes its own strong existence assertions. They are foundations *for* general category theory and for the rest of mathematics: "By 'foundation' we mean a single system of first order axioms in which all usual mathematical objects can be defined and all their usual properties proved" (Lawvere, 1966, p. 1). Feferman has never discussed any specific first order axioms for categorical foundations.

Hellman has lately discussed the specifics. As to ETCS he says:

> What remains problematic, however, regarding McLarty's reading of ETCS (which he attributes to Mac Lane), is its apparent commitment to a fixed, presumably maximal, real-world universe of sets, 'the category of sets.' This just strikes me as a convenient fiction. (Hellman, 2006, p. 154)

It is not my fiction. I have often rejected any fixed or maximal universe of sets. Just to quote one earlier reply to Hellman:

> Hellman warns against pretending to all embracing completeness in foundations. I agree. I only find it a less pressing issue because I would not know how to pretend to it if I tried. Clearly there are more sets, categories, smooth spaces, or whatever, than any given axiomatization can prove, and no axiomatization I have ever seen denies it. (McLarty, 2005a, p. 53n)

I not only attribute the same view to Mac Lane but quote him in that and other articles.[2] I have always endorsed his view: "any fixed foundation would preclude the novelty which might result from the discovery of new form" (Mac Lane, 1986, p. 455). Let us be clear that Mac Lane took this as a criticism of fixed foundations. He did not want to preclude novelty arising from progress in the future. And of course he knew that no theory of foundations could ever succeed at doing that in fact even if it was meant to. I endorse Lawvere saying "foundations come out of practice, and will change as practice develops" so that "pure foundations of mathematics are no foundations of mathematics at all" quoted in (McLarty, 1990, p. 370). We return to this below in connection with Awodey's anti-foundationalism.

The original ideas of categorical foundations were as fully open-ended as Hellman's own. Only they see this open-endedness in terms of mathematical forms and mathematical practice rather than by Hellman's preferred device of formal modal logic.

Hellman's pluralism has come to draw on largely the same examples as led Lawvere to his views in the 1960s and 1970s. It would be interesting to pursue more detailed comparison with that work. Elementary topos theory was born simultaneously with synthetic differential geometry, as described notably in Lawvere (1979). The original motives for both are mathematical rather than logical and are quite different from the motives offered for them by Hellman (2006, pp. 155ff.). At the same time Lawvere developed aspects of proof theory and constructive logic in this setting, as in Lawvere (1969), and the constructivist aspect was taken farther by Fourman and Scott (1979), Hyland (1982) and many others since. There is a large technical literature on this mathematics. Any one of these ideas can be axiomatized as an independent foundation, with more or less plausibility depending on the specific idea and the specific axiomatic take on it.

More attractive from a foundational viewpoint is to regard them all as living in CCAF, that is the category of categories as a foundation, as Lawvere intended from his earliest days as a student of category theory before he had offered any axioms for categorical foundations. The problem with CCAF according to Hellman is that the axioms do not "really mean" what they say:

---

[2] See especially (McLarty, 2005b; McLarty, 2007).

> when we speak of the "objects" and "arrows" of a metacategory of categories as categories and functors, respectively, what we really mean is "structures (or at least "interrelated things") satisfying the algebraic axioms of CT", i.e. we are using "satisfaction which is normally understood set-theoretically. That is not to say that there are no alternative ways of understanding "satisfaction"; second-order logic or a surrogate such as the combination of mereology and (monadic) plural quantification of modal structuralism would also suffice. But clearly there is some dependence on a background that explicates satisfaction of sentences by structures, and this background is not "category theory" itself, either as a schematic system of definitions or as a substantive theory of a metacategory of categories. (Hellman, 2006, p. 157)

In fact Lawvere (1966) and I (1991) really do mean our axioms as published. They are straightforward first order axioms. We really do not mean something else about structures, algebraic axioms, a set-theoretic satisfaction relation, mereology, or modal logic. It is unclear whether Hellman believes there are formal defects in our published first order proofs. He may merely believe that for some unstated philosophical reason we should mean something other than what we say. Until he specifies which of those he believes, I cannot reply directly.

I can say that people often make new mathematical ideas unnecessarily hard for themselves by insisting they are 'really' older ideas in disguise. This is no new issue for category theory:

> Students afflicted with this misunderstanding have trouble escaping the idea that objects are 'really' structured sets and arrows are 'really' structure preserving functions. So they keep looking for the truth 'behind' the category axioms instead of learning to use the axioms. They have trouble learning categorical definitions not because the definitions are too complex but because they believe the axioms must 'really mean' something other than what they say. (McLarty, 1990, p. 365)

We return to the notion of "assertory" axioms. Shapiro (2005, p. 65) puts it somewhat differently, invoking the Frege-Hilbert debate, asking whether "the terms in the proposed 'axioms' [...] have meaning beforehand." The answer published by Lawvere and Mac Lane is *yes*. We know what sets and categories are antecedently to axiomatization. More fully though, neither Lawvere nor Mac Lane claims to know a single objectively real and objectively complete category of absolutely *all* sets (or categories). Both know perfectly well that there *cannot* ever be a complete first order axiomatization of such a thing. Rather, each says "the" category of sets (or of categories) is ever-further discovered by mathematical experience through time. They differ somewhat on the epistemology of this process. Lawvere and Rosebrugh (2003) describes mathematical experience as producing mathematical truth and in fact Lawvere has a dialectical conception of this best described in his (1996). Mac Lane (1986) takes a kind of Popperian view where mathematics is not empirically falsifiable and thus neither true nor false, but correct or incorrect. A great many questions remain on both of those views, as indeed many questions remain on the conceptually parallel ideas of "coherence" in Shapiro (1997, 2000) and "possibility" in Putnam (1967) and Hellman (1989).

I agree with Mac Lane and Lawvere (not by coincidence) in holding that all assertions in science can be expected to evolve over time. Here my foundationalism comes closer to the anti-foundationalism of Awodey (1996, 2004):

> there is neither a once-and-for-all universe of all mathematical objects nor a once-and-for-all system of all mathematical inferences. (Awodey, 2004, p. 58)

On Awodey's view the mathematician makes provisional assertions about the objects at hand and the inferences to be recognized, and provisional assertions should not be called foundations. I hold that foundations are necessarily provisional in that they will necessarily change and advance with time.

I do not know if any supporters of categorical foundations believe that a foundation for mathematics can be fixed forever. Most categorists interested in foundations today learned ZFC set theory before they heard of categorical foundations, in some cases before categorical foundations existed, and of course in Lawvere's case before he invented categorical foundations. Categorical foundations came to us as a new alternative. If you have only ever learned one rigorous foundation for mathematics, say ZFC, you could believe it is the only one possible. But after you have learned two, say ZFC and ETCS, you cannot believe they are the only two.

Mac Lane is explicit that foundations for him are not a priori limitations on what mathematics can do, nor accounts of what mathematics has really been about from time immemorial. Foundations are "proposals for the organization of Mathematics" (1986, p. 406) and undoubtedly better organizations will yet arise if mathematics endures into the future. My foundationalism and Awodey's anti-foundationalism have real conceptual differences but are not worlds apart.

2 Feferman and Rao

Feferman (2006) makes two objections to categorical foundations:

> It is not clear what exactly is meant by categorical foundations for category theory and how it proposes to handle the problem of the category of all categories and that of arbitrary functor categories. There is also a specific mathematical objection that has been raised by Rao concerning the construction of localizations in homotopical algebra that make use of transfinite induction and recursion. As he says, "it is not clear how to formulate these in categorical terms ... Solving these problems [by such means] looks remote at the moment." (Feferman, 2006, p. 186)

In fact he applies that first objection to set theoretic foundations as well. Feferman states clearly "I am opposed to [set theory] on fundamental philosophical grounds" (2006, p. 185).

His mathematical objection, though, is meant to show that set theory has an advantage over category theory. Feferman of course means membership-based theories such as ZFC have an advantage over categorical set theory. Rao refers to some cutting edge methods of topology, and indeed finds it "not clear how to formulate [the central constructions of these methods] in categorical terms" (Rao, 2006, p. 278). But routine means of formulation were given decades ago when logicians proved ETCS has the same expressive power as ZFC.[3] Most mathematics is expressed the

[3] For full proofs see (Mitchell, 1972; Osius, 1974), beautifully organized in (Johnstone, 1977, §9.3), more recently (Mac Lane & Moerdijk, 1992, §VI.10). See (McLarty, 2004).

same way verbatim in both foundations, and everything expressible at all in one can be routinely translated into the other.

Equality of expressive power does not mean ZFC and ETCS are equal in proof-theoretic strength. In fact ZFC is much stronger. But it does mean some extension of ETCS has the strength of ZFC. (Lawvere, 1964, p. 34) gave one using a neatly stripped-down reflection principle R. All isomorphism invariant theorems of ZFC are expressed and proved the same way verbatim in ETCS+R, and absolutely all theorems and proofs in ETCS+R translate verbatim into ZFC. This covers essentially all of mathematics outside pure set theory.

Membership in ZFC is not isomorphism invariant: to say that $S \in T$ and T is isomorphic to U obviously does not imply $S \in U$. So ZFC membership does not translate verbatim to ETCS. But the *Mostowski collapsing theorem* (Jech, 2006, p. 69) shows how to restate every ZFC sentence as an isomorphism invariant statement about well-founded extensional relations within ZFC. That version of it translates verbatim into ETCS+R, as shown by both Mitchell (1972) and Osius (1974).

Such a brute force translation of Rao's mathematics into ETCS+R would be identical verbatim to the ZFC version at most points, and clunky at some points. Of course a practical categorical approach to these problems would not just use the brute force translation. It would adapt the approach to suit categorical terms. Perhaps Rao only meant to say that he does not see any positive advantage to using categorical terms. It would be interesting to hear more. But that is no question of adequacy of foundations, just a question of adaptation to his way of approaching his problem. It is well known that ZFC and ETCS+R are routinely intertranslatable.

3 The Differences

Categorical set theory and membership based set theories do not differ in their technical relations to metatheories, neither one formally requires the other, and neither one describes structures the other cannot. Arguments purporting to show they differ in those ways are simply mistaken.

Rather, ETCS and ZFC differ overtly: ZFC takes membership as primitive and ETCS takes composition of functions as primitive. Lawvere put it in both practical terms and tendentious ontological terms saying

> Philosophically, it may be said that these developments partially support the thesis that even in set theory and elementary mathematics it is also true as has long been felt in advanced algebra and topology, namely that the substance of mathematics resides not in Substance (as it is made to seem when $\in$ is the irreducible predicate, with the accompanying necessity of defining all concepts in terms of a rigid elementhood relation) but in Form (as is clear when the guiding notion is isomorphism-invariant structure, as defined, for example, by universal mapping properties). (Lawvere, 1965, p. 7)

So ETCS is closer to the working methods of mainstream mathematics than ZFC is, and like those methods it rests ontologically on form and structure rather than membership and substance. Lawvere and Rosebrugh (2003) and Lawvere and

Schanuel (2007) expand on both claims. Mac Lane (1986) goes into much more detail than Lawvere on the categorical practice of mainstream mathematics and offers somewhat different ontological conclusions from it.

Lawvere (1965) mentions advanced topology and algebra but every year more mathematics is done in explicitly categorical terms. As a textbook example the influential Lang (1965) gives many full, precise descriptions of algebraic structures up to isomorphism by their categorical properties, specifically their *universal mapping* properties. Then he gives brief and less explicit set-theoretic constructions to prove each kind of structure exists.[4] The set theory in that book is not precisely defined or axiomatized in any terms, but it is all on its face categorical set theory. It uses none of the characteristic features of ZFC. It uses no well-founding of sets, no transfinitely iterated membership, no extensionality on the level of sets. It frequently describes a subset $U \subseteq S$ of a given set S by saying which elements of S are in U. That is extensionality on the level of subsets and it is common to ETCS and ZFC. It never describes a set S by saying what sets are elements of it. That would be extensionality on the level of sets which ZFC has and ETCS does not. These are genuine differences between categorical set theory and membership-based set theory. They might be fruitfully debated by philosophers. But I think it would be hard to deny that mathematicians have chosen the categorical approach in practice (McLarty, 2008).

What about the ontological issues of Substance versus Form, or membership versus isomorphism invariant structure? These are the special province of structuralists and in fact Hellman finds the situation "encourages us to seek some kind of synthesis of CTS with MS" (2005, p. 560). The abbreviations stand for Category Theoretic Structuralism and Hellman's own Modal Structuralism. That topic should be pursued, and ought to benefit also from the greatest difference between categorical foundations and membership-based set theory, the one we have barely touched on. That is the wide range of alternatives.

Beyond the intrinsic pluralism of CCAF foundations themselves, there are other categorical foundations and frameworks quite unlike either ETCS or CCAF. All variants of membership based set theory, including ZFC and for example Aczel (1988) and Forster (1995), deal with essentially the same idea of a set as a discrete collection of elements. So does the first categorical foundation in print, ETCS. But the first categorical foundation conceived was *The Category of Categories as Foundation* taking categories themselves as fundamental (Lawvere, 1963, 1966). It includes discrete collections as a special case without taking them as fundamental. Another foundation yet takes smooth spaces as in differential geometry as basic (Bell, 1998; Lawvere, 1979; McLarty, 1988). On that foundation the points, curves, surfaces, and so on in any space are all equally aspects of its spatial organization. Points are not constituents out of which the curves and so on, or the space itself, are constituted. Another is the *free topos* as described by Lambek and Scott in this volume. A variety of approaches are described by Landry and Marquis (2005). No matter what you take as foundation, any of these ideas can serve as a framework for intrinsic description of structure. All are drawn from the current working tools of

[4] Lang (2005) is now in its third edition. Many other textbooks have taken up the categorical style. I cite Lang just because of its influence over many decades.

structural mathematics (McLarty, 2008). Structuralists ought to draw on that mathematical experience. The foundational claims of any of these alternatives can be more valuably debated when the basic issues which arise even for categorical set theory are understood.

References

Aczel, P. (1988) *Non-well-founded Sets*, Stanford, CA: Center for the Study of Language and Information.

Awodey, S. (1996) Structuralism in mathematics and logic, *Philosophia Mathematica* **4**, 209–237.

Awodey, S. (2004) An answer to G. Hellman's question: "does category theory provide a framework for mathematical structuralism?", *Philosophia Mathematica* **12**, 54–64.

Bell, J. (1998) *A Primer Of Infinitesimal Analysis*, Cambridge: Cambridge University Press.

Eilenberg, S. and Mac Lane, S. (1945) General theory of natural equivalences, *Transactions of the American Mathematical Society* **58**, 231–294.

Feferman, S. (1977) Categorical foundations and foundations of category theory, in Butts, R. and Hintikka, J., eds., *Logic, Foundations of Mathematics, and Computability Theory*, Dordrecht: D. Reidel, pp. 149–69.

Feferman, S. (2005) Predicativity, in Shapiro, S., ed., *The Oxford Handbook of Philosophy of Mathematics and Logic*, Oxford: Oxford University Press, pp. 590–562.

Feferman, S. (2006) Enriched stratified systems for the foundations of category theory, In Sica, G., editor, *What is category theory?*, Monza: Polimetrica, pp. 185–204.

Forster, T. (1995) *Set theory with a universal set*, New York: Oxford University Press.

Fourman, M. P. and Scott, D., eds. (1979) *Applications of sheaves (Durham, 1977)*, Number 753 in Lecture Notes in Mathematics, Springer-Verlag.

Hartshorne, R. (1977) *Algebraic Geometry*, Berlin: Springer.

Hellman, G. (1989) *Mathematics Without Numbers*, Oxford: Oxford University Press.

Hellman, G. (2003) Does category theory provide a framework for mathematical structuralism? *Philosophia Mathematica*, **11**, 129–157.

Hellman, G. (2005) Structuralism, in Shapiro, S., ed., *The Oxford Handbook of Philosophy of Mathematics and Logic*, Oxford: Oxford University Press, pp. 536–562.

Hellman, G. (2006) What is categorical structuralism? in J. van Benthem, G. Heinzmann, M. Rebuschi, H. Visser, eds., *The Age of Alternative Logics: Assessing Philosophy of Logic and Mathematics Today*, Berlin: Springer-Verlag, pp. 131–162.

Hyland, M. (1982) The effective topos, in Troelstra, A. S. and van Dalen, D., eds., *The L.E.J. Brouwer Centenary Symposium*, New York: North-Holland, pp. 165–216.

Jech, T. (2006) *Set Theory*, New York, NY: Academic Press.

Johnstone, P. (1977) *Topos Theory*, New York, NY: Academic Press.

Landry, E. and Marquis, J.-P. (2005) Categories in context: Historical, foundational, and philosophical, *Philosophia Mathematica* **13**, 1–43.

Lang, S. (1965) *Algebra*, Reading MA: Addison-Wesley.

Lang, S. (2005) *Algebra*, Reading MA: Addison-Wesley.

Lawvere, F. W. (1963) *Functorial Semantics of Algebraic Theories*, PhD thesis, Columbia University, Published with author commentary in: *Reprints in Theory and Applications of Categories*, No. 5 (2004) pp. 1–121, available on-line at http://138.73.27.39/tac/reprints/articles/5/tr5abs.html.

Lawvere, F. W. (1964) An elementary theory of the category of sets, *Proceedings of the National Academy of Science of the U.S.A* **52**, 1506–1511.

Lawvere, F. W. (1965) An elementary theory of the category of sets. Lecture notes of the Department of Mathematics, University of Chicago, Reprint with commentary by the author and Colin McLarty in: *Reprints in Theory and Applications of Categories*, No. 11 (2005) pp. 1–35, on-line at http://138.73.27.39/tac/reprints/articles/11/tr11abs.html.

Lawvere, F. W. (1966) The category of categories as a foundation for mathematics, in Eilenberg, S. et al., eds., *Proceedings of the Conference on Categorical Algebra, La Jolla, 1965*, Berlin: Springer, pp. 1–21.

Lawvere, F. W. (1969) Adjointness in foundations, *Dialectica*, **23**, 281–96.

Lawvere, F. W. (1979) Categorical dynamics, in *Topos Theoretic Methods in Geometry*, number 30 in Various Publications, Aarhus: Aarhus University, pp. 1–28.

Lawvere, F. W. (1996) Grassman's dialectics and category theory, in Schubring, G., ed., *Hermann Günther Grassmann (1809–1877): Visionary Mathematician, Scientist and Neohumanist Scholar*, Dordrecht: Kluwer Academic Publishers, pp. 255–264.

Lawvere, F. W. and Rosebrugh, R. (2003) *Sets for Mathematics*, Cambridge: Cambridge University Press.

Lawvere, W. and Schanuel, S. (2007) *Conceptual Mathematics: A First Introduction to Categories*, Second edition, Cambridge: Cambridge University Press.

Mac Lane, S. (1986) *Mathematics: Form and Function*, Berlin: Springer.

Mac Lane, S. (1998) *Categories for the Working Mathematician*, Second edition, New York: Springer.

Mac Lane, S. and Moerdijk, I. (1992) *Sheaves in Geometry and Logic*, Berlin: Springer-Verlag.

McLarty, C. (1988) Defining sets as sets of points of spaces, *Journal of Philosophical Logic* **17**, 75–90.

McLarty, C. (1990) The uses and abuses of the history of topos theory, *British Journal for the Philosophy of Science* **41**, 351–375.

McLarty, C. (1991) Axiomatizing a category of categories, *Journal of Symbolic Logic* **56**, 1243–1260.

McLarty, C. (2004) Exploring categorical structuralism, *Philosophia Mathematica*, **12**, 37–53.

McLarty, C. (2005a) Learning from questions on categorical foundations, *Philosophia Mathematica*, **13**, 44–60.

McLarty, C. (2005b) Saunders Mac Lane (1909–2005) his mathematical life and philosophical works, *Philosophia Mathematica* **13**, 237–251.

McLarty, C. (2006) Two aspects of constructivism in category theory, *Philosophia Scientiae*, Cahier Spécial **6**, 95–114.

McLarty, C. (2007) The last mathematician from Hilbert's Göttingen: Saunders Mac Lane as a philosopher of mathematics, *British Journal for the Philosophy of Science* **58**, 77–112.

McLarty, C. (2008) What structuralism achieves, in Mancosu, P., ed., *The Philosophy of Mathematical Practice*, Oxford: Oxford University Press, pp. 354–69.

Mitchell, W. (1972) Boolean topoi and the theory of sets, *Journal of Pure and Applied Algebra* **2**, 261–274.

Osius, G. (1974) Categorical set theory: A characterization of the category of sets, *Journal of Pure and Applied Algebra* **4**, 79–119.

Putnam, H. (1967) Mathematics without foundations, *Journal of Philosophy* **64**, 5–22.

Rao, V. (2006) On doing category theory within set theoretic foundations, In Sica, G., ed., *What is category theory?*, Monza: Polimetrica, pp. 275–289.

Shapiro, S. (1997) *Philosophy of Mathematics: Structure and Ontology*, Oxford: Oxford University Press.

Shapiro, S. (2000) *Thinking about Mathematics*, Oxford: Oxford University Press.

Shapiro, S. (2005) Categories, structures, and the Frege-Hilbert controversy: The status of metamathematics, *Philosophia Mathematica* **13**(1), 61–77.

Part III
Between Foundations of Classical and Foundations of Constructive Mathematics

The Axiom of Choice in the Foundations of Mathematics

John L. Bell

The principle of set theory known as the *Axiom of Choice* (**AC**) has been hailed as "probably the most interesting and, in spite of its late appearance, the most discussed axiom of mathematics, second only to Euclid's axiom of parallels which was introduced more than two thousand years ago".[1] It has been employed in countless mathematical papers, a number of monographs have been exclusively devoted to it, and it has long played a prominently role in discussions on the foundations of mathematics.

In 1904 Ernst Zermelo formulated the Axiom of Choice in terms of what he called *coverings* (Zermelo, 1904). He starts with an arbitrary set M and uses the symbol M' to denote an arbitrary nonempty subset of M, the collection of which he denotes by $\mathbf{M}$. He continues:

> Imagine that with every subset M' there is associated an arbitrary element m_1', that occurs in M' itself; let m_1' be called the "distinguished" element of M'. This yields a "covering" γ of the set M by certain elements of the set M. The number of these coverings is equal to the product [of the cardinalities of all the subsets M'] and is certainly different from 0.

The last sentence of this quotation—which asserts, in effect, that coverings always exist for the collection of nonempty subsets of any (nonempty) set—is Zermelo's first formulation of **AC**.[2] This is now usually stated in terms of *choice functions*: here a choice function on a collection $\mathcal{S}$ of nonempty sets is a map f with domain $\mathcal{S}$ such that $f(X) \in X$ for every $X \in \mathcal{S}$. Zermelo's first formulation of the Axiom of Choice then reads: **AC1** *Any collection of nonempty sets has a choice function.* **AC1** can also be reformulated in terms of relations, viz. **AC2** *For any relation R between sets A, B,*

$$\forall x \in A \exists y \in BR(x, y) \Rightarrow \exists f : A \to B \forall x \in AR(x, fx).$$

[1] Fraenkel, Bar-Hillel, Levy (1973), Section II.4.

[2] Zermelo does not actually give the principle an explicit name at this point, however. He does so only in his (1908), where he uses the term "postulate of choice".

G. Sommaruga (ed.), *Foundational Theories of Classical and Constructive Mathematics*, The Western Ontario Series in Philosophy of Science 76,
DOI 10.1007/978-94-007-0431-2_8,

In his (1908) Zermelo offered a formulation of **AC** couched in somewhat different terms from that given in his earlier paper. Let us call a *choice set* for a family of sets $\mathcal{S}$ any subset $T \subseteq \cup\mathcal{S}$ for which each intersection $T \cap X$ for $X \in \mathcal{S}$ has exactly one element. Zermelo's second formulation of **AC** amounts to the assertion[3] that any family of mutually disjoint nonempty sets has a choice.

Zermelo asserts that "the purely objective character" of this principle "is immediately evident". In making this assertion he meant to emphasize the fact that in this form the principle makes no appeal to the possibility of making "choices". It may also be that Zermelo had something like the following "combinatorial" justification of the principle in mind. Given a family $\mathcal{S}$ of mutually disjoint nonempty sets, call a subset $S \subseteq \cup\mathcal{S}$ a *selector* for $\mathcal{S}$ if $S \cap X \neq \emptyset$ for all $X \in \mathcal{S}$. Clearly selectors for $\mathcal{S}$ exist; $\cup\mathcal{S}$ itself is an example. Now one can imagine taking a selector S for $\mathcal{S}$ and "thinning out" each intersection $S \cap X$ for $X \in \mathcal{S}$ until it contains just a single element. The result[4] is a choice set for $\mathcal{S}$.

Let us call Zermelo's 1908 formulation the *combinatorial* axiom of choice: **CAC**[5] *Any collection of mutually disjoint nonempty sets has a choice set.* It is to be noted that **AC1** and **CAC** for *finite* collections of sets are both provable (by induction) in the usual set theories. As is well-known, Zermelo's original purpose in introducing **AC** was to establish a central principle of Cantor's set theory, namely, that every set admits a well-ordering and so can also be assigned a cardinal number. His introduction of the axiom, as well as the use to which he put it, provoked considerable criticism from the mathematicians of the day. The chief objection raised was to what some saw as its highly non-constructive, even idealist, character: while the axiom asserts the possibility of making a number of—perhaps even uncountably many—arbitrary "choices", it gives no indication whatsoever of how these latter are actually to be effected, of how, otherwise put, choice functions are to be *defined*. For this reason Bertrand Russell regarded the principle as doubtful at best. The French Empiricists Baire, Borel and Lebesgue, for whom a mathematical object could be asserted to exist only if it can be uniquely defined went further in explicitly repudiating the principle in the uncountable case.

On the other hand, a number of mathematicians came to regard the Axiom of Choice as being true a priori. These all broadly shared the view that for a mathematical entity to exist it was not necessary that it be uniquely definable. Zermelo himself calls **AC** a "logical principle" which "cannot... be reduced to a still simpler one" but which, nevertheless, "is applied without hesitation everywhere in mathematical deductions." Ramsey asserts that "the Multiplicative Axiom seems to me the

---

[3] Zermelo's formulation reads literally: *A set S that can be decomposed into a set of disjoint parts A,B,C..., each containing at least one element, possesses at least one subset S_1 having exactly one element with each of the parts A,B,C,..., considered.*

[4] This argument, suitably refined, yields a rigorous derivation of **AC** in this formulation from Zorn's lemma.

[5] It is this formulation of **AC** that Russell and others refer to as the *multiplicative axiom*, since it is easily seen to be equivalent to the assertion that the product of arbitrary nonzero cardinal numbers is nonzero.

most evident tautology".[6] Hilbert employed **AC** in his defence of classical mathematical reasoning against the attacks of the intuitionists: indeed his ε-operators are essentially just choice functions. For him, "the essential idea on which the axiom of choice is based constitutes a general logical principle which, even for the first elements of mathematical inference, is indispensable."[7]

A particularly interesting analysis of the axiom of choice was formulated by Paul Bernays.[8] He saw **AC** as the result of a natural extrapolation of what he terms "extensional logic", valid in the realm of the finite, to infinite totalities. He considers formulation **AC2**, with the two sets A and B identical. In the special case in which A contains just two (or, more generally, finitely many elements), **AC2** is essentially just the usual distributive law for $\wedge$ over $\vee$. Bernays now observes:

> The universal statement of the principle of choice is then nothing other than the extension of an elementary-logical law [i.e. the distributive law] for conjunction and disjunction to infinite totalities, and the principle of choice constitutes thus a completion of the logical rules that concerns the universal and the existential judgment, that is, of the rules of existential inference, whose application to infinite totalities also has the meaning that certain elementary laws for conjunction and disjunction are transferred to the infinite.

He goes on to remark that the principle of choice "is entitled to a special position only to the degree that the *concept of function* is required for its formulation." Most striking is his further assertion that the concept of function "in turn receives an adequate implicit characterization only through the principle of choice."

What Bernays seems to be saying here is that in asserting the antecedent of **AC2**, in this case $\forall x \in A \exists y \in A\ R(x, y)$, one is implicitly asserting the existence of a function $f : A \to A$ for which $R(x, fx)$ holds for all x—that is, the consequent of **AC2**. On the surface, this seems remarkably similar to the justification of **AC** under constructive interpretations of the quantifiers: indeed, under (some of) those interpretations (discussed further below), the assertability of an alternation of quantifiers $\forall x \exists y R(x, y)$ means *precisely* that one is given a function f for which $R(x, fx)$ holds for all x. However, Bernays goes on to draw the conclusion that, for the concept of function arising in this way, "the existence of a function with a [given] property in no way guarantees the existence of a concept-formation through which a determinate function with [that] property is uniquely fixed." In other words, the existence of a function may be asserted without the ability to provide it with an explicit definition.[9] This is incompatible with stronger versions of constructivism.

Bernays and the constructivists both affirm **AC2** through the claim that its antecedent and its consequent *have the same meaning*. The difference is that, while Bernays in essence agrees with the constructive interpretation in treating the quantifier block $\forall x \exists y$ as meaning $\exists f \forall x$, he interprets the existential quantifier in the latter

[6] Ramsey (1926).

[7] Quoted in Section 4.8 of Moore (1982).

[8] Bernays (1930), translated in Mancosu (1998).

[9] This fact, according to Bernays, renders the usual objections against the principle of choice invalid, since these latter are based on the misapprehension that the principle "claims the possibility of a choice"

classically, so that in affirming "there is a function" it is not necessary, as under the constructive interpretation, actually to be *given* such a function.

Per Martin-Löf has recently[10] contrasted the constructive affirmability of Zermelo's 1904 formulation of the axiom of choice—which we shall take in the version **AC2**, and which Martin-Löf terms the *intensional* axiom of choice—with Zermelo's 1908 formulation, the combinatorial axiom of choice **CAC**.

Martin-Löf's discussion takes place within a simplified version of *constructive (dependent) type theory* (CTT), the system of constructive mathematics, based on intuitionistic logic, he introduced some years ago and which has become standard.[11] In CTT the primitive relation of identity of objects (necessarily of the same type) is *intensional*. In set theory, on the other hand, the identity relation is treated extensionally since two sets are identified if they have the same elements (Axiom of Extensionality). In CTT a set in the usual set-theoretic sense corresponds to an *extensional* set, that is, a set carrying an equivalence relation representing "extensional" equality of its elements.

That being the case, it is natural to formulate within CTT a version of **AC** for extensional sets. Martin-Löf calls this the *extensional* axiom of choice (**EAC**). To state this we need to introduce the notion of an extensional function. Thus let A and B be two sets carrying equivalence relations $=_A$ and $=_B$ respectively. A function $f : A \to B$ is called *extensional*, Ext(f), if $\forall xx' \in A\,(x =_A x' \to fx =_B fx')$. Then **EAC** may be stated: for any relation R between A and B,

$$\forall x \in A \exists y \in B\, R(x, y) \Rightarrow \exists f : A \to B[\text{Ext}(f) \wedge \forall x \in A\, R(x, fx)].$$

Martin-Löf shows that, in CTT, **CAC** and **EAC** are equivalent.

Now the equivalence between **CAC** and **EAC**, is established within CTT where **AC2** is *already provable*.[12] There the equivalence between **CAC** and **EAC** is a nontrivial assertion. In *set theory*, on the other hand, not only are **CAC** and **EAC** equivalent, but they are themselves both equivalent to **AC2**. It becomes natural then to ask: can Martin-Löf's argument be presented within set theory without courting triviality?

I believe this can be done by noting that Martin-Löf also establishes the equivalence, in CTT, of **CAC** with the assertion that unique representatives can be picked from the equivalence classes of any given equivalence relation. Let us abbreviate this as **EQ**. In deriving **CAC** (actually the equivalent **EAC**, but no matter) from **EQ**, Martin-Löf employs **AC2**, so establishing, in CTT, the implication

$$\textbf{EQ} + \textbf{AC2} \Rightarrow \textbf{CAC}.$$

The problem thus boils down to giving a faithful version of the argument for this implication within set theory.

---

[10] Martin-Löf (2006).

[11] See Martin-Löf (1975, 1982, 1984).

[12] For a proof see, e.g., Tait (1994).

To do this, **AC2** must be furnished with a *constructively valid set-theoretical* formulation. This can be achieved by invoking the "propositions as types" doctrine (PAT)[13] underlying CTT. **CDTT** the central thesis of PAT is that each proposition is to be *identified* with the type, set, or assemblage of its proofs. As a result, such proof types, or sets of proofs, have to be accounted the *only* types, or sets. Strikingly, then, in the "propositions as types" doctrine, a type, or set, simply *is* the type, or set, of proofs of a proposition, and, reciprocally, a proposition *is* just the type, or set, of its proofs. In PAT logical operations on propositions are interpreted as certain mathematical operations on sets: in particular $\forall$ is interpreted as Cartesian product $\prod$ and $\exists$ as coproduct (disjoint union) $\coprod$.[14]

Under PAT, **AC2** may be taken to assert the existence, for any doubly-indexed family of sets $\{A_{ij} : i \in I, j \in J\}$, of a bijection

$$(+) \qquad \prod_{i\in I}\coprod_{j\in J} A_{ij} \cong \coprod_{f\in J^I}\prod_{i\in I} A_{if(i)}.$$

The requisite, indeed canonical, isomorphism is easily supplied in the form of the map

$$g \mapsto (\pi_1 \circ g, \pi_2 \circ g) = g^*,$$

where π_1, π_2 are the projections of ordered pairs onto their first and second coordinates.

Note that

(#) for $g \in \prod_{i\in I}\coprod_{j\in J} A_{ij}$, g^* is a pair of functions (e, f) with $f \in J^I$ and $e \in \prod_{i\in I} A_{if(i)}$.

Now **CAC** can be shown, in standard (intuitionistic) set theory, to be equivalent to the assertion that, for any doubly-indexed family of sets $\{A_{ij} : i \in I, j \in J\}$,

$$\prod_{i\in I}\bigcup_{j\in J} A_{ij} = \bigcup_{f\in J^I}\prod_{i\in I} A_{if(i)},$$

which is in turn equivalent to

$$(*) \qquad \prod_{i\in I}\bigcup_{j\in J} A_{ij} \subseteq \bigcup_{f\in J^I}\prod_{i\in I} A_{if(i)}.$$

I shall present a natural derivation within set theory of $(*)$ from $(+)$ and **EQ**, so providing what seems to me a purely set-theoretical formulation of Martin-Löf's argument.

First observe that there is a natural epimorphism

---

[13] See Tait (1994).

[14] Here $\coprod_{i\in I} A_i$ may be identified with $\bigcup_{i\in I}(A_i \times \{i\})$.

$$\prod_{i\in I}\coprod_{j\in J} A_{ij} \twoheadrightarrow \prod_{i\in I}\bigcup_{j\in J} A_{ij}$$

given by

$$g \mapsto \pi_1 \circ g.$$

Write $\approx$ for the equivalence relation on $\prod_{i\in I}\coprod_{j\in J} A_{ij}$ given by

$$g \approx h \Leftrightarrow \pi_1 \circ g = \pi_1 \circ h.$$

Each $k \in \prod_{i\in I}\bigcup_{j\in J} A_{ij}$ may be identified with the $\approx$-equivalence class $\{g : \pi_1 \circ g = k\} = \tilde{k}$. Using **EQ**, choose a system of unique representatives from the $\approx$-equivalence classes. This amounts to introducing a map

$$u : \prod_{i\in I}\bigcup_{j\in J} A_{ij} \to \prod_{i\in I}\coprod_{j\in J} A_{ij}$$

for which $u(k) \in \tilde{k}$, i.e.

(∗∗) $$\pi_1 \circ u(k) = k,$$

for all $k \in \prod_{i\in I}\bigcup_{j\in J} A_{ij}$.

Now to establish (∗), we take any $k \in \prod_{i\in I}\bigcup_{j\in J} A_{ij}$. Then under the natural bijection between $\prod_{i\in I}\coprod_{j\in J} A_{ij}$ and $\coprod_{f\in J^I}\prod_{i\in I} A_{if(i)}$ given in (+), $u(k)$ is correlated with the pair of maps

$$(\pi_1 \circ u(k),\ \pi_2 \circ u(k)),$$

i.e., using (∗∗), with

$$(k, \pi_2 \circ u(k)).$$

Writing $f = \pi_2 \circ u(k)$, it follows from (#) that

$$f \in J^I \text{ and } k \in \prod_{i\in I} A_{if(i)},$$

whence

$$k \in \bigcup_{f\in J^I}\prod_{i\in I} A_{if(i)}.$$

So we have derived (∗).

What is really going on here appears to be the following. Under the epimorphism

$$\prod_{i\in I}\coprod_{j\in J} A_{ij} \twoheadrightarrow \prod_{i\in I}\bigcup_{j\in J} A_{ij}$$

information is "lost", to wit, the identity , for a given member g of the domain of the epi, and an arbitrary $i \in I$, of the $j \in J$ for which $g(i) \in A_{ij}$. The map u furnished by **EQ** essentially resupplies that information. So starting with $k \in \prod_{i \in I} \bigcup_{j \in J} A_{ij}$, if one applies u to it, and then applies to the result the bijection given in (+), one winds up with a map $f \in J^I$ for which $k(i) \in A_{if(i)}$ for all $i \in I$. This is precisely what is demanded by (∗).

In an intensional constructive framework such as CTT, the axiom of choice is compatible with intuitionistic logic, that is, with the non-affirmation of the law of excluded middle. But in 1975 Diaconescu showed[15] that, in extensional frameworks such as topos theory or set theory, the usual formulations of the axiom of choice *imply the law of excluded middle*, so making logic classical. And Martin-Löf's analysis shows that, in CTT, the imposition of (a form of) extensionality on the axiom of choice will enable Diaconescu's theorem to become applicable, again yielding classical logic.[16] That extensionality in some form is required to derive Diaconescu's theorem can be observed in a number of different ways in addition to Martin-Löf's penetrating analysis. Here are three.

1. *Second-order logic*. Let $\mathcal{L}$ be a second-order language with individual variables $x, y, z, \ldots$, predicate variables $X, Y, Z, \ldots$ and second-order function variables $F, G, H, \ldots$. Here a second-order function variable F may be applied to a predicate variable X to yield an individual term FX. The scheme of sentences

AC* $$\forall X[\Phi(X) \rightarrow \exists x X(x)] \rightarrow \exists F \forall X[\Phi(X) \rightarrow X(FX)]$$

may be taken as the axiom of choice in $\mathcal{L}$.

We assume that the background logic of $\mathcal{L}$ is intuitionistic logic. Given certain mild further presuppositions, **AC*** can be shown to imply **LEM**, the law of excluded middle that, for any for any proposition A, $A \vee \neg A$. These mild further presuppositions latter may be stated:

Predicative Comprehension $\exists X \forall x[X(x) \leftrightarrow \phi(x)]$

Here ϕ is a formula not containing any bound predicate variables.

Extensionality of Functions $\forall X \forall Y \forall F[X \equiv Y \rightarrow FX = FY]$

Here $X \equiv Y$ is an abbreviation for $\forall x[X(x) \leftrightarrow Y(x)]$, that is, X and Y are *extensionally equivalent*.

In addition we assume the presence of two individuals 0 and 1. Their distinctness is expressed by means of the trivial presupposition $0 \neq 1$.

Now let A be a given proposition. By Predicative Comprehension, we may introduce predicate constants U, V together with the assertions

(1) $$\forall x[U(x) \leftrightarrow (A \vee x = 0)] \qquad \forall x[V(x) \leftrightarrow (A \vee x = 1)]$$

---

[15] Diaconescu (1975).

[16] Note, however, that if the axiom of choice is formulated within set theory or topos theory in the "harmless"—indeed mathematically useless—way (+), it is perfectly compatible with intuitionistic logic.

Let $\Phi(X)$ be the formula $X \equiv U \vee X \equiv V$. Then clearly we may assert $\forall X[\Phi(X) \to \exists x X(x)]$ so **AC*** may be invoked to assert $\exists F \forall X[\Phi(X) \to X(FX)]$. Now we can introduce a function constant K together with the assertion

$$(2) \qquad \forall X[\Phi(X) \to X(KX)]$$

Evidently we may assert $\Phi(U)$ and $\Phi(V)$, so it follows from (2) that we may assert $U(KU)$ and $V(KV)$, whence also, using (1),

$$[A \vee KU = 0] \wedge [A \vee KV = 1]$$

Using the distributive law (which holds in intuitionistic logic), it follows that we may assert

$$A \vee [KU = 0 \wedge KV = 1]$$

From the presupposition that $0 \neq 1$ it follows that

$$(3) \qquad A \vee KU \neq KV$$

is assertable. But it follows from (1) that we may assert $A \to U \equiv V$, and so also, using Extensionality of Functions, $A \to KU = KV$. This yields the assertability of $KU \neq KV \to \neg A$, which, together with (3) in turn yields the assertability of

$$A \vee \neg A$$

that is, **LEM**.

Note that in deriving **LEM** from version **AC*** essential use was made of the principles of Predicative Comprehension and Extensionality of Functions. It follows that, in systems of constructive mathematics affirming **AC** (but not **LEM**) *either the principle of Predicative Comprehension or the Principle of Extensionality of Functions must fail.* While the Principle of Predicative Comprehension can be given a constructive justification, no such justification can be provided for the principle of Extensionality of Functions. Functions on predicates are given intensionally, and satisfy just the corresponding Principle of Intensionality $\forall X \forall Y \forall F[X = Y \to FX = FY]$. The Principle of Extensionality can easily be made to fail by considering, for example, the predicates P: *rational featherless biped* and Q: *human being* and the function K on predicates which assigns to each predicate the number of words in its description. Then we can agree that $P \equiv Q$ but $KP = 3$ and $KQ = 2$.

2. *Hilbert's Epsilon Calculus.* In the logical calculus developed by Hilbert in the 1920s the Axiom of Choice appears in the form of a postulate he called the *logical ε-axiom* or the *transfinite axiom.* To formulate this postulate he introduced, for each formula $\alpha(x)$, a term (an *epsilon term*) $\varepsilon_x \alpha$ or simply ε_α which, intuitively, is intended to name an indeterminate object satisfying $\alpha(x)$. The ε-axiom then takes the form

$$(\varepsilon) \qquad \exists x \alpha(x) \to \alpha(\varepsilon_\alpha)$$

All that is known about ε_α is that, if anything satisfies α, it does.[17] Now since α may contain free variables other than x, the identity of ε_α depends, in general, on the values assigned to these variables. So ε_α may be regarded as the result of having chosen, for each assignment of values to these other variables, a value of x so that $\alpha(x)$ is satisfied. That is, ε_α may be construed as a choice function, and the ε-axiom accordingly seen as a version of **AC**.

An *ε-calculus* $\mathcal{P}_\varepsilon$ is obtained by starting with a system $\mathcal{P}$ of first-order predicate logic, augmenting it with epsilon terms, and adjoining as an axiom scheme the formulas (ε). It is known that when $\mathcal{P}$ is classical predicate logic, $\mathcal{P}_\varepsilon$ is *conservative* over $\mathcal{P}$, that is, each assertion of $\mathcal{P}$ demonstrable in $\mathcal{P}_\varepsilon$ is also demonstrable in $\mathcal{P}$: The move from $\mathcal{P}$ to $\mathcal{P}_\varepsilon$ does not enlarge the body of demonstrable assertions in $\mathcal{P}$. But for *intuitionistic* predicate logic the situation is otherwise.

In fact it is easy to see that, if $\mathcal{P}$ is taken to be intuitionistic predicate logic, then a number of first-order assertions undemonstrable within $\mathcal{P}$, for instance $\exists x(\exists x\alpha(x) \to \alpha(x))$, are provable within $\mathcal{P}_\varepsilon$. More interesting is the fact that certain purely *propositional* assertions undemonstrable within $\mathcal{P}$ are rendered provable within $\mathcal{P}_\varepsilon$.[18] These include Dummett's scheme $A \to B \vee B \to A$ and (hence) the intuitionistically invalid De Morgan law $\neg(A \wedge B) \to \neg A \vee \neg B$. But, curiously, the Law of Excluded Middle does *not* become demonstrable as a result of passing from intuitionistic $\mathcal{P}$ to $\mathcal{P}_\varepsilon$.

This is related to the fact (remarked on above) that in deriving **LEM** from **AC** one requires the principle of Extensionality of Functions. The analogous principle within the ε-calculus is the *Principle of Extensionality for ε-terms*:

(Ext) $$\forall x[\alpha(x) \leftrightarrow \beta(x)] \to \varepsilon_\alpha = \varepsilon_\beta.$$

An argument similar to the derivation of **LEM** from **AC** given above yields **LEM** from (Ext) within the intuitionistic ε-calculus.

It is interesting to note that the use of (Ext) can be avoided in deriving **LEM** in the intuitionistic ε-calculus if one employs *relative ε-terms*, that is, allows ε to act on *pairs* of formulas, each with a *single* free variable. Here, for each pair of formulas $\alpha(x), \beta(x)$ we introduce the "relativized" ε-term $\varepsilon_x\alpha/\beta$ and the "relativized" ε-axioms

$$(1)\quad \exists x\beta(x) \to \beta(\varepsilon_x\alpha/\beta) \qquad\qquad (2)\quad \exists x\,[\alpha(x) \wedge \beta(x)] \to \alpha(\varepsilon_x\alpha/\beta).$$

That is, $\varepsilon_x\alpha/\beta$ may be thought of as an individual that satisfies β if anything does, and which in addition satisfies α if anything satisfies both α and β. Notice that the usual ε-term $\varepsilon_x\alpha$ is then $\varepsilon_x\alpha/x = x$. In the classical ε-calculus $\varepsilon_x\alpha/\beta$ may be defined by taking

$$\varepsilon_x\alpha/\beta = \varepsilon_y[[y = \varepsilon_x(\alpha \wedge \beta) \wedge \exists x(\alpha \wedge \beta)] \vee [y = \varepsilon_x\beta \wedge \neg\exists x(\alpha \wedge \beta)]]$$

---

[17] David Devidi has had the happy inspiration of calling ε_α "the thing most likely to be α."

[18] Bell (1993a,b).

But the relativized ε-scheme is not derivable in the intuitionistic ε-calculus since it can be shown to imply **LEM**. To see this, given a formula γ define

$$\alpha(x) \equiv x = 1 \qquad \beta(x) \equiv x = 0 \vee \gamma$$

Write a for $\varepsilon_x \alpha/\beta$. Then we certainly have $\exists x\beta(x)$, so (1) gives $\beta(a)$, i.e.

(3) $$a = 0 \vee \gamma$$

Also $\exists x(\alpha \wedge \beta) \leftrightarrow \gamma$, so (2) gives $\gamma \to \alpha(a)$, i.e.

$$\gamma \to a = 1$$

whence

$$a \neq 1 \to \neg\gamma$$

so that

$$a = 0 \to \neg\gamma$$

And the conjunction of this with (3) gives $\gamma \vee \neg\gamma$, as claimed.

3. *Weak set theories lacking the axiom of extensionality.* In Bell (forthcoming) a first order *weak set theory* **WST** is introduced which lacks the axiom of extensionality and supports only minimal set-theoretic constructions. **WST** may be considered a fragment both of (intuitionistic) Δ_0-Zermelo set theory and Aczel's constructive set theory.[19] Like CTT, **WST** is too weak to allow the derivation of **LEM** from **AC**. But (again as with constructive type theories) beefing up **WST** with extensionality principles (even very moderate ones) enables the derivation to go through.

I end with some further thoughts on the status of the axiom of choice in constructive type theory and the "propositions as types" framework. We have observed above that **AC** interpreted à la "propositions as types" is (constructively) canonically true, while construed set- (or topos-) theoretically it is anything but, since so construed its affirmation yields classical logic. This prompts the question: what modification needs to be made to the "propositions-as-types" framework so as to yield the set- (or topos-) theoretic interpretation of **AC**? An answer (due to M.E. Maietti)[20] to this question can be furnished within the general framework of (variable) type theories through the use of so-called *monotypes* (or mono-objects), that is, types containing at most one entity or having at most one proof. In the category **Set** of ordinary sets, mono-objects are *singletons*, that is, sets containing at most one element.

Monotypes correspond to monic maps. This can be illustrated concretely by considering the categories **Indset** of *indexed* sets and $\mathbf{Set}^{\to}$ of *bivariant* sets. The objects of **Indset** are indexed sets of the form $M = \{\langle i, M_i \rangle : i \in I\}$ and those of $\mathbf{Set}^{\to}$ maps $A \to B$ in **Set**, with appropriately defined arrows in each case. It can be shown

[19] Aczel and Rathjen (2001).

[20] Maietti (2005).

that these two categories are equivalent. If we think of (the objects of) **Set** as representing simple or static types, then (the objects of) **Indset**, and hence also of $\mathbf{Set}^{\rightarrow}$, represent variable types. It is easily seen that a monotype, or object, in **Indset**, is precisely an object M for which each M_i has at most one element. Moreover, under the equivalence between **Indset** and $\mathbf{Set}^{\rightarrow}$, such an object corresponds to a monic map-object in $\mathbf{Set}^{\rightarrow}$.

Now consider $\mathbf{Set}^{\rightarrow}$ as a topos. Under the topos-theoretic interpretation in $\mathbf{Set}^{\rightarrow}$, formulas correspond to monic arrows, which in turn correspond to mono-objects in **Indset**. Carrying these correspondences over entirely to **Indset** yields the sought modification of the "propositions-as-types" framework to bring it into line with the topos-theoretic interpretation of formulas, namely, to take formulas or propositions to correspond to *mono*-objects, rather than to *arbitrary* objects. Let us call this the "formulas-as-monotypes" interpretation.

Finally let us reconsider **AC** under the "formulas-as-monotypes" interpretation within **Set**. In the "propositions-as-types" interpretation as applied to **Set**, the universal quantifier $\forall i \in I$ corresponds to the product $\prod_{i\in I}$ and the existential quantifier $\exists i \in I$ to the coproduct, or disjoint sum, $\coprod_{i\in I}$. Now in the "formulas-as-monotypes" interpretation, under which formulas correspond to singletons, $\forall i \in I$ continues to correspond to $\prod_{i\in I}$, since the product of singletons is still a singleton. But the interpretation of $\exists i \in I$ is changed. In fact, the interpretation of $\exists i \in I\, A_i$ (with each A_i a singleton) now becomes $\left[\coprod_{i\in I} A_i\right]$, where for each set X, $[X] = \{u : u = 0 \wedge \exists x.x \in X\}$ is the *canonical singleton* associated with X.

It follows that, under the "formulas-as-monotypes" interpretation, the proposition $\forall i \in I \exists j \in J A_{ij}$ is interpreted as the singleton

$$\prod_{i\in I}\left[\coprod_{j\in J} A_{ij}\right] \tag{1}$$

and the proposition $\exists f \in J^I \forall i \in I\, A_{if(i)}$ as the singleton

$$\left[\coprod_{f\in J^I}\prod_{i\in I} A_{if(i)}\right]. \tag{2}$$

Under the "formulas-as-monotypes" interpretation **AC** would be construed as asserting the existence of an isomorphism between (1) and (2).

Now it is readily seen that to give an element of (1) amounts to no more than affirming that, for every $i \in I$, $\bigcup_{j\in J} A_{ij}$ is nonempty. But to give an element of (2) amounts to specifying maps $f \in J^I$ and g with domain I such that $\forall i \in I\, g(i) \in A_{if(i)}$. It follows that to assert the existence of an isomorphism between (1) and (2), that is, to assert **AC** under the "formulas-as-monotypes" interpretation, is tantamount to asserting **AC** in its usual form, so leading in turn to classical logic. This is in sharp contrast with **AC** under the "propositions-as-types" interpretation, where its assertion is automatically correct and so has no nonconstructive consequences.

References

Aczel, P. and M. Rathjen (2001) Notes on Constructive Set Theory. Technical Report 40, Mittag-Leffler Institute, The Swedish Royal Academy of Sciences. Available on first author's webpage www.cs.man.ac.uk/petera/papers

Bell, J. L. (1993a) Hilbert's Epsilon-Operator and Classical Logic, *Journal of Philosophical Logic*, **22**, 1–18.

Bell, J. L. (1993b) Hilbert's Epsilon Operator in Intuitionistic Type Theories, *Mathematical Logic Quarterly*, **39**.

Bell, J. L. (2008) The Axiom of Choice and the Law of Excluded Middle in Weak Set Theories. *Mathematical Logic Quarterly*, **54**, 194–201.

Bell, J. L. (2009) *The Axiom of Choice*, London: College Publications.

Bernays, P. (1930–31) Die Philosophie der Mathematik und die Hilbertsche Beweistheorie, *Blätter für deutsche Philosophie* **4**, pp. 326–367, Translated in Mancosu, (1998).

Diaconescu, R. (1975) Axiom of Choice and Complementation. *Proceedings of the American Mathematical Society* **51**, 176–178.

Fraenkel, A., Y. Bar-Hillel and A. Levy (1973) *Foundations of Set Theory*, 2nd edition. Amsterdam: North-Holland.

Goodman, N. and Myhill, J. (1978) Choice Implies Excluded Middle, *Zeitschrift für Mathematische Logik und Grundlagen der Mathematik* **24**(5), 461.

Hilbert D. (1926) Über das Unendliche, *Mathematische Annalen* **95**. Translated in van Heijenoort, (ed.) *From Frege to Gödel: A Source Book in Mathematical Logic* 1879–1931, Harvard University Press, 1967, pp. 367–392.

Maietti, M.E. (2005) Modular Correspondence Between Dependent Type Theories and Categories Including Pretopoi and Topoi. *Mathematical Structures in Computer Science* **15 6**, 1089–1145, Cambridge MA: Harvard University Press.

Mancosu, P. (1998) *From Brouwer to Hilbert*, Oxford: Oxford University Press.

Martin-Löf, P. (1975) An Intuitionistic Theory of Types: predicative part, in Rose, H. E. and Shepherdson, J. C., (eds.), *Logic Colloquium* 73, Amsterdam: North-Holland, pp. 73–118.

Martin-Löf, P. (1982) Constructive Mathematics and Computer Programming, in Cohen, L. C. Los, J. Pfeiffer, H. and Podewski, K.P., (eds.), *Logic, Methodology and Philosophy of Science VI*, Amsterdam: North-Holland, pp. 153–179.

Martin-Löf, P. (1984) *Intuitionistic Type Theory*, Naples: Bibliopolis.

Martin-Löf, P. (2006) 100 Years of Zermelo's Axiom of Choice: What Was the Problem With It? *The Computer Journal* **49**(3), pp. 345–350.

Moore, G. H. (1982) *Zermelo's Axiom of Choice. Its Origins, Development and Influence*, Berlin: Springer.

Ramsey, F. P. (1926) The Foundations of Mathematics, *Proceedings of the London Mathematical Society* **25**, 338–384.

Tait, W. W. (1994) The Law of Excluded Middle and the Axiom of Choice, in George, A. (ed.) *Mathematics and Mind*, New York: Oxford University Press, pp. 45–70.

Zermelo, E. (1904) Neuer Beweis, dass jede Menge Wohlordnung werden kann (Aus einem an Herrn Hilbert gerichteten Briefe), *Mathematische Annalen* **59**, pp. 514–516. Translated in van Heijenoort, (ed.) *From Frege to Gödel: A Source Book in Mathematical Logic* 1879–1931, Cambridge, MA: Harvard University Press, 1967, pp. 139–141.

Zermelo, E. (1908) Neuer Beweis für die Möglichkeit einer Wohlordnung, *Mathematische Annalen* **65**, pp. 107–128. Translated in van Heijenoort, (ed.) *From Frege to Gödel: A Source Book in Mathematical Logic* 1879–1931, Cambridge, MA: Harvard University Press, 1967, pp. 183–198.

Reflections on the Categorical Foundations of Mathematics

Joachim Lambek and Philip J. Scott

1 Introduction

Most practicing mathematicians see no need for the foundations of their subject. But those who wish to place it on a solid ground usually pick set theory, an axiomatic treatment of the membership relation expressed in first order logic. Some of us feel that higher order logic is more appropriate and, since Russell and Whitehead's *Principia Mathematica*, such a system has been known as type theory (more precisely, classical *impredicative* type theory with Peano's axioms). Although type theory has been greatly simplified by works of Alonzo Church, Leon Henkin, and others, and despite its naturalness for expressing mathematics, it was unjustly neglected until quite recently.

An apparently different approach to foundations is via category theory, a subject that was introduced by Samuel Eilenberg and Saunders Mac Lane in 1945. In 1964, F. W. Lawvere proposed to found mathematics on the category of categories (Lawvere, 1966). When he lectured on this at an international conference in Jerusalem, Alfred Tarski objected: "But what is a category if not a set of objects together with a set of morphisms?" Lawvere replied by pointing out that set theory axiomatized the binary relation of membership, while category theory axiomatized the ternary relation of composition.

Later Lawvere returned from the category of categories to the category of sets. Trying to axiomatize the latter (e.g. Lawvere, 1964), he ended up with the notion of an elementary topos, which made its first public appearance in joint work with Myles Tierney (Lawvere, 1970; Tierney, 1972). Elementary toposes have the advantage of describing not only sets, but also sheaves (called "variable sets" by Lawvere). Quoting Lawvere (1972):

> This is the development on the basis of elementary (first-order) axioms of a theory of "toposes" just good enough to be applicable not only to sheaf theory, algebraic spaces, global spectrum etc. as originally envisaged by Grothendieck, Giraud, Verdier, and Hakim but also to Kripke semantics, abstract proof theory, and the Cohen-Scott-Solovay method for obtaining independence results in set theory.

G. Sommaruga (ed.), *Foundational Theories of Classical and Constructive Mathematics*, The Western Ontario Series in Philosophy of Science 76,
DOI 10.1007/978-94-007-0431-2_9, © Springer Science+Business Media B.V. 2011

Indeed, it was soon realized that an elementary topos has an associated "internal" logic which is essentially a version of (intuitionistic) type theory. In the second part of our book *Introduction to higher order categorical logic* (Lambek and Scott, 1986), we tried to exploit the close connections between higher order logic (better called "higher order arithmetic") and topos theory.

2 Type Theory

By a *type theory* (or *higher order arithmetic*) we understand a formulation of higher order logic with Peano's axioms. We shall follow our book (1986) and consider type theories based on equality. Thus the language contains both *types* and *terms* (of the indicated types) as follows:

$$\begin{array}{llllll} \text{Types:} & 1 & \Omega & N & \Omega^A & A \times B \\ \text{Terms:} & * & a = a' & 0 & \{x : A \mid \varphi(x)\} & \langle a, b \rangle \\ & & a \in \alpha & Sn & & \end{array}$$

It is assumed that $1, N, \Omega$ are types, and that the types are closed under the operations Ω^A and $A \times B$ for given types A and B. Here 1 denotes a one-point type, Ω the type of propositions (or truth values), N the type of natural numbers, Ω^A the "powerset" of A, and $A \times B$ the cartesian product of types A and B. Among the terms (not indicated above) there are infinitely many variables of each type. We assume that $*$ is a term of type 1, 0 is a term of type N, and that the terms are closed under the operations $=, \in, S, \{- \mid -\}$, and $\langle -, - \rangle$, as indicated above, where it is understood that a, a' are of the same type A, b is of type B, α is of type Ω^A, n is of type N and $\varphi(x)$ is of type Ω. We adopt the current convention of writing $t : A$ for "t is a term of type A".

We also assume that there is a collection of theorems which include the usual axioms and which is closed under the usual rules of inference (i.e. equality, pairing, comprehension, extensionality, and Peano's axioms). The familiar logical symbols are now definable as follows (see also Lawvere, 1972)

$$\begin{array}{lll} \top & := * = * & \\ p \wedge q & := \langle p, q \rangle = \langle \top, \top \rangle & \text{where } p, q : \Omega \\ p \Rightarrow q & := p \wedge q = p & \\ \forall_{x:A} \phi(x) & := \{x : A \mid \phi(x)\} = \{x : A \mid \top\} & \text{where } \phi(x) : \Omega \\ \perp & := \forall_{x:\Omega} x & \\ \neg p & := \forall_{x:\Omega}(p \Rightarrow x) & \\ p \vee q & := \forall_{x:\Omega}(((p \Rightarrow x) \wedge (q \Rightarrow x)) \Rightarrow x) & \\ \exists_{x:A} \varphi(x) & := \forall_{y:\Omega}(\forall_{x:A}((\phi(x) \Rightarrow y) \Rightarrow y)) & \end{array}$$

The usual properties of these logical connectives can now be proved (see Lambek and Scott, 1986).

We will call a type theory *analytic* if it contains no types and terms other than the ones it must contain according to the above definition. Thus, an analytic type theory does not contain the type of humans or the type of vegetables, nor does it contain terms denoting the binary relations of loving or eating. Even the internal language of a topos (see below) is not analytic, since it admits as types all sets (a *set* in a topos being a morphism $1 \to \Omega^A$, for some object A).

Pure type theory $\mathcal{L}_0$ is the analytic type theory containing no theorems other than those following from the above inductive definition. Every analytic type theory has the form $\mathcal{L}_0/\theta$, where θ is a set of propositions (i.e. terms of type Ω) now considered as additional *nonlogical axioms*. We may even take θ to be the set of all theorems.

3 Elementary Toposes

A *topos*, according to Lawvere, is a cartesian closed category (ccc) with pullbacks, a subobject classifier Ω and a natural numbers object N. By a ccc we mean a category with a terminal object 1, cartesian products $A \times B$ and exponentiation C^B, together with a canonical bijection between arrows $(A \times B) \to C$ and arrows $A \to C^B$. As Lawvere himself pointed out (Lawvere, 1969), the prime example of a ccc is the proof theory of the positive intuitionistic propositional calculus, with

$$1 = \mathsf{T}\,, \quad A \times B = A \wedge B\,, \quad C^B = B \Rightarrow C$$

According to the so-called Curry-Howard isomorphism, the associated proof theory can also be described by the typed lambda calculus (with surjective pairing); hence it is quite natural that ccc's, typed lambda calculi, and the proof theory of positive intuitionistic propositional calculi turn out to be equivalent (see Lambek and Scott, 1986).

A *subobject classifier* in a ccc with pullbacks is an object Ω together with a canonical (monic) arrow $\mathsf{T} : 1 \to \Omega$ and a canonical bijection between subobjects B of A and their characteristic morphisms $\chi_B : A \to \Omega$.[1] This generalizes the familiar set-theoretic bijection between subsets of a set A and characteristic functions $A \to \Omega$, where Ω is a two-element set. Viewed as usual classical sets, Ω and the powerset Ω^A are Boolean algebras, whereas in toposes, Ω and Ω^A carry the more general structure of a Heyting algebra. It is therefore not surprising that the "internal logic" of a topos is in general intuitionistic.

According to Lawvere, a *Natural Numbers Object* (NNO) in a ccc is an object N, together with arrows $0 : 1 \to N$ and $S : N \to N$ such that, given arrows $a : 1 \to A$ and $f : N \to N$, there is a unique arrow $h : N \to A$ making the following diagram commute:

[1] This bijection is induced by pulling back the arrow $\mathsf{T} : 1 \to \Omega$ along $\chi_B : A \to \Omega$.

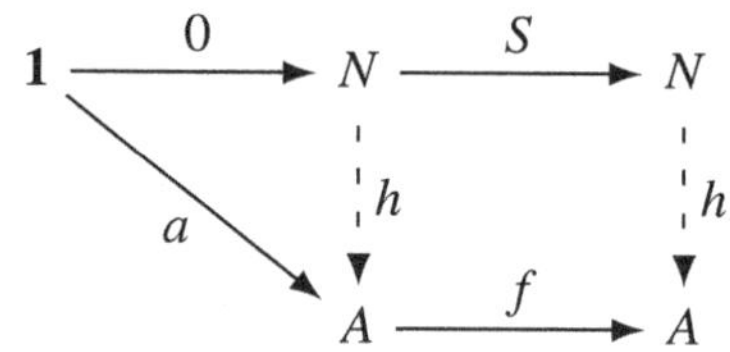

In the case of a topos, this yields Lawvere's categorical formulation of the well-known Peano axioms for set theory (Lawvere, 1964), which is seen here by putting $h(n) = f^n(a)$. In the case of cartesian closed categories (and their equivalent typed lambda calculi), this leads to notions of higher-type "iteration" arising in proof theory, recursive function theory, and theoretical computer science.

4 Comparing Type Theories and Toposes

In Lambek and Scott (1986) we compared two categories: the category of type theories, by which we mean intuitionistic type theories with axiom of infinity or, equivalently, Peano's axioms, and the category of toposes, which we understand to be elementary toposes with natural numbers object. As morphisms in the former we took "translations" between type theories, and in the latter, so-called "logical morphisms" between toposes (we ignored the alternative "geometric morphisms" arising from the Grothendieck tradition; for that, see Mac Lane and Moerdijk, 1992; Makkai and Reyes, 1977). We introduced functors between the two categories as follows. One functor L assigns to any topos $\mathcal{T}$ its "internal language" $L(\mathcal{T})$ (an intuitionistic type theory); the other functor T assigns to any type theory $\mathcal{L}$, the topos $T(\mathcal{L})$ "generated" by it, a kind of Lindenbaum-Tarski category constructed from the language. Let us briefly recall these two constructions.

The types of $L(\mathcal{T})$ are the objects of $\mathcal{T}$ and the closed terms of type A in $L(\mathcal{T})$ are the arrows $a : 1 \to A$ in $\mathcal{T}$. In particular, propositions of $L(\mathcal{T})$ are the arrows $p : 1 \to \Omega$ in $\mathcal{T}$. We say that *p holds in $\mathcal{T}$*, *p is true* in $\mathcal{T}$, or *p is a theorem* of the type theory $L(\mathcal{T})$, if and only if $p = \top$; that is, if p equals the distinguished arrow $\top : 1 \to \Omega$. Thus $L(\mathcal{T})$ has a "semantic" definition of theorem; it differs from logicians' more familiar (freely generated) type theories, in which terms are defined inductively from a small set of primitives, and in which "theorems" are introduced with the help of a recursive proof predicate. The internal language of a topos has some interesting properties. For example, $L(\mathcal{T})$ satisfies the *unique existence property*: if $\exists !_{x:A}\phi(x)$ holds in $\mathcal{T}$, then there is a closed term of type A, namely an arrow $a : 1 \to A$ in $\mathcal{T}$, such that $\phi(a)$ holds in $\mathcal{T}$. As Bertand Russell would have said: "a is the unique $x : A$ such that $\phi(x)$". We sometimes denote such a unique a by $\iota_{x:A}.\phi(x)$.

The topos $T(\mathcal{L})$ generated by the type theory $\mathcal{L}$ has as objects closed terms α of type Ω^A (modulo provable equality), and as morphisms $\alpha \to \beta$, where $\alpha : \Omega^A$ and $\beta : \Omega^B$, we choose those binary "relations" (closed terms) $\phi : \Omega^{A\times B}$ (again, modulo provable equality) such that the following is provable in $\mathcal{L}$:

$$\forall_{x:A}(x \in \alpha \Rightarrow \exists !_{y:B}(y \in \beta \wedge (x,y) \in \phi))$$

Intuitively, $T(\mathcal{L})$ is the category of "sets" and "functions" formally definable within the higher-order logic $\mathcal{L}$: its objects are the "sets" $\alpha, \beta, \ldots$ in $\mathcal{L}$ and its morphisms are the "provably functional relations" ϕ in $\mathcal{L}$ between such objects, all modulo provable equality.

We proved quite formally that there are two natural transformations $\varepsilon : LT \rightarrow id$ and $\eta : id \rightarrow TL$ rendering the functor T to be left adjoint to L. Moreover, we showed that η was an isomorphism, so that every topos is equivalent to the topos generated by its internal language. We pointed out in an exercise that a slight tightening of the definition of translation would also make ε an isomorphism; this was carried out by Lavendhomme and Lucas (1989). However, returning to our more natural notion of translation, we showed that, for any type theory $\mathcal{L}$, the translation $\mathcal{L} \rightarrow LT(\mathcal{L})$ is a conservative extension.

A type theory $\mathcal{L}$ may be interpreted in a topos $\mathcal{T}$ by means of a translation of languages $\mathcal{L} \rightarrow L(\mathcal{T})$ or, equivalently, by a logical morphism $T(\mathcal{L}) \rightarrow \mathcal{T}$, recalling that T is left adjoint to L. In some sense, every such interpretation may be viewed as a "model" of $\mathcal{L}$ in $\mathcal{T}$. By abuse of language, one often refers to $\mathcal{T}$ itself as the model. In particular, this view is justified for models of *pure type theory* $\mathcal{L}_0$, the initial object in the category of type theories and translations. For in this case, there is a unique translation from $\mathcal{L}_0$ to any type theory. In particular, for any topos $\mathcal{T}$, there is a unique translation $\mathcal{L}_o \rightarrow L(\mathcal{T})$, thus a unique logical morphism $T(\mathcal{L}_0) \rightarrow \mathcal{T}$. $\mathcal{F} = T(\mathcal{L}_0)$ is thus initial in the category of toposes and logical morphisms and is known as the *free topos*. Hence any elementary topos (with Natural Numbers Object) serves as a model of $\mathcal{L}_0$.

5 Models and Completeness

Following Leon Henkin's presentation of classical type theory (Henkin, 1950), we adopt a more restrictive notion of model. A *model* of a type theory $\mathcal{L}$ is a topos $\mathcal{T}$ satisfying three properties (for formulas in $L(\mathcal{T})$):

(a) **consistency**: $\perp$ is not true ;
(b) **disjunction property**: if $p \vee q$ is true in $\mathcal{T}$, then so is p or q;
(c) **existence property**: if $\exists_{x:A}\phi(x)$ is true in $\mathcal{T}$, then so is $\phi(a)$ for some closed term a of type A in $L(\mathcal{T})$, that is, for some morphism $a : 1 \rightarrow A$ in $\mathcal{T}$.

Following Alexander Grothendieck, we now call such a topos a *local topos*.

Peter Freyd observed that the above three linguistic properties can be expressed categorically as follows:

(a) the terminal object 1 is not initial;
(b) 1 is indecomposable;
(c) 1 is projective.

Local toposes of interest also have another property:

(d) all numerals are *standard*; that is, all the arrows $1 \to N$ have the form $S^n 0$ for some natural number n.

As we mentioned earlier, Russell and Whitehead (as well as Gödel and Henkin) dealt with *classical* type theory. This theory differs from intuitionistic type theory by the addition of a single axiom β (the *law of excluded middle*), which we may write as:

$$\forall_{x:\Omega}(\neg\neg x \Rightarrow x)$$

or, equivalently,

$$\forall_{x:\Omega}(x \vee \neg x)$$

A topos is said to be *Boolean* if its internal language is classical. In particular, this implies that $\Omega \cong 1 + 1$ (the coproduct). Boolean local toposes may be characterized as follows (see Seldin and Hindley, 1980):

Proposition 5.1 *A topos $\mathcal{T}$ is Boolean local iff it satisfies*

I Consistency: $\mathsf{T} \neq \mathsf{F} : 1 \to \Omega$.

II Universal Property: If $\phi(x)$ is a formula in $L(\mathcal{T})$ such that $\phi(a)$ holds in $\mathcal{T}$ for all closed terms $a : A$ in $L(\mathcal{T})$, then $\forall_{x:A}\phi(x)$ holds in $\mathcal{T}$.

The second property can be expressed in categorical language by saying that the terminal object 1 of $\mathcal{T}$ is a *generator : if $f, g : A \to B$ and $fa = ga$ for all $a : 1 \to A$, then $f = g$.*

Proof Assuming that $\mathcal{T}$ is Boolean and local, the universal property follows from the existence property, using negation.

Conversely, assume that $\mathcal{T}$ satisfies properties (i) and (ii) above. Among the subobjects of 1 are the (isomorphism classes of) monomorphisms $0 \hookrightarrow 1$ and $1 \hookrightarrow 1$, with characteristic morphisms $\mathsf{F} : 1 \to \Omega$ and $\mathsf{T} : 1 \to \Omega$, respectively. Here 0 is the initial object of $\mathcal{T}$. By (i), these are distinct subobjects of 1. We claim there are no others. For let $m : A \hookrightarrow 1$ be any subobject of 1. If there is an arrow $a : 1 \to A$, clearly $ma = 1_1$, hence $mam = m1_A$, so $am = 1_A$ and m is an isomorphism $A \cong 1$. If there is no arrow $1 \to A$, we claim $A \cong 0$. For trivially $\phi(a)$ then holds in $\mathcal{T}$ for all closed terms a of type A; hence $\forall_{x:A}\phi(x)$ holds in $\mathcal{T}$, whatever formula $\phi(x)$ we take. In particular, for any object B, let $\phi(x)$ be the formula $\exists!_{y:B}\psi(x, y)$, where $\psi(x, y)$ is, for example, $x \neq x \wedge y = y$. Then clearly ψ defines an arrow $A \to B$. Since 1 is a generator, there is at most one such arrow $A \to B$, and thus A is an initial object.

Therefore 1 has exactly two subobjects, and so there are exactly two arrows $\mathsf{T}, \mathsf{F} : 1 \to \Omega$. Thus the topos $\mathcal{T}$ is *two-valued*. Hence for all arrows $p : 1 \to \Omega$, $\neg\neg p = p$, hence $\neg\neg = 1_\Omega$, since 1 is a generator. Hence the formula $\forall_{x:A}\neg\neg x = x$ holds in $\mathcal{T}$, so $\mathcal{T}$ is Boolean.

Once $\mathcal{T}$ is Boolean, the universal property gives rise to the existence property (by negation). Similarly the conjunction property (which holds in any topos) gives rise to the disjunction property by de Morgan's law. Thus $\mathcal{T}$ is local.

Remark 5.2 In general in toposes, Boolean does not imply 2-valued; however it does in the presence of the disjunction property. Conversely, 2-valued does not imply Boolean, but it does if 1 is a generator.

Gödel's completeness theorem was originally enunciated for classical first order logic, but was extended by Henkin to higher order as follows (in our terminology):

> A proposition holds in $T(\mathcal{L})$, the topos generated by a classical type theory $\mathcal{L}$, if and only if it is true in all models of $\mathcal{L}$, i.e. in all Boolean local toposes.

Of course, if $\mathcal{L}$ is inductively generated, such propositions are usually called *provable*, and the Completeness Theorem asserts the equivalence between provability in $\mathcal{L}$ and truth in all models.

What about Gödel's more famous Incompleteness Theorem, which he himself had originally stated for classical type theory? An examination of its proof in our setting (carried out in the next section) shows it actually asserts the following:

> In a consistent analytic type theory $\mathcal{L}$ whose theorems are recursively enumerable, in order to characterize provability in $\mathcal{L}$, it is *not* sufficient to look only at local toposes which also satisfy the *ω-property:* if $\phi(S^n 0)$ is true for all natural numbers n, then $\forall_{x:N}\phi(x)$ is also true in the model.

The crucial role of the ω-property was first pointed out by David Hilbert. Classically, though not intuitionistically, it is equivalent to what we call the ω^* *property*: if $\exists_{x:N}\phi(x)$ is true in the model, then so is $\phi(S^n 0)$ for some natural number n. For a local topos, the ω^* property follows from the existence property, provided we assume that all numerals are standard. In fact, for a local topos, the ω^*-property is equivalent to the condition that all numerals are standard.

6 Gödel's Incompleteness Theorem

In any analytic type theory $\mathcal{L}_0/\theta$, we may effectively enumerate all terms of a given type. This may be done with the help of the well-known method of Gödel numbering, or even just by putting the terms into alphabetical order. In particular, let p_n be the nth proposition (closed term of type Ω) and α_n be the nth numerical predicate (closed term of type Ω^N).

The analytic type theories we are usually interested in also possess a recursive proof predicate, ensuring that the set of theorems is recursively enumerable.[2] If θ contains all the axioms and is closed under the rules of inference, θ is the set of theorems of $\mathcal{L}_0/\theta$ and is recursively enumerable by a primitive recursive function g. Thus $p_{g(n)}$ denotes the nth theorem of $\mathcal{L}_0/\theta$. Note that the set of numerical predicates in the internal language of the "usual" category of sets cannot be enumerated, recursively or otherwise, as follows from Cantor's theorem. This serves as an inspiration for the theorems of Gödel and Tarski, as we shall see.

Here is our formulation of Gödel's incompleteness theorem, which includes both the classical and intuitionist cases.

---

[2] The set of theorems (of an analytic type theory) is the set of propositions formally provable from the logical and nonlogical axioms, using the rules of inference of $\mathcal{L}_0$.

Theorem 6.1 *In a consistent analytic type theory $\mathcal{L}$ whose theorems are recursively enumerable, there is a proposition q which does not hold in any model in which all numerals are standard, yet its negation $\neg q$ is not provable. Thus, $\neg q$ must hold in every Boolean model in which all numerals are standard. Hence, if $\mathcal{L}$ has at least one model in which the numerals are standard, neither q nor $\neg q$ is a theorem.*

Proof For a type theory $\mathcal{L}$, we write $\vdash_{\mathcal{L}}$ to denote provability in $\mathcal{L}$. Recall that any primitive recursive function f can be *numeralwise represented* by a formula $\phi(x, y)$ in $\mathcal{L}_0$ such that

$$\text{for all } m \in \mathbb{N}, \qquad \vdash_{\mathcal{L}_0} S^{f(m)}0 = \imath_{y:N}.\phi(S^m 0, y)$$

where $\imath_{y:N}.\phi(x, y)$ denotes "the unique $y : N$ such that $\phi(x, y)$" (Russell's definite description operator, which we can introduce as an abbreviation in $\mathcal{L}_0$). Recall that provability in $\mathcal{L}_0$ implies provability in any type theory. The representability of the primitive recursive functions in $\mathcal{L}_0$ is shown in our book (Lambek and Scott, 1986, Remark 3.6, p. 266).

Consider the two primitive recursive functions f and g, represented by ϕ and ψ, respectively, where f enumerates the propositions $S^m 0 \notin \alpha_m$ (already considered by Cantor) and g enumerates the theorems of $\mathcal{L}$. Thus, for any $m \in \mathbb{N}$,

$$\vdash_{\mathcal{L}} S^m 0 \notin \alpha_m \quad \text{iff} \quad \text{for some } n \in \mathbb{N}, \;\; f(m) = g(n) \quad . \tag{1}$$

Putting $\chi = \phi \wedge \psi$, we may write the RHS of (1) as:

$$\text{for some } n \in \mathbb{N}, \quad \vdash_{\mathcal{L}} \chi(S^m 0, S^n 0), \tag{2}$$

which implies

$$\vdash_{\mathcal{L}} \exists_{y:N} \chi(S^m 0, y), \tag{3}$$

that is,

$$\vdash_{\mathcal{L}} S^m 0 \in \alpha_k, \;\; \text{where} \;\; \alpha_k := \{x : N \mid \exists_{y:N} \chi(x, y)\}. \tag{4}$$

Therefore

$$\vdash_{\mathcal{L}} S^m 0 \notin \alpha_m \quad \text{implies} \quad \vdash_{\mathcal{L}} S^m 0 \in \alpha_k$$

Putting $m = k$, we infer by consistency that $S^k 0 \notin \alpha_k$ is not a theorem of $\mathcal{L}$.

Let us try to reverse the above reasoning. Clearly (4)⇒(3). The implication (2)⇒(3) may be reversed if we pass to the internal language $\mathcal{L}'$ of any model of $\mathcal{L}$ in which all numerals are standard, which thus inherits the existence and disjunction properties. We thus obtain

$$\vdash_{\mathcal{L}'} S^m0 \in \alpha_k \text{ implies } \vdash_{\mathcal{L}'} \exists_{y:N}\chi(S^m0, y)$$

$$\text{implies for some } n \in \mathbb{N} \quad \vdash_{\mathcal{L}'} \chi(S^m0, S^n0)$$

$$\text{by the Existence Property in } \mathcal{L}'$$

$$\text{implies for some } n \in \mathbb{N} \ f(m) = g(n),$$

$$\text{hence } \vdash_{\mathcal{L}} S^m0 \notin \alpha_m \quad \text{by (1).}$$

Again, putting $m = k$ and recalling that $S^k0 \notin \alpha_k$ is not a theorem of $\mathcal{L}$, we infer that not $\vdash_{\mathcal{L}'} S^k0 \in \alpha_k$, hence $S^k0 \in \alpha_k$ does not hold in any model where the numerals are standard.

The theorem follows if we take q to be $(S^k0 \in \alpha_k)$.

Corollary 6.2 *Assuming that the "usual" category of sets $\mathcal{S}$ is a Boolean local topos in which all numerals are standard, the set of propositions of $\mathcal{L}_0$ which hold in $\mathcal{S}$ is not recursively enumerable. Hence $\mathcal{S}$ cannot be construed as the topos generated by an analytic type theory whose theorems are recursively enumerable.*

Remark 6.3 The assumption that all numerals are standard is redundant, if we define "standard numerals" to be the arrows $1 \to N$ in the "usual" category $\mathcal{S}$ of sets. Thus Gödel's proposition $\neg q$ is true in $\mathcal{S}$ but not provable.

7 Reconciling Foundations

7.1 Constructive Nominalism

Gödel's incompleteness theorem seemed to show that Formalism and Platonism are mutually incompatible philosophies of mathematics. Indeed, this is what Gödel himself had in mind. He believed that the ω-property must hold in the Platonic universe of mathematics, later to be called "the model in the sky" by William Tait (1986). The contradiction disappears if one abandons classical mathematics for a moderate form of Intuitionism. According to the Brouwer-Heyting-Kolmogorov interpretation of formal intuitionistic arithmetic, the validity of a universal statement $\forall_{x:N}\phi(x)$ does not follow from the collection of its numerical instances $\phi(S^n0)$, for each $n \in \mathbb{N}$, unless the validity of all these instances has been established in a *uniform* way. For all we know, a proof of $\phi(S^n0)$ may increase in length and complexity with n. No such objection applies to the ω^*- property.

Although the formulation of Gödel's incompleteness theorem remains valid for intuitionistic higher order logic, this is no longer the case if the ω-property is replaced by the ω^*-property. In fact, a statement in pure intuitionistic higher order logic is provable if and only if it holds in $\mathcal{F} = T(\mathcal{L}_0)$, the *free topos*. Recall, this is the initial object in the category of all toposes and logical morphisms, and is constructed linguistically as the topos generated from pure intuitionistic type theory

$\mathcal{L}_0$. As has been pointed out repeatedly (Lambek, 2004; Lambek and Scott, 1986), the free topos should satisfy moderate adherents of various traditional philosophical schools in the foundations of mathematics:

- Platonists, because as an initial object it is unique up to isomorphism;
- Formalists, or even nominalists, because of its linguistic construction;
- Constructivists, or moderate intuitionists, because the underlying type theory is intuitionistic;
- Logicists, because this type theory is a form of higher order logic, although it must be complemented by an axiom of infinity, say in the form of Peano's axioms.

This eclectic point of view has been called "constructive nominalism" in Couture and Lambek (1991).

Proofs that the free topos is local have been obtained by Boileau and Joyal (Boileau, 1975; Boileau and Joyal, 1981), and by us (Lambek and Scott, 1980, 1986). Our ultimate proof was based on what is called the *Freyd Cover*, obtained by "glueing" $\mathcal{F}$ into the "usual" category of sets. Freyd showed that every locally small topos $\mathcal{T}$ gives rise to a local topos $\widehat{\mathcal{T}}$ in which all numerals are standard, together with a logical functor $G : \widehat{\mathcal{T}} \to \mathcal{T}$. The condition that $\mathcal{T}$ is *locally small* ensures that each set of arrows $Hom_{\mathcal{T}}(A, B)$ lives (as an object) in this category $\mathcal{S}$ of sets; the latter is presumed to be local and such that all numerals are standard. But what is this "usual" category of sets? We shall return to this question; for now, the reader may have to use her intuition to identify $\mathcal{S}$.

Returning to Freyd's argument (Freyd, 1978), let $\mathcal{T} = \mathcal{F}$, the free topos. Then, by initiality, there is a unique logical functor $F : \mathcal{F} \to \widehat{\mathcal{F}}$. Thus we obtain a logical functor $GF : \mathcal{F} \to \mathcal{F}$, which must equal $id_{\mathcal{F}}$, again by initiality. It follows that $\mathcal{F}$ inherits (from $\widehat{\mathcal{F}}$, hence from $\mathcal{S}$) the properties of being local and that all numerals are standard.

7.2 What Is the Category of Sets?

We saw above that we were able to construct a local topos in which all numerals are standard, which should satisfy moderate intuitionists. Unfortunately, Freyd's proof assumed the existence of the "usual" category of sets $\mathcal{S}$, which is itself assumed to be a local topos in which all numerals are standard. The category $\mathcal{S}$ may be said to live, if not in the world of mathematics, then in the world of metamathematics. If the metamathematician is herself an intuitionist, she might believe that this category of sets could be the free topos itself. But then we reach a circularity: to prove the free topos is a model in which numerals are standard, we must assume an ambient category of sets $\mathcal{S}$ which itself has that property, and of course we cannot then postulate that to be the free topos. One is reminded of Lewis Carroll (1895).

What if the metamathematician believes in classical logic? In that case, she must assume the existence of a model topos $\mathcal{S}$ in which the terminal object is a generator, and in which all numerals are standard, the "usual" category of sets. While the existence of such model toposes can be shown with the help of the axiom of choice (Lambek and Scott, 1986), can even a single one be "constructed"? For example, consider *classical* type theory $\mathcal{L}_1 = \mathcal{L}_0/\beta$, where the formula $\beta = \forall_{x:\Omega}(\neg\neg x \Rightarrow x)$ is added to $\mathcal{L}_0$ as a new axiom. The Boolean topos $T(\mathcal{L}_1)$ generated by $\mathcal{L}_1$ is the initial object in the category of all Boolean toposes. Is $T(\mathcal{L}_1)$ a model? Unfortunately it is not local, by the Incompleteness Theorem for $\mathcal{L}_1$. Indeed, the disjunction property fails for any undecidable sentence q, since we can prove in $\mathcal{L}_1$ that $\vdash q \vee \neg q$. Indeed, we conjecture that no such classical model topos can be constructed, at least if we require it to satisfy reasonable properties.

8 What Is Truth?

What is truth? This question, once raised by Pilate, received different answers from different mathematicians.

Hilbert famously proposed the problem of showing that mathematical statements are true if and only if they can be proved. Like all of us, he assumed the set of proofs to be recursive.

Brouwer once asserted that mathematical statements are true if and only if they are known. In retrospect, he should have said "can be known", if truth is to be independent of time.

Gödel believed that a (classical) mathematical statement is true if and only if it holds in some kind of Platonic universe, which we take to be a Boolean local topos in which numerals are standard.

It follows from Gödel's incompleteness theorem that Hilbert's position is incompatible with the assumption that the Platonic universe is classical. However, if we assume that this universe is intuitionistic (the free topos), there is no contradiction. Moreover, Brouwer's modified position is vindicated if we interpret "knowable" as "provable".

Tarski defines truth differently. He said "p is true" instead of asserting p. By abuse of language, we often abbreviate *p is true* by p, ignoring quotation marks, like most mathematicians. Tarski then said that a numerical predicate τ *defines* truth (for a language $\mathcal{L}$) provided

$$\text{for all } n \in \mathbb{N}, \quad \vdash_{\mathcal{L}} (p_n \Leftrightarrow S^n 0 \in \tau).$$

where we use the same conventions of Gödel numbering as in Gödel's theorem. Here is our formulation of Tarski's undefinability theorem.

Theorem 8.1 *In any consistent analytic type theory, truth (in Tarski's sense) is not definable by a numerical predicate.*

Proof As in the proof of Gödel's theorem, suppose there were such a τ, and let

$$p_{f(n)} := S^n 0 \notin \alpha_n.$$

Then

$$\vdash_{\mathcal{L}} (S^{f(n)}0 = \iota_{y:N}.\phi(S^n 0, y))$$

where ϕ represents f. Put

$$\alpha_k := \{x : N \mid \exists_{y:N}(y \in \tau \wedge \phi(x, y))\}$$

Then we have the following provable equivalences:

$$\begin{aligned} \vdash_{\mathcal{L}} (S^k 0 \in \alpha_k &\Leftrightarrow S^{f(k)}0 \in \tau \\ &\Leftrightarrow p_{f(k)} \\ &\Leftrightarrow S^k 0 \notin \alpha_k) \end{aligned}$$

which contradicts consistency.

We will attempt to briefly compare the different notions of truth. Let $\vdash_{\mathcal{L}_0}$ stand for provability in $\mathcal{L}_0$, hence for truth in the free topos $\mathcal{F}$. We would like to interpret this as truth in Brouwer's sense. Comparing this with Tarski's notion of meta-truth, we would believe that, for all propositions p, $(\vdash_{\mathcal{L}_0} p) \Leftrightarrow p$. In particular, soundness corresponds to the entailment ($(\vdash_{\mathcal{L}_0} p) \Rightarrow p$).

Most post-Gödel mathematicians still believe in soundness. However soundness already implies consistency; for by soundness

$$(\vdash_{\mathcal{L}_0} p \wedge \vdash_{\mathcal{L}_0} \neg p) \Rightarrow (p \wedge \neg p) \ .$$

Yet, Gödel's second incompleteness theorem (not treated here) shows that consistency of $\mathcal{L}_0$ cannot be proved in $\mathcal{L}_0$; hence (using the encoding methods of the second Gödel theorem) soundness in the above sense cannot be formally proved either.

We may also ask whether Gödel's notion of truth (in a classical Platonic universe $\mathcal{S}$) implies Tarski's notion of meta-truth, i.e. whether ($(\mathcal{S} \models p) \Rightarrow p$). Now this implies ($(\vdash_{\mathcal{L}_1} p) \Rightarrow p$), where $\mathcal{L}_1 = \mathcal{L}_0/\beta$ is pure classical type theory, which is initial among all classical type theories (including the internal language of $\mathcal{S}$). Thus Tarski's notion of meta-truth implies soundness of $\mathcal{L}_1$, which again cannot be proved in $\mathcal{L}_1$ (when suitably encoded), by the same argument as above.

9 Continuously Variable Sets

It would appear that metamathematics is an attempt by mathematicians to lift themselves up by their own bootstraps. This had already been noted by Lewis Carroll (1895), in connection with the rule of modus ponens. It is also evident to anyone who looks at Gentzen style deductive systems, which derive the meaning of logical connectives from that of the meta-logical ones.

If we cannot single out a distinguished Boolean local topos as a candidate for the classical category of sets, we may be forced to look at the totality of all such models.

From an algebraic point of view, Gödel's completeness theorem asserts

Every topos is a subtopos of a direct product of local toposes.

This is analogous to the familiar assertion:

Every commutative ring is a subring of a direct product of local rings.

However, the latter statement can be improved to one that plays a crucial rôle in modern algebraic geometry (see Grothendieck and Dieudonné, 1960)

Every commutative ring is the ring of continuous global sections of a sheaf of local rings.

It has been realized for some time that Gödel's completeness theorem can be improved analogously:

Every topos is equivalent to the topos of global sections of a sheaf of local toposes .

It had also been clear that the models of any type theory, including those of the internal language of a topos, are the points of a topological space and that the truth of a proposition varies continuously from point to point. With any proposition in $\mathcal{L}_0$ one associates a *basic open set* consisting of all models in which the proposition is true. After various starts towards the sheaf representation of toposes, the result was ultimately established by Awodey (2000).[3]

10 Some Intuitionistic Principles

The fact that the free topos is local (and has only standard numerals) may be exploited to prove a number of intuitionistic principles for pure intuitionistic type theory $\mathcal{L}_0$, as we showed in our book (Lambek and Scott, 1986):

Consistency: not ($\vdash \bot$) .
Disjunction Property: If $\vdash_{\mathcal{L}_0} p \vee q$, then $\vdash_{\mathcal{L}_0} p$ or $\vdash_{\mathcal{L}_0} q$.
Existence Property: If $\vdash_{\mathcal{L}_0} \exists_{x:A}\phi(x)$ then $\vdash_{\mathcal{L}_0} \phi(a)$ for some closed term a of type A.

[3] Having observed that the truth of a proposition varies continuously from point to point, one of the present authors was led to announce the sheaf representation at conferences in Sussex and Amsterdam, but he made a bad choice of the basic open sets and used a definition of "local" which employed only the disjunction property. The first fault was rectified in a joint paper with Moerdijk (Lambek, 1994), further expository development occurred in our book (Lambek and Scott, 1986), and the second fault was rectified in (Lambek and Moerdijk, 1982), in which the author introduced a large number of "Henkin constants" to witness existential statements. This was shown to be unnecessary in a more recent article of Awodey (2000), who replaced the earlier logical proofs, based on definition by cases, by a purely categorical one. Similar ideas had also been pursued by Peter Freyd.

Troelstra's Uniformity principle for $A = \Omega^C$: If $\vdash_{\mathcal{L}_0} \forall_{x:A}\exists_{y:N}\phi(x,y)$ then $\vdash_{\mathcal{L}_0} \exists_{y:N}\forall_{x:A}\phi(x,y)$.

In the free topos $\mathcal{F}$, the uniformity principle says the arrows $\Omega^C \to N$ are constant (i.e. factor through some standard numeral).

Independence of premisses: If $\vdash_{\mathcal{L}_0} \neg p \Rightarrow \exists_{x:A}\phi(x)$ then $\vdash_{\mathcal{L}_0} \exists_{x:A}(\neg p \Rightarrow \phi(x))$.

Markov's Rule: If $\vdash_{\mathcal{L}_0} \forall_{x:A}(\phi(x) \vee \neg\phi(x))$ and $\vdash_{\mathcal{L}_0} \neg\forall_{x:A}\neg\phi(x)$, then $\vdash_{\mathcal{L}_0} \exists_{x:A}\phi(x)$.

This says: if in pure intuitionist type theory $\mathcal{L}_0$ we have that ϕ is provably decidable, and if there is a proof of $\neg\forall_{x:A}\neg\phi(x)$, then there must also be a constructive proof of $\exists_{x:A}\phi(x)$, i.e. (by the existence property) a proof in $\mathcal{L}_0$ of $\phi(a)$, for some closed term a of type A.

The Existence Property with a parameter of type $A = \Omega^C$: If $\vdash_{\mathcal{L}_0} \forall_{x:A}\exists_{y:B}\phi(x,y)$ then $\vdash_{\mathcal{L}_0} \forall_{x:A}\phi(x,\psi(x))$, where $\psi(x)$ is some term of type B.

A similar statement for the disjunction property with a parameter of type A is also provable. The disjunction property and already the unique existence property fail for parameters of type N, but hold in the internal language $L(\mathcal{F}(x))$, where $\mathcal{F}(x)$ is the free topos with an indeterminate $x : 1 \to N$ adjoined. The existence property in this case amounts to showing that the slice topos $\mathcal{F}/N$ is local, hence that N is projective in $\mathcal{F}$. This is equivalent to closure of the logical system under a rule of countable choice. For second order arithmetic, there is a proof due to A. Troelstra (1973, Theorem 4.5.12) based on methods of S. Hayashi (1977) which is proof-theoretic in nature. There is apparently not yet a clean categorical proof of such results.

11 Concluding Remarks

Aside from the historical discussion of our categorical approach to the foundations of mathematics, our formulation of the proof of Gödel's Incompleteness Theorem exploits the struggle between two primitive recursive functions. One enumerates all theorems and the other enumerates the Cantorian formulas which exclude the nth numeral from the nth numerical predicate. In our view, Gödel's theorem does not assert that provability fails to capture the notion of absolute truth in *the* Platonic universe. Rather, it asserts that other models of set theory are required than those which resemble the alleged Platonic universe. In fact, we have some doubts about the constructive existence of a Platonic universe, except in the context of intuitionistic (higher-order) arithmetic. Even there, the proof that our candidate, the free topos, is a model depends on the metamathematical assumption that a model of set theory exists.

References

Awodey, S. (2000) Sheaf Representation for Topoi, *Journal of Pure and Applied Algebra* **145**, 107–121.

Barwise, J. ed. (1977) *Handbook of Mathematical Logic*, Amsterdam: North-Holland, Elsevier.

Boileau, A. (1975) Types vs. Topos, Phd thesis, University of Montreal.

Boileau, A. and Joyal, A. (1981) La logique des topos, *Journal of Symbolic Logic* **46**, 6–16.

Carroll, L. (1895) What The Tortoise Said To Achilles, *Mind* **4**, No. 14, 278–280.

Church, A. (1940) A Foundation for the Simple Theory of Types, *Journal of Symbolic Logic* **5**, 56–68.

Couture, J. (1997) Analyse logique et analyticité: de Carnap à Gödel, *Dialectica* **51**(2), 95–117.

Couture, J. and Lambek, J. (1991) Philosophical Reflections on the Foundations of Mathematics, *Erkenntnis* **34**, 187–209.

Eilenberg, S. and Mac Lane, S. (1945) General Theory of Natural Equivalences, *Transactions of the American Mathematical Society* **58**, 231–294.

Feferman, S. (1977) Theories of Finite Type Related to Mathematical Practice, in: Barwise (1977), 913–971.

Freyd, P. (1978) On proving that 1 is an indecomposible projective in various free categories, manuscript.

Gödel, K. (1931) Über formal unentscheidbare Sätze der Principia Mathematica und verwandter Systeme I. *Monatshefte für Mathematische Physik* **38**, 173–198.

Gödel, K. (1958) Über eine bisher noch nicht benützte Erweiterung des finiten Standpunktes. *Dialectica* **12**, 280–287.

Grothendieck, A. and Dieudonné, J. (1960) *Eléments de géométrie algébrique, tome I, le langage des schémas*, I.H.E.S. Publ. Math **4**, Paris, 1960. Second Edition published in *Die Grundlehren der math. Wissenschaften*, **166** (1971) Springer-Verlag.

Hayashi, S. (1977) On Derived Rules of Intuitionistic Second Order Arithmetic, *Commentarii Mathematici Universitatis Sancti Pauli* **26–1**, 77–103.

Henkin, L.A. (1950) Completeness in the Theory of Types, *Journal of Symbolic Logic* **15**, 81–91.

Howard, W.A. (1980) The Formulas-as-types Notion of Construction, in Seldin and Hindley eds. (1980), 479–490.

Lambek, J. (1994) Are the Traditional Philosophies of Mathematics Really Incompatible? *The Mathematical Intelligencer* **16**, 56–62.

Lambek, J. (2004) What is the World of Mathematics? *Annals Pure and Applied Logic* **126**, 149–158.

Lambek, J. and Moerdijk, I. (1982) Two Sheaf Representations of Elementary Toposes, in Troelstra, A.S. and van Dalen, D., eds. *The L. E. J. Brouwer Centenary Symposium* Vol 110, Studies in Logic and the Foundations of Mathematics, Amsterdam: North-Holland, pp. 275–295.

Lambek, J. and Scott, P.J. (1980) Intuitionist Type Theory and the Free Topos, *Journal of Pure and Applied Algebra* **19**, 215–257.

Lambek, J. and Scott P.J. (1986) *Introduction to Higher Order Categorical Logic*, Cambridge Studies in Advanced Mathematics 7, Cambridge: Cambridge University Press.

Lavendhomme, R. and Lucas, T. (1989) Toposes and Intuitionistic Set Theories, *Journal of Pure and Applied Algebra* **57**, 141–157.

Lawvere, F.W. (1964) An Elementary Theory of the Category of Sets. *Proceedings of the National Academy of Sciences USA* **50**, 869–872.

Lawvere, F.W. (1966) The Category of Categories as a Foundation for Mathematics, in Eilenberg, S. et. al., eds. *Proceedings of the Conference on Categorical Algebra, La Jolla 1965*, Berlin, Heidelberg and New York: Springer, pp. 1–20.

Lawvere, F.W. (1969) Adjointness in Foundations, *Dialectica* **23**, 281–296.

Lawvere, F.W. (1970) Quantifiers and Sheaves, in *Actes du Congrès International des Mathématiques*, Nice 1970, tome I, Paris: Gauthier-Villars, pp. 329–334.

Lawvere, F.W. (1972) Introduction *to Toposes, Algebraic Geometry, and Logic LNM* **274**, New York: Springer, pp. 1–12.
Mac Lane, S. and Moerdijk, I. (1992) *Sheaves in Geometry and Logic*, Universitext, New York: Springer.
Makkai, M. and Reyes, G. (1977) *First Order Categorical Logic LNM* **611**, New York: Springer.
McLarty, C. (1992) *Elementary Categories, Elementary Toposes*, Oxford: Clarendon Press.
Seldin, J.P. and Hindley, J.R. eds. (1980) *To H. B. Curry: Essays on Combinatory Logic, Lambda Calculus, and Formalism*, London: Academic Press.
Smorynski, C. (1977) The Incompleteness Theorems, in Barwise (1977), 821–865.
Tait, W.W. (1986) Truth and Proof: The Platonism of Mathematics, *Synthèse* **69**, 341–370.
Tarski, A. (1996) *Logic, Semantics, Metamathematics*, Oxford: Clarendon Press.
Tierney, M. (1972) Sheaf Theory and the Continuum Hypothesis, in *Toposes, Algebraic Geometry, and Logic LNM* **274**, Berlin: Springer, pp.13–42.
Troelstra, A. S. (ed) (1973) *Metamathematical Investigation of Intuitionistic Arithmetic and Analysis* New York: Springer *LNM* **344**. (For the correction of Theorem 4.5.12 in the text, based on Hayashi's work (1977)) see Corrections to the volume, on the webpage *http://staff.science.uva.nl/~ anne*

Part IV
Foundations of Constructive Mathematics

Local Constructive Set Theory and Inductive Definitions

Peter Aczel

1 Introduction

Local Constructive Set Theory (**LCST**) is intended to be a local version of constructive set theory (**CST**). Constructive Set Theory is an open-ended set theoretical setting for constructive mathematics that is not committed to any particular brand of constructive mathematics and, by avoiding any built-in choice principles, is also acceptable in topos mathematics, the mathematics that can be carried out in an arbitrary topos with a natural numbers object. We refer the reader to Aczel and Rathjen (2000/01) for any details, not explained in this paper, concerning **CST** and the specific **CST** axiom systems **CZF** and $\mathbf{CZF}^+ \equiv \mathbf{CZF} + \mathbf{REA}$.

CST provides a global set theoretical setting in the sense that there is a single universe of all the mathematical objects that are in the range of the variables. By contrast a local set theory avoids the use of any global universe but instead is formulated in a many-sorted language that has various forms of sort including, for each sort α a power-sort $\mathcal{P}\alpha$, the sort of all sets of elements of sort α. For each sort α there is a binary infix relation $\in_\alpha$ that takes two arguments, the first of sort α and the second of sort $\mathcal{P}\alpha$. For each formula ϕ and each variable x of sort α, there is a comprehension term $\{x : \alpha \mid \phi\}$ of sort $\mathcal{P}\alpha$ for which the following scheme holds.

Comprehension: $(\forall y : \alpha)[\ y \in_\alpha \{x : \alpha \mid \phi\} \quad \leftrightarrow \quad \phi[y/x]\ ]$.

Here we use the notation $\phi[a/x]$ for the result of substituting a term a for free occurences of the variable x in the formula ϕ, relabelling bound variables in the standard way to avoid variable clashes.

Our use of the terminology *local* for a version of a set theory has its origin in the use of that term by John Bell in his book (Bell, 1988). His notion of a *local set theory* is a certain kind of syntactic version of the category theoretic notion of a topos. Each of his local set theories uses a local language that has type symbols built up from ground type symbols. The type symbols have various forms including

G. Sommaruga (ed.), *Foundational Theories of Classical and Constructive Mathematics*, The Western Ontario Series in Philosophy of Science 76,
DOI 10.1007/978-94-007-0431-2_10,

the form of a power type **PA**, where **A** is a type. There are terms of each type and the set-like terms of the local language are the terms of some power type. So, in a local set theory, there is no global universe of sets, but each set has to be understood as local to some power type. Here we will keep to this general idea but will not be using the precise details of the formulations in Bell's book. In particular we prefer to use the word *sort* rather than *type*. Our first example of a local set theory will be what we will call Local Intuitionistic Zermelo (**LIZ**). This is essentially a variant of what Bell has called at the end of chapter 7 of Bell (1988), the free naturalised local set theory; i.e. the local set theory for the free topos with a natural numbers object. It is also natural to describe it as a version of intuitionistic higher order arithmetic.

There are several reasons for our interest in the setting up of a local version of **CST**. One reason is in connection with the formulation of predicative and generalised predicative versions of the notion of an elementary topos. We expect that a local (generalised) predicative axiom system for **CST** will have as its category theoretic models categories that are (generalised) predicative toposes, according to some suitable weakening of the notion of an elementary topos. Some category theorists dislike global set theories because they claim that the focus of global theories on the structure of the binary membership relation on the universe of sets is irrelevant to mainstream mathematics. So a local approach to **CST** may be more appealing to a category theorist interested in constructive mathematics and the carrying over of the beautiful, but fully impredicative apparatus, of topos theory to the generalised predicative context.

Another reason for our interest in the development of a local version of **CST** is to do with the dependent type theoretical setting for constructive mathematics initiated by Per Martin-Löf. That setting aims to provide a philosophically motivated foundational framework for constructive mathematics that makes explicit the fundamental notions. It is the natural translation of the **CST** axiom systems such as **CZF** and **CZF**$^+$ into formulations of the type theoretic setting that have been used to justify the claim that those axiom systems are constructively acceptable. Although the translation is indeed natural it is technically somewhat complicated due to the transfinitely iterative nature of the global universe. Some of that complication can be avoided when directly interpreting the language of local set theory into type theory. We consider this important in connection with our third reason for our interest in a local version of **CST**.

In recent years there has been a growing interest in the development of old and new areas of constructive mathematics and there have been competing settings for this such as the Bishop style approach and the type theoretical, set theoretical and category theoretical approaches. Each has its advantages. The Bishop style approach is informal and works directly with the intensional constructive notions. The type theoretical approach is a more formal philosophically motivated approach. The set theoretical approach is fully extensional and is close to the mainstream set theoretical approach to classical mathematics. The category theoretic approach is more conceptual with its focus on the algebraic structure of the fundamental mathematical notions. Definitions and results formulated in one approach should carry over to the other approaches. But this is not always a straightforward matter.

For example let us focus on the relationship between type theoretical and set theoretical constructive mathematics. In some presentations of work in constructive mathematics definitions and results are given in an ambiguous style, intended to be understood in both the type theoretical and the set theoretical setting. There is a danger that such a style leads to a lack of rigour. So we advocate another approach. Develop constructive mathematics so that it can be straightforwardly formalised in suitable axiom systems for local **CST**. As local **CST** has a straightforward interpretation in global **CST** and a fairly straightforward direct type theoretic interpretation we get a simple rigorous approach to having definitions and results simultaneously in both settings.

The details of the direct type theoretic interpretation requires more type-theoretic treatment than seems appropriate for this paper and so has been left for another occasion. Suffice it to state here that each sort α is interpreted as a setoid $[[\alpha]]$; i.e. a type, with an equivalence relation $=_{[[\alpha]]}$ and each proposition $(a_1 =_{[[\alpha]]} a_2)$, for $a_1, a_2 : [[\alpha]]$ is required to be small; i.e. a value in a type universe U. Also each set-term of sort $\mathcal{P}\alpha$ will be interpreted as an object $\mathsf{pair}(A, f) : (\Sigma X : U)(X \to [[\alpha]])$.

We will want our local version of **CST** to have a straightforward interpretation in global **CST**. But some care is needed in setting up the language. Classical set theory has the powerset axiom and the full separation scheme. So an interpretation of a local set theory in classical set theory has each sort α interpreted as a set $[[\alpha]]$ with the sort $\mathcal{P}\alpha$ interpreted as the powerset $Pow([[\alpha]])$ and each comprehension term $\{x : \alpha \mid \phi\}$ of sort $\mathcal{P}\alpha$ interpreted as a subset of $[[\alpha]]$.

But a key feature of **CST** is that the powerset axiom and the full separation scheme are not available, as these are too impredicative. So, instead of interpreting each sort as a set, the interpretation of our local version of **CST** will interpret each sort as a class, with $[[\mathcal{P}\alpha]]$ interpreted as the powerclass of $[[\alpha]]$; i.e. the class $Pow([[\alpha]])$ of all subsets of the class $[[\alpha]]$. Also, each comprehension term $\{x : \alpha \mid \phi\}$ of sort $\mathcal{P}\alpha$ will now be interpreted as a subclass of the class $[[\alpha]]$ which, in general, may not be a set in $[[\mathcal{P}\alpha]]$.

Class terminology and notation provides a useful device when working in classical axiomatic set theory. It proves to be even more useful when working in **CST** when many comprehension terms that represent sets in classical set theory can only be taken to be classes in **CST**. To treat classes in a set theory in a flexible way, without making them values in the range of bound variables, it is convenient to formulate a set theory in a suitable free logic. In general, a free logic allows the use of terms which may not represent values in the range of the variables.

There are a variety of approaches to the setting up of a free logic. For example some approaches such as those of Beeson (1985), are intended for use with function symbols that may be interpreted as partial functions, so that terms may be undefined in an interpretation. In such an approach it is natural to require equality and other relation symbols to be strict in the sense that they are only intended to hold for arguments that are in the range of the bound variables. In our approach to free logic we will be more liberal. We do not want equality to be strict, as equality between classes has a natural extensional treatment. Also the membership relation should only be strict in its first argument.

A key axiom scheme of **CZF** is the set induction scheme. It is a suitably constructive version of the classical foundation axiom that expresses that all sets are well-founded. More specifically the scheme states that the universe is inductively generated as the smallest class such that every subset of the class is an element of the class. By making essential use of the scheme we have a class induction metatheorem for **CZF**. The metatheorem expresses that a general kind of inductive definition can be used to inductively generate a class as the smallest class satisfying closure conditions specified by the inductive definition. Moreover, by making essential use of the axiom **REA** of **CZF**$^+$, for certain inductive definitions the inductively generated class will be a set. In addition, using **REA** again, a useful set-compactness result concerning set inductive definitions can be obtained.

These inductive definition results can play an important role in the development of constructive mathematics in **CST**. See, for example Aczel (2006), where it is shown using set compactness that every inductively generated formal topology is set-presentable. So we would like to have these inductive definition results available in our local **CST**. But there is a problem with the proofs of these results in local **CST**. The results can be formulated in local **CST**. But the proofs of the results use the set induction scheme and the axiom **REA**, a scheme and axiom that are global and do not have direct local formulations. The other axioms and schemes of **CZF** and **CZF**$^+$ do have local formulations. So we will introduce new axiom systems **CZFI** and **CZF*** that have axioms and schemes that directly express the inductive definition results of **CZF** and **CZF**$^+$ respectively. They are both extensions of the axiom system **CZF**$^-$ obtained from a formulation of **CZF** by leaving out the set induction scheme. We will see that both **CZFI** and **CZF*** have local versions.

We review the **CST** axiom systems **CZF**, **CZF**$^+$ and **CZF**$^-$ in Section 2. We also discuss the inductive definition results that can be proved and formulate the axiom systems **CZFI** and **CZF***. In Section 3 we introduce our free logic and the free versions of the **CST** axiom systems. We go on, in Section 5, to formulate our local versions of these axiom systems. But before that we introduce the ideas of local set theory by formulating a local version of intuitionistic Zermelo set theory. We give an application of inductive definitions to well-founded trees in Section 6.

2 Inductive Definitions in CST

In this section we review results concerning inductive definitions that have been obtained in the **CST** axiom systems **CZF** and **CZF**$^+$.

2.1 Inductive Definitions in CZF

The axiom system **CZF** is formulated in the usual first order language of axiomatic set theory with equality and membership as the two relation symbols. It has the axioms and rules of inference for intuitionistic predicate logic with equality and uses the non-logical axioms and schemes of Extensionality, Pairing, Union, Infinity, Restricted Separation, Strong Collection, Subset Collection and Set Induction. See Aczel and Rathjen (2000/01) for the details of these axioms and schemes. Alternatively the reader may get a good enough idea by looking at the presentation of the axiom system $\mathbf{CZF}_{\mathfrak{f}}$ in the next section. Here we just consider the Infinity axiom, which states that there is an ω-inductive set, where we define a class A to be *ω-inductive* if $\emptyset \in A$ and $(\forall x \in A)[\ x \cup \{x\} \in A\ ]$.

Let $\mathbf{CZF}^-$ be obtained from **CZF** first, by leaving out from the axioms and schemes the set induction scheme, and second by strengthening the axiom of Infinity to the axiom of Strong Infinity and adding the Mathematical Induction Scheme. Strong Infinity states that there is a smallest ω-inductive set, while Mathematical Induction states that the smallest ω-inductive set is a subset of each ω-inductive class. Note that both Strong Infinity and Mathematical Induction can be derived in **CZF**.

The axiom system $\mathbf{CZF}^-$ is fully predicative. It is the set induction scheme that gives **CZF** its logical strength. That scheme expresses that the universe of sets is the smallest class such that every subset of the class is an element of the class. Although the scheme is not predicative in the traditional sense it is not fully impredicative either, as it does not imply the powerset axiom or the full separation scheme. It is natural to call it generalised predicative, as it is predicative relative to certain kinds of inductive definition which may be infinitary rather than the finitary inductive definitions which are acceptable in predicative mathematics.

It will not be difficult to formulate a local version of $\mathbf{CZF}^-$. But the set induction scheme is a global property of the universe of sets and there seems to be no direct way to formulate a local version of that scheme so as to obtain a local version of the whole of **CZF**. An important metatheorem about **CZF**, which would seem to express the logical strength of **CZF**, states that class inductive definitions of classes hold for **CZF**. We shall see, in Section 5, that the metatheorem can be formulated and derived for an extension of the local version of $\mathbf{CZF}^-$ and we will take that extension as our local version of **CZF**. The additional axioms of the extension will directly express that set inductive definitions of classes hold. But in this section we consider inductive definitions in the global context.

We think of an inductive definition as an abstract axiom system having (inference) steps X/a consisting of a (possibly infinite) set X of premisses and a conclusion a. The theorems of the axiom system form the smallest class closed under the inference steps; i.e. for each step X/y, if the premisses are in the class then so is the conclusion. Any class Φ can be viewed as a class inductive definition whose steps X/a are the ordered pairs (X, a) in Φ. A class is defined to be Φ-closed if, for each step X/a of Φ, if every element of X is in the class then so is a. The class inductively defined by Φ, if it exists, is the smallest Φ-closed class.

Definition: 2.1 *A set theoretical axiom system T has the* class induction property *if, for each class Φ of T there is a smallest Φ-closed class $I(\Phi)$ of T; i.e. there is a class I such that the following are derivable in T.*

1. *I is Φ-closed, and*
2. *if the class A is Φ-closed then $I \subseteq A$.*

The metatheorem may now be formulated as follows.

Theorem 2.2 (Class Induction Metatheorem for CZF) *The theory* **CZF** *has the class induction property.*

Note that the Set Induction Scheme may be restated as $V = I(\Phi)$, where Φ is the class of all pairs (X, X). The scheme is clearly a global property about the universe V.

It will be straightforward to formulate a local version of the predicative system **CZF**$^-$. We now formulate an extension **CZFI** of **CZF**$^-$ by adding a new binary infix relation symbol $\vdash$ to the language satisfying the following axioms and scheme where, for each class Φ,

$$I(\Phi) = \{x \mid \Phi_0 \vdash x \text{ for some subset } \Phi_0 \text{ of } \Phi\}.$$

ind0: $\vdash$ is monotone in its first argument; i.e. for all sets Φ, Φ',

$$\Phi \subseteq \Phi' \;\Rightarrow\; (\forall x)[\; \Phi \vdash x \;\Rightarrow\; \Phi' \vdash x \;].$$

ind1: For all sets Φ the class $I(\Phi)$ is Φ-closed.
ind2: For each class A, if Φ is any set such that A is Φ-closed then $I(\Phi) \subseteq A$.

Note that, because of **ind0**, $I(\Phi) = \{x \mid \Phi \vdash x\}$ for each set Φ. Also **ind1** and **ind2** combine to state that for each set Φ, the class $I(\Phi)$ is the smallest Φ-closed class. The next result states that, for **CZFI**, $I(\Phi)$ is the smallest Φ-closed class, even when Φ is a class that may not be a set.

Theorem 2.3 (Class Induction Metatheorem for CZFI) *The axiom system* **CZFI** *has the class induction property.*

Proof Let Φ be a class. We first show that $I(\Phi)$ is Φ-closed. So let X/a be a step of Φ such that $X \subseteq I(\Phi)$. We must show that $a \in I(\Phi)$. By our assumption,

$$(\forall x \in X)(\exists \Phi_0 \in Pow(\Phi))\; \Phi_0 \vdash x.$$

So, by Strong Collection, there is a subset $\mathcal{Y}$ of $Pow(\Phi)$ such that

$$(\forall x \in X)(\exists \Phi_0 \in \mathcal{Y})\; \Phi_0 \vdash x.$$

Now let $\Phi_1 = \{(X, a)\} \cup \bigcup \mathcal{Y}$ and observe that Φ_1 is a subset of Φ having X/a as a step, with $X \subseteq I(\Phi_1)$ so that, by **ind1**, $a \in I(\Phi_1) \subseteq I(\Phi)$ as the operator I is monotone.

It remains to show that $I(\Phi)$ is a subclass of each Φ-closed class A. So let $a \in I(\Phi)$; i.e. $\Phi_0 \vdash a$ for some subset Φ_0 of Φ. If A is Φ-closed then it is Φ_0-closed so that, by **ind2**, $a \in I(\Phi_0) \subseteq A$, as desired. ■

2.2 Inductive Definitions in CZF$^+$

A useful strengthening of **CZF** is the axiom system $\mathbf{CZF}^+ \equiv \mathbf{CZF} + \mathbf{REA}$. Here **REA** is the regular extension axiom, which states that every set is a subset of a regular set. A *regular set* is an inhabited transitive set[1] A, such that for each $a \in A$, if $R \subseteq a \times A$ such that $(\forall x \in a)(\exists y \in A)\ (x, y) \in R$ then there is $b \in A$ such that both $(\forall x \in a)(\exists y \in b)\ (x, y) \in R$ and $(\forall y \in b)(\exists x \in a)\ (x, y) \in R$. The following results about inductive definitions may be derived in **CZF**$^+$, see Aczel and Rathjen (2000/01, 2008).

A set B is a *set bound for a class* Φ if, for each step Y/z of Φ, there is $b \in B$ and a surjective $f : b \to Y$. The class Φ is defined to be *bounded* if it has a set bound and for each set Y the class $\{z \mid (Y, z) \in \Phi\}$ is a set. Note that, in **CZF**, each set is bounded. For each set Φ_0 let $I_0(\Phi_0)$ be the intersection of all Φ_0-closed sets; i.e. the class $\bigcap\{Y \mid Y \text{ is a } \Phi_0\text{-closed set}\}$. Also, for each class Φ let

$$I(\Phi) = \bigcup\{I_0(\Phi_0) \mid \Phi_0 \in Pow(\Phi)\}.$$

Bounded Induction Scheme (BIS): *For each class Φ, the class $I(\Phi)$ is a subclass of each Φ-closed class and hence is the smallest Φ-closed class. Moreover if Φ is bounded then $I(\Phi)$ is a set and so $I(\Phi) = I_0(\Phi)$.*

Another useful result of **CZF**$^+$ is the Set Compactness property for set inductive definitions. See Aczel and Rathjen (2000/01) for the original result and Aczel and Rathjen (2008) for a proof of the more recent improvement, **SSC**.

Strong Set Compactness Property (SSC): *For every set Φ there is a set B of subsets of Φ such that for every subset Φ' of Φ every element of $I_0(\Phi')$ is in $I_0(\Phi_0)$ for some subset Φ_0 of Φ' that is in B.*

Theorem 2.4 (CZF$^+$) *Each instance of* **BIS** *can be derived as can the statement* **SSC**.

Let $\mathbf{CZF}^* \equiv \mathbf{CZF} + \mathbf{BIS} + \mathbf{SSC}$. As each set is bounded we get the following consequence of **BIS**.

Corollary 2.5 (CZF*) *For each set Φ the class $I_0(\Phi)$ is a set and so $I_0(\Phi) = I(\Phi)$, the smallest Φ-closed class.*

For classes Φ and A let $I(\Phi, A) = I(\Phi_A)$, where $\Phi_A = \Phi \cup \{(\emptyset, x) \mid x \in A\})$. Note that $I(\Phi, A)$ is the smallest Φ-closed class that includes A. It immediately follows from corollary 2.5 that if Φ is a set then $I(\Phi, A)$ is a set for each set A. We have the following consequence of **SSC**.

Corollary 2.6 (Set Compactness for CZF*) *For all sets Φ, A there is a set B of subsets of A such that, for all sets $A' \subseteq A$, each element of $I(\Phi, A')$ is an element of $I(\Phi, A_0)$ for some subset A_0 of A' that is in B.*

[1] That is a set that has an element and is such that it is a subset of its powerset.

3 The Free Version of CST

3.1 A Free Logic

We present our free version of intuitionistic predicate logic with equality. We assume that formulae are generated from the atomic formulae in the usual way using the logical constants $\bot$ and $\top$, the binary connectives $\wedge, \vee$ and $\rightarrow$ and the quantifiers $(\forall x)$ and $(\exists x)$ for each variable x. Abbreviations for $\neg$ and $\leftrightarrow$ are defined as usual.

We first consider the following Hilbert style axiomatisation of Intuitionistic logic with equality and then present our free modification. A standard set of axioms together with the rule of modus ponens will do for intuitionistic propositional logic. We just focus on the axioms and rules for equality and the quantifiers. The equality axioms are as follows, where a, a_1, a_2 are terms and x is a variable.

$$a = a \qquad\qquad (a_1 = a_2) \wedge \phi[a_1/x] \rightarrow \phi[a_2/x]$$

We consider the following standard quantifier rules and axioms, where x is a variable not free in the formula θ and a is any term.

$$\frac{\theta \rightarrow \phi}{\theta \rightarrow (\forall x)\phi} \qquad\qquad (\forall x)\phi \rightarrow \phi[a/x]$$

$$\phi[a/x] \rightarrow (\exists x)\phi \qquad\qquad \frac{\phi \rightarrow \theta}{(\exists x)\phi \rightarrow \theta}$$

We use $a\!\downarrow$ to abbreviate $(\exists x)[a = x]$, where the variable x is chosen not to occur free in the term a. This expresses that the term a is in the range of the variables. For our free logic we modify the two quantifier axioms as follows.

$$a\!\downarrow \wedge\, (\forall x)\phi \rightarrow \phi[a/x] \qquad\qquad a\!\downarrow \wedge\, \phi[a/x] \rightarrow (\exists x)\phi$$

We also need the axiom, $y\!\downarrow$, for each variable y.

3.2 The Axiom System $\mathbf{CZF_f}$

The axiom system **CZF** for constructive set theory has usually been axiomatised in the first order language with equality and one binary infix relation symbol $\in$. Here we give an axiomatisation in our free logic, where we allow the formation of comprehension terms[2] $\{x \mid \phi\}$ for arbitrary formulae ϕ and have the following general non-logical axioms and schemes. Further axioms and schemes will be given after we have introduced some abbreviations. In the following ϕ is any formula and a, b are terms.

---

[2] Free occurrences of x in ϕ become bound in $\{x \mid \phi\}$.

Comprehension: $a \in \{x \mid \phi\} \leftrightarrow a\downarrow \wedge \phi[a/x]$.
Extensionality: $(\forall x)(x \in a \leftrightarrow x \in b) \rightarrow a = b$.
Elements are Sets: $(a \in b) \rightarrow a\downarrow$.

Some Abbreviations

In these abbreviations a, a_1, a_2, b, c, r are terms; i.e. either variables or comprehension terms. There may be a standard constraint that a variable is not allowed to occur free in a formula or term. For example in the abbreviation for $\{x \in a \mid \phi\}$ the variable x should not occur free in the term a and in the abbreviation for $(\exists! y)\phi$ the variable x should not occur free in ϕ.

$$\begin{aligned}
(\forall x \in a)\phi &\equiv (\forall x)[x \in a \rightarrow \phi] \\
(\exists x \in a)\phi &\equiv (\exists x)[x \in a \wedge \phi] \\
\{x \in a \mid \phi\} &\equiv \{x \mid x \in a \wedge \phi\} \\
\{b \mid x \in a\} &\equiv \{y \mid (\exists x \in a)\, y = b\} \\
a_1 \cap a_2 &\equiv \{x \mid x \in a_1 \wedge x \in a_2\} \\
a_1 \cup a_2 &\equiv \{x \mid x \in a_1 \vee x \in a_2\} \\
a_1 \subseteq a_2 &\equiv (\forall x \in a_1)\, x \in a_2 \\
\bigcap c &\equiv \{x \mid (\forall y \in c)\, x \in y\} \\
\bigcup c &\equiv \{x \mid (\exists y \in c)\, x \in y\} \\
Pow(a) &\equiv \{y \mid y \subseteq a\} \\
\{a_1, a_2\} &\equiv \{x \mid x = a_1 \ \vee x = a_2\} \\
\{a\} &\equiv \{a, a\} \\
(a_1, a_2) &\equiv \{\{a_1\}, \{a_1, a_2\}\} \\
(\exists! y)\, \phi &\equiv (\exists x)(\forall y)[(y = x) \leftrightarrow \phi] \\
a \times b &\equiv \{z \mid (\exists x \in a)(\exists y \in b)[z = (x, y)]\} \\
V &\equiv \{x \mid \top\} \\
\emptyset &\equiv \{x \mid \bot\}
\end{aligned}$$

$$\begin{aligned}
r : a \succ &\equiv (\forall x \in a)(\exists y)[(x, y) \in r] \\
r : a \succ b &\equiv (\forall x \in a)(\exists y \in b)[(x, y) \in r] \\
r : a \succ\!\!\prec b &\equiv r : a \succ b \wedge (\forall y \in b)(\exists x \in a)[(x, y) \in r] \\
r : a \rightarrow b &\equiv r \in Pow(a \times b) \wedge (\forall x \in a)(\exists\, !y)\ (x, y) \in r \\
r : a \twoheadrightarrow b &\equiv r : a \rightarrow b \wedge (\forall y \in b)(\exists x \in a)\ (x, y) \in r \\
\mathbf{mv}(a, b) &\equiv \{z \in Pow(a \times b) \mid z : a \succ b\} \\
\mathbf{exp}(a, b) &\equiv \{z \in Pow(a \times b) \mid z : a \rightarrow b\}
\end{aligned}$$

For our set theoretic representation of the natural numbers we use the following abbreviations.

$$\begin{aligned}
0 &\equiv \emptyset \\
a^+ &\equiv a \cup \{a\} \\
IND_\omega(a) &\equiv 0 \in a \ \wedge \ (\forall y \in a)[\, y^+ \in a\,] \\
\mathbb{N} &\equiv \bigcap \{x \mid IND_\omega(x)\}
\end{aligned}$$

Set Induction Scheme: For each term a,

$$Pow(a) \subseteq a \;\rightarrow\; V \subseteq a.$$

Mathematical Induction Scheme: For each term a,

$$IND_{\omega}(a) \rightarrow \mathbb{N} \subseteq a.$$

The Set Existence Axioms and Schemes

Pairing: $(\forall x_1, x_2)\ \{x_1, x_2\}\downarrow$.
Union: $(\forall z)\ \bigcup z\downarrow$.
Restricted Separation Scheme: $(\forall y)\ \{x \in y \mid \theta\}\downarrow$,
for each restricted formula θ; i.e. formula θ in which each quantifier occurs in one of the forms $(\forall u \in v)$ or $(\exists u \in v)$.
Strong Infinity: $\mathbb{N}\downarrow$.
Strong Collection Scheme: $(\forall x)[\ r : x \succ\ \rightarrow\ (\exists y)\ r : x \succ\!\prec y\ ]$,
for each term r.
Fullness: $(\forall x)(\forall y)(\exists z \in Pow(\mathbf{mv}(x,y)))(\forall u \in \mathbf{mv}(x,y))(\exists u_0 \in z)[u_0 \subseteq u]$.

Remarks

1. Using the Strong Collection Scheme both the Mathematical Induction Scheme and the Strong Infinity axiom can be derived using the following axiom.

   Infinity: $(\exists x)\ IND_{\omega}(x)$.

2. The original formulation of **CZF** used a certain scheme, the Subset Collection Scheme, instead of the Fullness axiom. That scheme and the Fullness axiom are equivalent, given the other axioms and schemes.
3. Easy consequences of the Strong Collection Scheme and the Fullness axiom are the following scheme and axiom, respectively.

   Replacement Scheme: $a\downarrow \wedge (\forall x \in a)(\exists\, !y)\phi\ \rightarrow\ \{y \mid (\exists x \in a)\phi\}\downarrow$, for each term a.
   Exponentiation: $a\downarrow \wedge\ b\downarrow \rightarrow \mathbf{exp}(a,b)\downarrow$, for arbitrary terms a, b.

We call this free formulation of **CZF**, $\mathbf{CZF_f}$.

Theorem 3.1 $\mathbf{CZF_f}$ *is a conservative extension of* **CZF**.

Proof We only sketch the idea for the straightforward proof. We can define a translation of the language of $\mathbf{CZF_f}$ into that of **CZF** by systematically eliminating comprehension terms. This can be done by repeatedly replacing each equality $(a = b)$ by $(\forall x)[\ x \in a\ \leftrightarrow\ x \in b\ ]$, each formula $(a \in b)$, where a is a comprehension term by $(\exists x)[(\forall y)(y \in a \leftrightarrow y \in x) \wedge x \in b]$ and each formula $(x \in \{y \mid \phi\})$ by $\phi[x/y]$. Eventually all comprehension terms will be eliminated from a formula ϕ giving a formula $\phi^\#$ such that $\phi \leftrightarrow \phi^\#$ is provable in $\mathbf{CZF_f}$ and if ϕ is provable in $\mathbf{CZF_f}$ then $\phi^\#$ is provable in **CZF**. If ϕ already is in the language of **CZF** then $\phi^\# \equiv \phi$ so that if it is provable in $\mathbf{CZF_f}$ then it is provable in **CZF**. ■

3.3 *The Axiom Systems* $\mathbf{CZF_f}^-$, $\mathbf{CZF_fI}$ *and* $\mathbf{CZF_f}^*$

By leaving out the Set Induction Scheme from $\mathbf{CZF_f}$ we get the free version $\mathbf{CZF_f}^-$ of $\mathbf{CZF}^-$.

We get the free version $\mathbf{CZF_fI}$ of **CZFI** by adding the binary relation symbol $\vdash$ to the language of $\mathbf{CZF_f}^-$ and adding the following axioms and scheme to $\mathbf{CZF_f}^-$ where, for terms c, b,

$$\begin{aligned} Closed(c,b) &\equiv (\forall y \in Pow(b))(\forall x)[(y,x) \in c \rightarrow x \in b], \\ I(c) &\equiv \{x \mid (\exists z \in Pow(c))\ z \vdash x\}. \end{aligned}$$

ind0: $(\forall z, z')[\ z \subseteq z' \rightarrow (\forall x)(z \vdash x \rightarrow z' \vdash x)\ ]$,
ind1: $\forall z\ Closed(z, I(z))$,
ind2: $\forall z\ [Closed(z,b) \rightarrow I(z) \subseteq b]$, for each term b.

We next formulate the free version of $\mathbf{CZF}^*$. For each term c let

$$\begin{aligned} I_0(c) &\equiv \bigcap\{y \mid Closed(c,y)\} \\ I(c) &\equiv \bigcup\{I_0(z) \mid z \in Pow(c)\} \end{aligned}$$

Note that $Closed(I(c), c)$ is a theorem of $\mathbf{CZF_f}^-$.
Let

$$Bounded(c) \equiv \forall y\ \{x \mid (y,x) \in c\}\!\downarrow\ \wedge\ (\exists z)(\forall y \in dom(c))(\exists v \in z)(\exists w)\ w : v \twoheadrightarrow y,$$

where $dom(c) \equiv \{y \mid \exists x\ (y,x) \in c\}$.

Bounded Induction Scheme (BIS): For terms c, a,

$$[Closed(c,a) \rightarrow I(c) \subseteq a] \wedge [Bounded(c) \rightarrow I(c)\!\downarrow]$$

Strong Set Compactness (SSC):

$$(\forall z)(\exists y \in Pow(Pow(z)))(\forall z' \in Pow(z))[\ I_0(z') = \bigcup\{I_0(z_0) \mid z_0 \in Pow(z') \cap y\}\ ]$$

Finally we let $\mathbf{CZF_f}^* \equiv \mathbf{CZF_f}^- + \mathbf{BIS} + \mathbf{SSC}$.

4 Local Intuitionistic Zermelo Set Theory

We outline a formal system **LIZ** which is a local version of Intuitionistic Zermelo set theory in which there is a one element sort $\mathsf{1}$, an infinite ground sort $\mathbb{N}$, product sorts $\alpha\times\beta$ and power sorts $\mathcal{P}\alpha$. The sorts have a natural interpretation in a set theory such as **IZ**, Intuitionistic Zermelo Set Theory, each sort α being interpreted as a set $[[\alpha]]$, with $[[\mathsf{1}]] = \{\emptyset\}$, $[[\mathbb{N}]]$ any set with $\emptyset \in [[\mathbb{N}]]$ that is closed under $n \mapsto n \cup \{n\}$, and $[[\alpha \times \beta]] = [[\alpha]] \times [[\beta]]$, using the standard definition of the cartesian product of two sets, and $[[\mathcal{P}\alpha]] = Pow([[\alpha]])$, where Pow is the powerset operation on sets. With obvious interpretations of the constants $*, 0$ and function symbols $(\ )^+, (\ ,\ )$, used below in forming terms, the axioms of **LIZ**, given below, are easily seen to be theorems of **IZ**.

There is an unlimited supply of variables of each sort. The terms of each sort and the formulae are simultaneously inductively generated. Here are the rules for generating terms.

1. Every variable of sort α is a term of sort α.
2. $*$ is a term of sort $\mathsf{1}$.
3. 0 is a term of sort $\mathbb{N}$.
4. If a is a term of sort $\mathbb{N}$ then a^+ is also a term of sort $\mathbb{N}$.
5. If a, b are terms of sorts α, β respectively then (a, b) is a term of sort $\alpha \times \beta$.
6. If ϕ is a formula and x is a variable of sort α then $\{x : \alpha \mid \phi\}$ is a term of sort $\mathcal{P}\alpha$.

We call terms of the form $\{x : \alpha \mid \phi\}$ *α-classes* and write $a : \alpha$ when a is an α-term; i.e. a term of sort α. The atomic formulae have one of the forms $(a_1 =_\alpha a_2)$ for α-terms a_1, a_2 or $(a \in_\alpha b)$ when a is an α-term and b is a $\mathcal{P}\alpha$-term. We will usually supress the subscript α.

Formulae are generated from the atomic formulae in the usual way using the logical constants $\bot$ and $\top$, the binary connectives $\wedge, \vee$ and $\rightarrow$ and the quantifiers $(\forall x : \alpha)$ and $(\exists x : \alpha)$ for each variable x of sort α. Abbreviations for $\neg$ and $\leftrightarrow$ are defined as usual.

We use the following axiomatisation of Intuitionistic many sorted logic with equality. Any set of axioms and rules of inference for intuitionistic propositional logic will do. We just focus on the axioms and rules for equality and the quantifiers. The equality axioms are as follows, where a, a_1, a_2 are α terms and x is a variable of sort α.

$$a = a \qquad\qquad (a_1 = a_2) \wedge \phi[a_1/x] \rightarrow \phi[a_2/x]$$

We use the following quantifier rules and axioms, where x is a variable of sort α not free in the formula θ and a is an α-term.

$$\frac{\theta \rightarrow \phi}{\theta \rightarrow (\forall x : \alpha)\phi} \qquad\qquad (\forall x : \alpha)\phi \rightarrow \phi[a/x]$$

$$\phi[a/x] \rightarrow (\exists x : \alpha)\phi \qquad\qquad \frac{\phi \rightarrow \theta}{(\exists x : \alpha)\phi \rightarrow \theta}$$

We assume the following non-logical axioms and scheme, where a_1, a_2, a are α-terms, b_1, b_2 are β-terms, c is a $\alpha \times \beta$-term and d_1, d_2 are $\mathcal{P}\alpha$-terms.

1. $(a_1, b_1) = (a_2, b_2) \;\rightarrow\; (a_1 = a_2) \wedge (b_1 = b_2)$
2. $(\exists x : \alpha)(\exists y : \beta)[c = (x, y)]$
3. $(\forall z : 1)[z = *]$
4. $(\forall x : \alpha)(x \in d_1 \;\leftrightarrow\; x \in d_2) \;\rightarrow\; d_1 = d_2$
5. $(\forall y : \alpha)[\, y \in \{x : \alpha \mid \phi\} \;\leftrightarrow\; \phi[y/x]\,]$, for each formula ϕ.
6. $(\forall x : \mathbb{N})\, \neg(x^+ = 0)$
7. $(\forall x_1, x_2 : \mathbb{N})[\, (x_1^+ = x_2^+) \;\rightarrow\; (x_1 = x_2)\,]$

Some Abbreviations

It is convenient to introduce the following abbreviations where a, a_1, a_2 are $\mathcal{P}\alpha$-terms, b is a β-term and c is a $\mathcal{P}\mathcal{P}\alpha$-term.

$$
\begin{aligned}
(\forall x \in a)\phi &\equiv (\forall x : \alpha)[x \in a \;\rightarrow\; \phi] \\
(\exists x \in a)\phi &\equiv (\exists x : \alpha)[x \in a \;\wedge\; \phi] \\
\{x \in a \mid \phi\} &\equiv \{x : \alpha \mid x \in a \;\wedge\; \phi\} \\
\{b \mid x \in a\} &\equiv \{y : \beta \mid (\exists x \in a)\; y = b\} \\
Pow(a) &\equiv \{y : \mathcal{P}\alpha \mid y \subseteq a\} \\
a_1 \cap a_2 &\equiv \{x : \alpha \mid x \in a_1 \;\wedge\; x \in a_2\} \\
a_1 \cup a_2 &\equiv \{x : \alpha \mid x \in a_1 \;\vee\; x \in a_2\} \\
a_1 \subseteq a_2 &\equiv (\forall x \in a_1)\; x \in a_2 \\
\bigcap c &\equiv \{x : \alpha \mid (\forall y \in c)\; x \in y\} \\
\bigcup c &\equiv \{x : \alpha \mid (\exists y \in c)\; x \in y\}
\end{aligned}
$$

If a is a $\mathcal{P}\alpha$-term, b is a $\mathcal{P}\beta$-term we use the following abbreviations.

$$
\begin{aligned}
a \times b &\equiv \{z : \alpha \times \beta \mid (\exists x \in a)(\exists y \in b)[z = (x, y)]\} \\
(\exists\, !y \in b)\; \phi &\equiv (\exists x \in b)(\forall y \in b)[(y = x) \;\leftrightarrow\; \phi]
\end{aligned}
$$

As in the previous section we use the following abbreviations where now a is a $\mathcal{P}\alpha$-term, b is a $\mathcal{P}\beta$-term and r is a $\mathcal{P}(\alpha \times \beta)$-term.

$$
\begin{aligned}
r : a \succ\!\!- \;&\equiv (\forall x \in a)(\exists y)[(x, y) \in r] \\
r : a \succ\!\!- b &\equiv (\forall x \in a)(\exists y \in b)[(x, y) \in r] \\
r : a \succ\!\!\prec b &\equiv r : a \succ\!\!- b \wedge (\forall y \in b)(\exists x \in a)[(x, y) \in r] \\
r : a \rightarrow b &\equiv r \in Pow(a \times b) \wedge (\forall x \in a)(\exists\, !y \in b)\; (x, y) \in r \\
r : a \twoheadrightarrow b &\equiv r : a \rightarrow b \wedge (\forall y \in b)(\exists x \in a)\; (x, y) \in r \\
\mathbf{mv}(a, b) &\equiv \{z \in Pow(a \times b) \mid z : a \succ\!\!- b\} \\
\mathbf{exp}(a, b) &\equiv \{z \in Pow(a \times b) \mid z : a \rightarrow b\}
\end{aligned}
$$

The Natural Numbers

The natural numbers are here represented using the following abbreviations where c is a $\mathcal{P}\mathbb{N}$-term.

$$IND_\omega(c) \equiv 0 \in c \ \wedge \ (\forall x \in c)\ x^+ \in c$$
$$\mathbb{N} \equiv \bigcap\{z : \mathcal{P}\mathbb{N} \mid IND_\omega(z)\}$$

We have not included any mathematical induction scheme for $\mathbb{N}$, as we can show that $(\mathbb{N}, 0, (\ )^+)$ satisfies the Dedekind-Peano axioms, including the mathematical induction axiom:

$$(\forall z \in Pow(\mathbb{N}))[\ IND_\omega(z) \ \rightarrow \ \mathbb{N} \subseteq z\ ].$$

Inductive Definitions in LIZ

By an α-inductive definition we simply mean a term $c : \mathcal{P}(\mathcal{P}\alpha \times \alpha)$ for some sort α. If $b : \mathcal{P}\alpha$ let $I_0(c) \equiv \bigcap\{y : \mathcal{P}\alpha \mid Closed(c, y)\}$, where

$$Closed(c, b) \ \equiv \ (\forall y \in Pow(b))(\forall x : \alpha)[(y, x) \in c \ \rightarrow \ x \in b].$$

Because of the full impredicativity of **LIZ** the set $I_0(c)$ is provably the smallest c-closed set; i.e. the following sentences can be derived in **LIZ** for each term c of sort $\mathcal{P}(\mathcal{P}\alpha \times \alpha)$.

$\mathbf{ind_0 1}$: $(\forall z : \mathcal{P}(\mathcal{P}\alpha \times \alpha))\ Closed(z, I_0(z))$,
$\mathbf{ind_0 2}$: $(\forall z : \mathcal{P}(\mathcal{P}\alpha \times \alpha))\ (\forall y : \mathcal{P}\alpha)[Closed(z, y) \rightarrow I_0(z) \subseteq y]$.

5 Some Axiom Systems for Local CST

The formal system **LIZ** is a thoroughly impredicative axiom system, each set having its powerset. We wish to have a predicative version of it, which will be a local version, $\mathbf{LCZF_f}^-$, of $\mathbf{CZF_f}^-$, to which we can add the forms of inductive definition of sets and classes that are available in constructive set theory, thereby giving us local versions $\mathbf{LCZF_fI}$ and $\mathbf{LCZF_f}^*$ of $\mathbf{CZF_fI}$ and $\mathbf{CZF_f}^*$ respectively.

5.1 Many-Sorted Free Logic

The Intuitionistic many-sorted logic with equality, as described in section 4, is modified to become a free logic by modifying the quantifier axioms as follows for each α-term a.

$$a\downarrow \wedge (\forall x : \alpha)\phi \rightarrow \phi[a/x] \qquad a\downarrow \wedge \phi[a/x] \rightarrow (\exists x : \alpha)\phi$$

where $a\downarrow$ abbreviates the formula $(\exists x : \alpha)(x = a)$, with x chosen not free in a. Also, for each variable y, we add the axiom

$$y\downarrow .$$

5.2 *The Axiom System* $\mathbf{LCZF_f}^-$

We keep the same terms and formulae of **LIZ**, but use the many-sorted free logic where the sorts are intended to be interpreted as classes of constructive set theory which are generally not accepted as sets. In particular the sort $\mathcal{P}\alpha$ is intended to be interpreted as the class of all subsets of the class interpreting the sort α. So the comprehension terms of sort $\mathcal{P}\alpha$ can represent classes of values of sort α while not necessarily representing values in the range of the variables of sort $\mathcal{P}\alpha$, such variables only ranging over the sets of values of sort α.

We need the following axioms for the sorts $\mathsf{1}, \mathbb{N}$ and $\alpha\times\beta$, where a, b, c are terms of sorts $\alpha, \beta, \mathbb{N}$ respectively.

$$*\downarrow \qquad a\downarrow \wedge b\downarrow \;\leftrightarrow\; (a,b)\downarrow$$

$$0\downarrow \qquad c\downarrow \;\leftrightarrow\; c^+\downarrow$$

We also need, for each sort α, the axioms $(a \in b) \rightarrow a\downarrow$, for terms a, b of sorts $\alpha, \mathcal{P}\alpha$ respectively. The axiomatization of $\mathbf{LCZF_f}^-$ is completed by adding the following Mathematical Induction Scheme and Set Existence Axioms.

Mathematical Induction Scheme

$IND_\omega(a) \;\rightarrow\; (\mathbb{N} \subseteq a)$, for each $\mathbb{N}$-class a.

Set Existence Axioms

It remains to consider when we want $a\downarrow$ for $\mathcal{P}\alpha$-terms $a \equiv \{x : \alpha \mid \phi\}$. For this we have set existence axioms based on those for **CZF**.

Pairing: $(\forall x_1, x_2 : \alpha)\ \{x_1, x_2\}\downarrow$.
Union: $(\forall z : \mathcal{P}\mathcal{P}\alpha)\ \bigcup z\downarrow$.
Restricted Separation Scheme: $(\forall y : \mathcal{P}\alpha)\ \{x \in y \mid \theta\}\downarrow$,
for each restricted formula θ; i.e. formula θ in which each quantifier occurs in one of the forms $(\forall u \in v)$ or $(\exists u \in v)$, where v is a variable.
Strong Infinity: $\mathbb{N}\downarrow$.

Strong Collection Scheme: $(\forall x : \mathcal{P}\alpha)[\ r : x \succ\!\!- \ \rightarrow (\exists y : \mathcal{P}\beta) r : x \succ\!\!\prec y\ ]$, for each $(\alpha \times \beta)$-class r.

Fullness:

$$(\forall x : \mathcal{P}\alpha)(\forall y : \mathcal{P}\beta)(\exists z \in Pow(\mathbf{mv}(x, y)))(\forall u \in \mathbf{mv}(x, y))(\exists u_0 \in z)[u_0 \subseteq u].$$

We have now described our local version, $\mathbf{LCZF_f}^-$, of $\mathbf{CZF_f}^-$.

5.3 *The Axiom System* LCZF$_\mathbf{f}$I

In order to have class inductive definitions in **LCST** we formulate a local version, $\mathbf{LCZF_fI}$, of $\mathbf{CZF_fI}$ by adding to the language of $\mathbf{LCZF_f}^-$ a binary infix relation symbol $\vdash_\alpha$, for each sort α. This relation symbol takes a first argument of sort $\mathcal{P}(\mathcal{P}\alpha \times \alpha)$ and a second argument of sort α. We use the abbreviation for $Closed(c, b)$ as in the previous section. We add the following axioms and scheme, where for each term $c : \mathcal{P}(\mathcal{P}\alpha \times \alpha)$,

$$\begin{aligned} I_o(c) &\equiv \{x : \alpha \mid c \vdash x\}, \\ I(c) &\equiv \bigcup\{I_0(z) \mid z \in Pow(c)\}. \end{aligned}$$

ind0 : $(\forall z, z' : \mathcal{P}\alpha)[\ z \subseteq z' \rightarrow (\forall x : \alpha)(z \vdash_\alpha x \rightarrow z' \vdash_\alpha x)\ ]$,
ind1 : $(\forall z : \mathcal{P}(\mathcal{P}\alpha \times \alpha))\ Closed(z, I(z))$,
ind2 : $(\forall z : \mathcal{P}(\mathcal{P}\alpha \times \alpha))\ [Closed(z, b) \rightarrow I(z) \subseteq b]$, for each term b.

5.4 *The Axiom System* LCZF$_\mathbf{f}^*$

For each term $c : \mathcal{P}(\mathcal{P}\alpha \times \alpha)$, let

$$\begin{aligned} Bounded(c) \ \equiv\ & (\forall y : \mathcal{P}\alpha)\ \{x : \alpha \mid (y, x) \in c\}\!\downarrow \\ & \wedge (\exists z : \mathcal{P}\mathcal{P}\beta)(\forall y \in dom(c))(\exists v \in z)\ (\exists w : \mathcal{P}(\beta \times \alpha))\ w : v \twoheadrightarrow y, \end{aligned}$$

where $dom(c) \equiv \{y : \beta \mid (\exists x : \alpha)\ (y, x) \in c\}$.

Bounded Induction Scheme (BIS): For terms $c : \mathcal{P}(\mathcal{P}\alpha \times \alpha)$ and $a : \mathcal{P}\alpha$,

$$[Closed(c) \rightarrow I(c) \subseteq a] \wedge [Bounded(c) \rightarrow I(c)\!\downarrow]$$

Strong Set Compactness (SSC):

$$\begin{aligned} &(\forall z : \mathcal{P}(\mathcal{P}\alpha \times \alpha))(\exists y \in Pow(Pow(z))) \\ &\qquad (\forall z' \in Pow(z))[\ I_0(z') = \bigcup\{I_0(z_0) \mid z_0 \in Pow(z') \cap y\}\ ] \end{aligned}$$

We let $\mathbf{LCZF_f}^* \equiv \mathbf{LCZF_f}^- + \mathbf{BIS} + \mathbf{SSC}$.

6 Well-Founded Trees in Local CST

In constructive mathematics the inductive definitions used to generate well-founded trees are particularly important. In the context of Martin-Löf's constructive type theory Martin-Löf introduced the inductively defined W-types $(Wx : A)B(x)$, of well-founded trees, where B is a family of types indexed by a type A. The set theoretic version of W-types is naturally given as follows.

If A, B are classes such that $B(x) = \{y \mid (x, y) \in B\}$ is a set for each $x \in A$ then we inductively define a class of well-founded trees, where at each node, the branching of the tree is indexed by one of the sets $B(a)$, for $a \in A$. So if $a \in A$ and $p(y)$ is a tree in the class for each $y \in B(a)$ then a tree $sup(a, p)$ should be in the class having, as immediate subtrees, the trees $p(y)$, for $y \in B(a)$. In global set theory it is natural to represent $sup(a, p)$ as simply the ordered pair (a, p). This leads to the inductive definition of $W_{x \in A} B(x)$ as the smallest class W such that $\Gamma W \subseteq W$ where, for each class X,

$$\Gamma X = \sum_{x \in A} X^{B(x)} = \{(x, p) \mid x \in A \ \& \ p : B(x) \to X\}.$$

So $W_{x \in A} B(x) = I(\Phi)$ where Φ is the inductive definition having the steps

$$ran(p)/(a, p)$$

for $(a, p) \in \Gamma V$.

This approach to representing W-types in **CST** is global. We would like to have a local version. We first need a suitable way to represent finite sequences in local **CST**. The following lemma gives us what we need.

Lemma: 6.1 *For each class X there is a class $\widehat{X}$, an element $<> \in \widehat{X}$ and an injective function* $- : - : X \times \widehat{X} \to (\widehat{X} - \{<>\})$.

Proof It is enough to let $\widehat{X} = Pow(\mathbb{N} \times X)$, $<> = \emptyset$ and, for $a \in X$ and $\sigma \in \widehat{X}$,

$$a : \sigma \ = \ \{(0, a)\} \cup \{(n^+, x) \mid (n, x) \in \sigma\}.$$

■

Note that a finite sequence $a_0, a_1, \ldots, a_{n-1}$ of elements of X is represented as the element of $\widehat{X}$

$$a_0 : (a_1 : (\cdots (a_{n-1} : <>) \cdots))$$

which is just the set $\{(0, a_0), (1, a_1), \ldots, (n-1, a_{n-1})\}$.

We will get a local version to the above global approach to representing W-types when we assume that each set $B(x)$ is a subset of some class B_0 and use $S : \Gamma C \to C$ given by the following result of local **CST** to represent the sup operation.

Theorem 6.2 (LCZF$_{\mathbf{f}}$I) *There is a class C and an injective class function $S : \Gamma C \to C$.*

Proof We define $C = Pow(D)$ where $D = A \times \widehat{B_0 \times A}$. For $d = (a, \sigma) \in D$ and $b \in B_0$ let

$$b * d = ((b, a) : \sigma) \in \widehat{B_0 \times A}.$$

For $(a, p) \in \Gamma C$ let

$$S(a, p) = \{a\} \times (\{<>\} \cup \bigcup_{b \in B(a)} p^*(b)) \in C$$

where, for $b \in B(a)$,

$$p^*(b) = \{b * d \mid d \in p(b)\} \in Pow(\widehat{B_0 \times A}).$$

To show that $S : \Gamma C \to C$ is injective, let $(a_1, p_1), (a_2, p_2) \in \Gamma C$ such that $S(a_1, p_1) = S(a_2, p_2)$. Then, for $i = 1, 2$ and $x \in A$,

$$x = a_i \iff (x, <>) \in S(a_i, p_i)$$

so that $a_1 = a_2$. Let $a = a_1 = a_2$ and $b \in B(a)$.

Claim: $p_1(b) = p_2(b)$.

Proof By symmetry it suffices to show that every element n of $p_1(b)$ is an element of $p_2(b)$. So let $d_1 = (a'_1, \sigma_1) \in p_1(b)$. Then $((b, a'_1) : \sigma_1) = b * d_1 \in p_1^*(b)$ so that $(a, ((b, a'_1) : \sigma_1)) \in S(a, \sigma_1) = S(a, \sigma_2)$ and hence

$$(a, ((b, a'_1) : \sigma_1)) = (a, b_2 * d_2)$$

for some $b_2 \in B(a)$ and some $d_2 = (a'_2, \sigma_2) \in p_2(b_2)$. So

$$((b, a'_1) : \sigma_1) = b_2 * d_2 = ((b_2, a'_2) : \sigma_2)$$

so that $b = b_2, a'_1 = a'_2$ and $\sigma_1 = \sigma_2$ and hence

$$d_1 = (a'_1, \sigma_1) = (a'_2, \sigma_2) = d_2 \in p_2(b_2) = p_2(b);$$

i.e. $d_1 \in p_2(b)$. ∎

By the claim $p_1(b) = p_2(b)$ for all $b \in B(a)$ so that $p_1 = p_2$ and hence $(a_1, p_1) = (a_2, p_2)$, proving the theorem.

Using S let $W'_{x \in A} B(x) = I(\Phi')$ where Φ' is the inductive definition having the steps

$$ran(p)/S(a, p)$$

for $(a, p) \in \Gamma C$. In order to show that this definition gives an adequate representation of W-types we need to show that the set theoretic version of the elimination rule for W-types can be proved; i.e. we need the following result, where $W = W'_{x \in A} B(x)$.

Theorem 6.3 (LCZF$_f$I) *If C' is a class and $S' : \Gamma C' \to C'$ then there is a unique class function $K : W \to C'$ such that, for $(a, p) \in \Gamma W$,*

$$K(S(a, p)) = S'(a, K \circ p).$$

Proof We inductively define K to be the smallest class such that if $(a,p) \in \Gamma C$ and $(a,q) \in \Gamma C'$ such that

$$\{(p(b), q(b)) \mid b \in B(a)\} \subseteq K$$

then $(S(a,p), S'(a,q)) \in K$. The theorem is a consequence of the following claims.

Claim 1: If $(w,z) \in K$ then $w \in W$.
Claim 2: If $w \in W$ then there is a unique $z \in C'$ such that $(w,z) \in K$. By the previous claims $K : W \to C'$.
Claim 3: If $(a,p) \in \Gamma W$ then $K(S(a,p)) = S'(a, K \circ p)$.
Claim 4: If $K' : W \to C'$ such that $K'(S(a,p)) = S'(a, K' \circ p)$ for all $(a,p) \in \Gamma W$ then $K' = K$.

Claim 1 is proved by induction following the inductive definition of K while Claims 2 and 4 are proved by induction following the inductive definition of W. ■

Theorem 6.4 ($\mathbf{LCZF_f}^*$) *If A is a set then so is W.*

Proof Assume that A is a set and observe that the inductive definition Φ' of W is bounded with set bound $\{B(a) \mid a \in A\}$. This is because, for each set X, if $X/S(a,p)$ is a step of Φ' then p maps $B(a)$ onto X, and the class of all $S(a,p)$ such that $X/S(a,p)$ is a step of Φ' is the class

$$\bigcup_{b \in B(a)} \{S(a,p) \mid p : B(a) \to X\},$$

which is a set using Exponentiation, Replacement and Union. So, using the Bounded Induction Scheme we see that W is a set. ■

Acknowledgments The final stages of writing this paper were carried out at SCAS, the Scandinavian Collegium for Advanced Study, Uppsala University. I am very grateful for the excellent working environment provided by SCAS.

References

Aczel, P. (2006) Aspects of General Topology in Constructive Set Theory, *Annals of Pure and Applied Logic*, **137**, 3–29.
Aczel, P. and Rathjen, M. (2000/01) *Notes on Constructive Set Theory*, Institut Mittag-Leffler, Report No. 40.
Aczel, P. and Rathjen, M. (2008) *Notes on Constructive Set Theory*, available at `http://www.mims.manchester.ac.uk/logic/mathlogaps/workshop/CST-book-June-08.pdf`, Manchester: Mathlogaps workshop.
Beeson, M.J. (1985) *Foundations of Constructive Analysis*, Heidelberg: Springer.
Bell, J.L. (1988) *Toposes and Local Set Theories; An Introduction*, Oxford Logic Guides, Oxford: Clarendon Press.

Proofs and Constructions

Charles McCarty

A critical examination of notions associated with mathematical intuitionism, with attention to mathematical practice, internal vs. external negations, constructive meaning, weak counterexamples, and theories of constructions. The article includes a theory of constructions based on nonstandard realizability.

1 Preamble

Today, intuitionism is so overgrown with clinging philosophy and history that intuitionistic mathematics has become largely invisible. Erstwhile friends of the subject think that the central mathematical edifice has crumbled under the weight of moss and creeper. Some look only to quarry the ruins and scavenge the remains. Worse, others confuse tendril with brick, mistaking the assorted philosophical ruminations of Egbertus Brouwer, Arend Heyting, and Michael Dummett as intuitionism itself. They overlook the falsity and weirdness of these worthies' remarks, treating them as necessarily valid expressions of an intuitionistic *Weltanschauung*. They prize those dark sayings as intuitionistically true, as if analytically so, because of their provenance—voiced by the holy men, anointed masters of the intuitionist trade. It is high time to rip from mathematical intuitionism the philosophical weeds and other encroachments, and to cast them onto the fires.

2 Brouwer, Hilbert and Mathematical Practice

Faced with philosophers' efforts to undermine the status of intuitionism by appeal to some arcanum they denominate 'mathematical practice,' I am always skeptical, if not derisory. Naturally, I too hold that professional mathematicians do mathematical things and, at times, practice doing them. The mathematicians I know sit relatively

G. Sommaruga (ed.), *Foundational Theories of Classical and Constructive Mathematics*, The Western Ontario Series in Philosophy of Science 76,
DOI 10.1007/978-94-007-0431-2_11,

motionless for longish periods and sometimes converse among themselves. A while back, they spent their working days scribbling on sheets of paper and across chalkboards. Now they hammer on keyboards and squint into orthicon tubes. But one cannot draw from the truism that mathematicians do apposite things the conclusion that there is some normative whatnot rightly termed 'mathematical practice' to which one can cogently appeal to settle entrenched philosophical disputes. This would be akin to inferring, from a casual observation that old crones can be halitosis-ridden or cranky, the conclusion that there are witches. There is no such thing as univocal, self-justifying mathematical practice and, even if there were, further argument would be needed to demonstrate that it is a lodestone for guiding those wayward intuitionists back onto the conventional straight and narrow.

So far as one can tell, the bulk of what philosophers assume to be mathematical practice consists in homebrewed synopses of (a) the amateur-philosophical *obiter dicta* of mathematicians, and (b) reports, unsupported by even the slenderest thread of statistical evidence, on what mathematicians seem to opine nowadays. Therefore, critical or norm-setting appeals to so-called data of these two genres are, at best, manifestations of *argumentum ad populum*, the sophomoric fallacy perpetrated by maintaining, e.g., that the mathematical outlook held by the historical Brouwer was false simply because a plurality of today's mathematicians would reject it. (The pressing question, "Do they *in fact* reject it?" will engage us shortly.)

Contemporary mathematicians are no more seers than were their predecessors of yore. It is likely that what the silent mathematical majority now regard as bedrock-solid foundation under their current mathematical construction-zone is rejected with derision by batches of minority-opinion mathematicians alive among us who may yet gain their ascendancy in the days ahead. I might guess that the preponderance of mathematicians currently doing anything at all serious in set theory are convinced both that there is a single, well-defined universe of sets and that truth in set theory is determinate therein and absolute. However, if adherents to the category-theoretic doctrines of Eilenberg, Mac Lane and Lawvere ever have their way, then this catechetical view of the set theorists will be deemed mere stuff and nonsense. Mathematicians demanding that there is a single universe of sets will be treated like undergraduates insisting that all groups are cyclic. For the disciples of Lawvere and company, there is no univocal, all-embracing hierarchy of classes ordered combinatorially by their own internal structures, but only one or another equivocal interpretation of the formal principles of elementary topoi, only loose panoplies of arrows-linking-dots.

What mathematicians pronounce, with all signs of full conviction, may be as ephemeral as the zephyrs. What one generation banks on as golden, the next rejects as counterfeit. Members of our math departments now spurn fluxions, indivisibles, higher-order infinites à la du Bois-Reymond, variable numbers, Dedekindian *Systeme*, and Fregean *Wertverläufe*. They insist that the geometric line is composed of sharply delineated, individual points, not of fuzzy segments of infinitesimal length. It astounds me that some still harbor the conceit that mathematical knowledge is cumulative, an idea foisted upon the public by lowbrow philosophy, grammatical convention, and discriminatory anthologizing of historical mathematical texts. Mathematics is a realm of vast intellectual agon and scribal variegation. Like the Bible or the screeds of the deified Wittgenstein, it will never tell you a straight story.

At this juncture, I am required to note that contemporary mathematicians seem more likely to accept views associated with the historical Brouwer than the contrary beliefs held by Brouwer's presumptive nemesis, David Hilbert. I am not here committing the fallacy *ad populum* just decried. I look to set the record straight, to correct a widespread goof that 'mathematical practice,' the *doxa* of modern mathematicians, sides largely with Hilbert and shows him to be the laurel-crowned, heavyweight champion of the old-time *Grundlagenstreit*. For example, current mathematicians seem to treat the axioms of set theory as verdicts of an intuition surprisingly Brouwerian: fundamental axioms are deemed true thanks to exercises of insight, imagination and description that are neither strictly logical nor wholly demonstrative. (A distinguished mathematician once assured me that Zermelo's Axiom of Unions is true because "If you have a bunch of sets, a bunch of boxes, you can always find a single box to dump all their contents into!") Those axioms are not now thought to be, as Hilbert might have wished, implicit definitions of the terms 'set' and 'element,' but are taken to be contentful claims with their own well-individuated, denotative meanings. As Brouwer insisted, the higher-order assertions of mathematicians are as individually meaningful and capable of truth (or falsity) as are the lowliest truisms of elementary mathematics, like Kant's $7 + 5 = 12$. Famously, Hilbert did not want things that way.

If they are honest, knowledgeable mathematicians and philosophers should allow that Brouwer was often and importantly right on other issues where Hilbert was just plain wrong. Hilbert thought it best that mathematics march triumphantly in lockstep with the hard sciences. Brouwer had it that pure mathematics is effectively conducted in relative isolation from physics, astronomy, and its other quantitative handmaids. Hilbert staunchly maintained that there are no unsolvable problems in mathematics, and hooted loud, publicly, and long at Emil and Paul du Bois-Reymond's insight, taken up by Brouwer (and later by Gödel) that mathematics is chock-full of *Ignorabimus*. Brouwer insisted that mathematics is radically incomplete and that there is, in it, a lordly treasure of unsolvable problems: perfectly sensible mathematical questions incapable of definitive mathematical answers. Hilbert devised complicated schemes for proving the consistency of formalized higher-order mathematics. Brouwer demurred, echoing Poincaré, and maintained that any bid to certify the consistency of mathematics on a large scale would be question-begging, relying upon presuppositions the epistemic credentials of which are at least as shady as those on trial for consistency.

For the great discoverers, mathematicians whose breakthroughs are hot news to thousands of lesser analytical lights, mathematics is a business of amazing plasticity. When a Newton or a Gauss writes or lectures on mathematics, he is changing, in large or little, that subject, and thereby changing what a mathematician does and is. I think that serious students of the discipline realize, ecstatic pronouncements of Hermann Weyl notwithstanding, the extent to which Hilbert was—both in reality and in his own mind—as much a revolutionary, as much a reformer, of mathematics as Brouwer ever hoped to be. Like Cantor, Hilbert explicitly rejected real analysis with infinitesimal quantities. And, of course, he rejected the intuitionistic mathematics of Brouwer, Poincaré, and du Bois-Reymond. It was in goodly part under the influence of Hilbert's Göttingen that point-set topology and analysis took on the forms they did in the early 20th Century, forms they held throughout that era. Predominantly, they were set-theoretic forms, in the idiom of Dedekind and Cantor, inflected by axioms and strict definitions, in stark contrast to those forms shunned by Hilbert: the calculatory forms demanded by Leopold Kronecker and the intuitional forms once preferred by du Bois-Reymond and, after him, Émile Borel.

3 Internal and External Negations

With that prolegomenon under our belts, I come now to a main point. I think that Brouwer, Heyting, Dummett, and kindred worthies of the guild were at times wrong, even monstrously so, but not because their opinions offended against that so chimerical mathematical practice. Rather, their views offend against theorems of intuitionistic mathematics. In other words, their views are sometimes demonstrably false, necessarily false, or as close to that as philosophical fancies ever come when dragged out into harsh daylight. In this writing, I have chosen to wrangle with Heyting and, to a lesser extent, with Brouwer, leaving Dummett for another glad day.

Clearly, there were moments when the great Heyting approved the idea that "$2 + 3 = 5$" means something like "I have effected the constructions associated with performing the addition operation on 2 and 3 and judged the result to be the same as 5." He even put it in writing:

> Indeed intuitionistic assertions must seem dogmatic to those who read them as assertions about facts, but they are not meant in this sense. Intuitionistic mathematics consists, as I have explained already to Mr. Class, in mental constructions; a mathematical theorem expresses a purely empirical fact, namely the success of a certain construction. "2+2 = 3+1" must be read as an abbreviation for the statement: "I have effected the mental constructions indicated by "2+2" and by "3+1" and have found that they lead to the same result." (Heyting, 1956, 8)

I pass entirely over the contradiction explicit in the above, and lightly over the oddity of thinking such numerical banalities as $2 + 2 = 3 + 1$ empirical. Even were they interpretable as Heyting wished, they would be no more empirical than my announcements "I remember having breakfast" and "I don't like parsnips." That aside, Heyting's idea is as off-target as one can be in such matters. First, were Heyting right, "$2 + 3 \neq 5$" could well mean, "I have not effected the constructions associated with performing the addition operation on 2 and 3 and judged the result to be the same as 5," an assertion of personal history to which a perfectly rational reply would not be, "No, it's not. The sum of 2 and 3 is 5," but "It grieves me that your education has been so deprived." Second, were Heyting right, I could make the false conditional "If $34 + 17 = 51$, then $39 + 18 = 47$" true merely by refraining from the bit of mental arithmetic Heyting requires for the truth of the antecedent. You can make the same point at home by substituting, for "$34+17 = 51$" in the conditional, any equation expressing a correct addition you have yet to perform.

At this stage, I imagine my ever-helpful readers responding that, in excoriating Heyting's views in this way, I have not made a crucial distinction, one required when speaking or writing sensibly about intuitionism. The distinction they have in mind is that between so-called internal and so-called external negations. Presumably, external negation, which I symbolize $\sim$ rather than $\neg$, is to be used when speaking 'from outside intuitionism' about what is provable, disprovable, constructible, true or false in intuitionistic mathematics. External negation is imagined to be classical in its logic, that is, discrete or decidable logically. In other words, $\sim$ always obeys the *tertium non datur* or **TND**:

$$\forall p\,(p \vee \sim p)$$

or so we are told.

This fictive external negation is imagined to stand in clear contrast to a stronger negation $\neg$, supposed to be employed by intuitionists exclusively and used only when conducting their mathematical selves in strictly intuitionistic company. It is believed to bear a special, proprietary intuitionistic meaning in virtue of which $\neg$ never obeys the general **TND** as a matter of logic, for the intuitionist holds that

$$\neg\forall p\,(p \vee \neg p).$$

When I wish to convey intuitionistic mathematical theorems or conjectures, I turn to $\neg$ for all my negating needs. When I refer, as if in the metatheory, to what the intuitionist can or cannot get up to, I am required to fall back upon the weaker $\sim$. At least that is the proposal of my helpful readers.

But it is pure moonshine. As an intuitionist, whether chatting about the theorems of the day with a fellow true believer, or at home with the missus, the word "not" in my mouth means "not." Neither I nor Brouwer can change the meanings of logical signs by fiat, even when driving under the influence of some 200-proof philosophy. Besides, it is a simple theorem of intuitionistic mathematics that there is only one operation of negation. This is as provable a fact as anything in mathematics, love or life can be.

4 There Is Only One Negation

To start with, we warm up the cerebral muscles by demonstrating that there is no such thing as classical negation, hence no operation on truth-values like the $\sim$ operation my interlocutor has cooked up. It is a verity of kindergarten set theory that the set of truth-values is in bijective correspondence with the powerset $\mathfrak{P}$ of $\{0\}$. Officially, I think of myself as working over the latter set, while writing informally either of subsets of $\{0\}$ or of truth-values, as the spirit moves me.

Over the collection $\mathfrak{P}$, all assignments of numbers are uniform: if I can attach to every subset of {0} a natural number, then there is some number I have attached to all of them. In proper symbolese, that is

$$\forall p \in \mathfrak{P}\ \exists n \in \mathbb{N}.\ R(p, n) \rightarrow \exists n \in \mathbb{N}\ \forall p \in \mathfrak{P}.\ R(p, n).$$

Following Troelstra (1980), intuitionists refer to the foregoing as the Uniformity Principle, or **UP** for short.

Now I put **UP** to work. Let F stand for the empty set, the least truth-value, and T for {0}, the greatest. If a classical negation like ~ exists, it must map truth-values into truth-values:

$$\sim: \mathfrak{P} \rightarrow \mathfrak{P},$$

so that ~ is not a constant function, ~ maps no truth-value into itself, and it satisfies

$$\forall p \in \mathfrak{P}.\ ((\sim p) = \mathrm{T}\ \vee (\sim p) = \mathrm{F}).$$

For the sake of easy reading, I have here enclosed the sign for the result of applying the ~ operation to truth-value p in parentheses, viz., ($\sim p$). The last expression above displayed means that the result of applying the presumptive classical operation ~ to an arbitrary truth-value is either true or false. This is required by the assumption that ~ satisfies laws of conventional logic.

I now define a relation R on $\mathfrak{P} \times \mathbb{N}$ so that

$$R(p, n) \leftrightarrow [(n = 0\ \wedge (\sim p) = \mathrm{F}) \vee (n = 1\ \wedge (\sim p) = \mathrm{T})].$$

The assumption that ~ is classical shows that R is a binary relation between truth-values and numbers that is totally defined over the truth-values:

$$\forall p \in \mathfrak{P}\ \exists n \in \mathbb{N}.\ R(p, n).$$

Thanks to **UP**, we know that some number is related by R to all truth-values:

$$\exists n \in \mathbb{N} \forall p \in \mathfrak{P}\ R(p, n).$$

But, unless ~ is a constant function, this is impossible. Therefore, we see that the idea that classical operation ~ exists is mistaken.

My helpful interlocutors cannot reply that the argument I have just given is fallacious, since ~ was never meant to apply to all truth-values, but only to those that report on the doings of intuitionistic mathematicians. For example, it was meant to attach to such Heyting-isms as "I have effected the construction corresponding to the sum of 2 and 1." It was not intended for "2 + 1 = 3" *simpliciter*.

This idea is easily proven nonsensical as well. There are no disjoint subsets X and Y of $\mathfrak{P}$, each of which contains a truth-value, such that ~ is defined on only one of them. Assume that X is the subset on which ~ is defined, ~ mapping X into X. It is easy to see—without exploiting uniformity—that

$$\forall p, q \in X \, \neg\neg(p = q).$$

Hence,

$$\forall p \in X \, \neg\neg(p = (\sim p)).$$

This contradicts the assumption that $\sim$ maps no truth-value into itself.

At last, I am ready to prove that $\neg$ is the sole negation-like operator on $\mathfrak{P}$. Assume that f is a function on $\mathfrak{P}$ that obeys the (intuitionistic) logical laws of negation: that $f(\mathrm{F}) = \mathrm{T}$ and that, for any truth-value p, there is no way that p and $f(p)$ can be true together, i.e.,

$$\forall p \, (p \wedge f(p)) = \mathrm{F}.$$

Since any truth-value is a subset of $\{0\}$, I can investigate the extensional identity of truth-values by tracking membership, in them, of the number 0. On the two assumptions just made, we see that

$$0 \in f(p) \leftrightarrow f(p) = \mathrm{T}.$$

Since the intersection of p and $f(p)$ is always empty and $f(\mathrm{F}) = \mathrm{T}$,

$$f(p) = \mathrm{T} \leftrightarrow p = \mathrm{F}.$$

The same properties hold of $\neg$ as an operation on truth-values. Therefore, it is true that

$$p = \mathrm{F} \leftrightarrow (\neg p) = \mathrm{T} \leftrightarrow 0 \in (\neg p).$$

Putting it all together, I conclude that

$$0 \in f(p) \leftrightarrow 0 \in (\neg p).$$

This means that, as operations on $\mathfrak{P}$, f and $\neg$ are identical. Hence, there is no other negation than $\neg$, no other unary function on truth-values that has the requisite intuitionistic properties. Please note that, for this outcome, I have not assumed that f is a classical operation obeying fallacious conventional logic. Nor, this time, have I assumed **UP**.

In the face of these necessary truths, my interlocutors cannot hazard a rescue for the idea of discriminable internal and external negations by distinguishing 'internal' from 'external' truth-values, the first of which is our friend $\mathfrak{P}$, the proper domain of $\neg$, and the second some different domain, the stomping ground of the imaginary $\sim$. Were it demonstrable, this suggestion would sink the prospects of intuitionism at the jetty. It amounts not only to admitting but also to enforcing the notion that the mathematics that is intuitionism is not all the mathematics, all the mathematical logic, all the clear mathematical thinking, that there is, and allowing that there is some other, nonintuitionistic way that a mathematical statement can be true. In other words, it would etch into stone tablets the falsehood that intuitionistic mathematics, in the beginning and in the end, depends upon an external, classical theory for its viability. This is a Quinean/Carnapian counsel of despair, and utterly mistaken, as anyone who has read and understood some intuitionistic mathematics will know.

5 Intuitionism and Meaning

This business about negation is the incestuous bedfellow of a wackier—because grander—idea: that the intuitionist, when he comports himself mathematically, not only decides, in the manner of Humpty-Dumpty, to alter the meanings of a few words like "not" and "exists," but also legislates wholesale semantic change. At his workplace, the intuitionist is supposed to speak an utterly oddball mathematical lingo, construed exclusively by fellow intuitionists and others who have graduated *summa* from the intuitionistic school. At the end of the working day, when the intuitionist clocks out, he is believed to stop squawking his special patois, and to resume ordinary English (or German or Hindi) for the sake of friends and family. This notion can certainly be seen lurking in the pages of Carnap and Quine, and, perhaps, in the writings of the sainted Wittgenstein (McCarty, 2008).

A limited experience in reading, learning, or teaching intuitionism more than suffices for overturning this canard. Even Heyting withheld full treatment of his famous 'explanations' of the logical words of intuitionism until the penultimate chapter of his justly famous *Intuitionism: an Introduction* (Heyting, 1956). After an entertaining dialogic introductory entitled *Disputation*, the reader plunges with Heyting straight into the book's mathematical thicket—and none the worse for that—with little or no semantic ado. Immediately, the reader sees his or her way about in this new intuitionistic forest thanks to an everyday grasp of natural, rational and real numbers, and of the familiar words, including logical words, that we—and Heyting—always use to convey our familiar thoughts about them. Without implicitly calling upon those meanings grasped jointly by Heyting and his entire adult readership, how would one explain the ease with which savants picked up his thoughts? How would it be possible to start studying this work of Heyting, to engage with him in his fictional *Disputation*, if we did not share with him a fullscale mathematical vocabulary, including logical signs, and its understanding? Heyting's assertions, as with my own throughout this article, are in no new language. They are simply new truths in an old language. And, like many new truths, are often denounced, even jeered at, by the *hoi polloi*.

6 Fatally Weak Counterexamples

My critics now object that the distinction between internal and external negations seems absolutely required for making sense of and for crediting Brouwerian weak counterexamples, apparent refutations of presumptive principles on the grounds that they reduce to unacceptable instances of the quantified **TND**. For example (here, I follow the pellucid exposition of Troelstra and van Dalen (1988, 13ff)) assume that the truth or falsity of Goldbach's conjecture has not been sorted out with finality earlier today. No one now knows whether every even number greater than two is the sum of two primes or there is a nonGoldbach number, an even number that is not the sum of two primes. The latter, if it exists, will have to be larger than 10^{17} (Oliveira e Silva, 2009).

For n a natural number, I define a $\{0, 1\}$-valued natural number function g so that $g(n) = 0$ if all even numbers less than or equal to n are Goldbach, and $g(n) = 1$ if there is a nonGoldbach number which is less than or equal to n. Since g's defining predicate is algorithmically decidable, g is a nondecreasing total function of numbers. It's very simple: g runs through the natural numbers checking for an exception to the conjecture, outputting 0 if one has yet to be found, and popping up with a 1 as soon as an exception is located. By almost any man's logic, if

$$\forall n.\, g(n) = 0,$$

then Goldbach's conjecture is true. If

$$\exists n.\, g(n) \neq 0,$$

then Goldbach erred.

All that is unexceptionable. The dubious reasoning behind the associated Brouwerian weak counterexample now proceeds as follows. I enclose my expression of that reasoning in scare quotes because I do not wish to endorse it. 'This shows that the statement

$$\forall n.\, g(n) = 0 \vee \exists n.\, g(n) \neq 0$$

is not intuitionistically true. Were it true, the reductive argument, rehearsed in the preceding paragraph, would entail that either Goldbach's conjecture or its negation is intuitionistically true. Unfortunately, we do not yet have a proof of that conjecture. And we do not yet have a proof of its negation. Therefore, the statement just displayed is not intuitionistically true.' By the way, my imagined interlocutors and critics mean the negative particle in the sentence immediately preceding to be an instance of external negation; they do not imagine that negation to commute with the expression "is intuitionistically true."

If this sample is at all characteristic of the genre, Brouwer's weak counterexamples are fallacious. Whether I or anyone can prove or disprove Goldbach's conjecture is a fact of absolutely no relevance when it comes to determining the validity or invalidity of a presumptive logical law like **TND** or the truth or falsity of such putative mathematical principles as

$$\forall g[\forall n.\, g(n) = 0 \vee \exists n.\, g(n) \neq 0].$$

Once we know that Goldbach's conjecture is competently proven, we know thereby that it is true. (I postpone indefinitely a question that the purveyors of these kinds of fallacies studiously avoid, viz., "In Brouwer's weak counterexample reasoning, what kind of proof is under discussion there, classical or intuitionistic?") However, it is a blatant, old-fashioned fallacy *ad ignorantiam*, painfully familiar to myriads of college sophomores, to argue that, since Goldbach's conjecture is yet unproved, it fails of truth, intuitionistic or other.

As the informed reader will know, masking this fallacy requires of those who deify Brouwer some fancy semantical footwork, a fair sample of which is the doctrine of external negation. You can see right off why my interlocutors wish to stem the incoming tide of proper critical thinking by distinguishing inappropriately between "it is not intuitionistically true that p" and "it is intuitionistically true that $\neg p$," where p is a truth-value, thus enshrining the heretical catechism of $\sim$. For example, take this instance of the **TND**:

$$\mathbf{GC} \vee \neg\mathbf{GC} \tag{1}$$

where **GC** is short for a statement of Goldbach's Conjecture. Call that instance q. My interlocutor wants to maintain that q is not intuitionistically true. However, at the same time, he does not wish to maintain that q is (intuitionistically) false, on the grounds $\neg q$ is logically refutable. After all,

$$\neg\neg(p \vee \neg p)$$

is a theorem of intuitionistic propositional logic. By my lights, both q and "it is (intuitionistically) true that q" mark the truth-values of mathematical statements. Indeed, they mark the *very same truth-value*. Moreover, as we saw *infra*, they live under the rule and power of one sole and absolute negation, $\neg$. In fact, no instance of **TND** is intuitionistically false. **TND** fails because it is not logically true, that is,

$$\neg\forall p(p \vee \neg p),$$

as is easily seen to follow from **UP**.

Unless reformulated, weak counterexamples are plain fallacies. Even so, they are intelligent fallacies, and can be rescued in any number of ways. In effect, the rescue relies upon a nice observation: although

$$\neg\neg(p \vee \neg p)$$

is a theorem of the logic, the schemes freshmen might think its first-order and second-order quantificational analogues, specifically,

$$\neg\neg\forall x\,(P(x) \vee \neg P(x))$$

$$\neg\neg\forall P\forall x\,(P(x) \vee \neg P(x)).$$

are not.

Intuitionistic Church's Thesis **CT** is the statement that every total relation over the natural numbers is uniformized by a Turing computable function. **CT** is consistent with the full intuitionistic set theory **IZF**, intuitionistic Zermelo-Fraenkel, if the latter is consistent. From **CT** it follows immediately that

$$\neg\forall n(\exists m.T(n,n,m) \vee \neg\exists m.T(n,n,m)),$$

where $T(x,y,z)$ is the standard Kleene T-predicate. Hence, **CT** proves that

$$\neg\forall P\forall x\,(P(x) \vee \neg P(x)).$$

Now, let n index a Turing machine $\{n\}$. Define the function g so that $g(m) = 0$ provided that $\{n\}$ fails to halt, given input n, at any computation stage up to and including m, and $g(m) = 1$ provided that $\{n\}$ on n halts at or before stage m. As noted above, the initial, reductive portion of the weak counterexample reasoning is unimpeachable: we are assured that

$$\text{if } \forall k.\ g(k) = 0, \text{then } \neg\exists m.T(n, n, m),$$

and

$$\text{if } \exists k.\ g(k) \neq 0, \text{then } \exists m.T(n, n, m).$$

Therefore,

$$\forall g : \mathrm{N} \to \{0, 1\}\ \ (\forall n.\ g(n) = 0\ \vee\ \exists n.\ g(n) \neq 0)$$

implies

$$\forall n\,(\exists m.T(n, n, m) \vee \neg\exists m.T(n, n, m)).$$

Since the latter is false (or falsified by **CT**), and provably, so is the former. Hence, we know that

$$\neg\forall g : \mathrm{N} \to \{0, 1\}\ \ (\forall n.\ g(n) = 0\ \vee\ \exists n.\ g(n) \neq 0),$$

and the reasoning, elementary as it is, leading to this conclusion is strong and cogent, not weak and faulty.

7 Proofs and Constructions

Loitering in the long shadow cast by Brouwer's fallacious weak counterexamples is the presumption that, when a proof exists, its existence and relevant evidentiary properties are glaringly obvious. According to the reasoning behind weak counterexamples, when there is a proof of the Goldbach Conjecture, we are supposed to be apprised of the existence of such a proof and to be convinced by it. The weak counterexample does not seem to admit any gap between there being a proof of a conjecture and our being willing to assent, on its basis, to that conjecture. This idea is a close friend, or perhaps an ancestor, of the notion, fostered by Heyting, that the epistemically important properties of proofs, or of constructions generally, are patent when instantiated and that statements reporting the instantiation need no justification once grasped. (I shall leave it to right-thinking epistemologists among my readers to comment critically on the difference between those psychological processes by which one grasps and comes to credit a mathematical method, and the provision of a mathematical justification for believing the method successful.) Here is Heyting on this subject, from Heyting (1956):

> A mathematical construction ought to be so immediate to the mind and its results so clear that it needs no foundation whatsoever (Heyting, 1956, 12).

As advice to mathematical writers, the exhortation, "Make your constructions as plain as a pikestaff," may be all to the good. Exhortation is not, however, what Heyting had in mind. He felt that, when c is a construction that proves p, no further proof is required that c exists or has the requisite proof-theoretic properties. That Heyting was too optimistic on this point is easy to confirm. For the sake of definiteness, I draw this example from S.C. Kleene's number realizability interpretation of intuitionistic arithmetic (Kleene, 1945). It is a model, or as near as we come to it, of Heyting's general conception of what it takes for a construction to prove something. For the sake of that example, I ignore the extra calculatory effort required to check that the partial recursive functions here described are defined when they need to be.

Let construction e be a Heyting-style proof—in this case, a realizability witness—for a specific instance of first-order mathematical induction, say

$$(A(0) \wedge \forall n.\ (A(n) \to A(n+1))) \to \forall n.\ A(n).$$

A lucky number e that proves or realizes the displayed formula will have to satisfy these two recursion equations, wherein m, n, p and q are natural numbers:

$$\{\{e\}(\langle p, q\rangle)\}(0) \;=\; p$$

while

$$\{\{e\}(\langle p, q\rangle)\}(n+1) \;=\; \{\{q\}(\{\{e\}(\langle p, q\rangle)\}(n))\}(n).$$

In this, an expression of the form $\{x\}(y)$ names the result, if it exists, of applying the recursive function with index x to input y, and $\langle x, y\rangle$ is a primitive recursive pair of x and y. An e that satisfies these two equations can be all-purpose and realize as well the second-order induction principle, if the realizability is that of Kreisel and Troelstra (Troelstra, 1973). The pertinent facts relating to the existence and recursion-theoretic powers of e are ones we could hardly claim to discern convincingly in a way that is immediate, clear and in no need of proof. To be assured, in general, that such an e exists would require proofs of the Second Recursion Theorem and *s-m-n* Theorem of Kleene (Rogers, 1987), or at least exemplars of their proof schemes suited to this particular case. On top of that, there is the not-so-minor matter of proving that an e, so constructed, does the trick: that it proves (i.e., realizes) the induction scheme. Commonly, this proof will call for a use of mathematical induction with a meta-induction formula at least as complex as $A(x)$. Reflection on examples like this strongly suggests that Heyting was wrong to think that the existence and significant proof-theoretic properties of constructions are, when they obtain, obvious to the Brouwerian mathematician.

Again, I warn readers against confusing, on the one hand, a psychological process undergone by an expert in recursion theory who could see, in a flash, that this kind of e exists with, on the other hand, a justification that such a number e truly exists.

Perhaps Heyting's greatest error was to think that the (nonexistent) special constructive meanings of statements made by intuitionists are defined by a privileged relation between sentences and proofs, real proofs. Such a relation is described in (Troelstra and van Dalen, 1988, 9), where we find such claims as

> a proof of $A \rightarrow B$ is a construction that transforms a proof of A into a proof of B
>
> a proof of $A \vee B$ is given by presenting either a proof of A or a proof of B.

Incidentally, I find it curious that Troelstra and van Dalen refer to this definition as a "stipulation." A defense attorney sometimes has the power to stipulate that a presumptively accredited document count as evidence during a court case, but no one has the power to stipulate what the word "and" means and to make it stick. No one has the power to stipulate that collections of statements, constructions, and proofs exist that truly satisfy constraints such as the two clauses above.

To be sure, the proofs Troelstra and van Dalen have in mind are not proofs from classical mathematics. The latter do not commute with disjunction: a conventional proof of $(A \vee B)$ is not generally a conventional proof either of A or of B. One does not normally go about proving logical truths, but something like a classical proof could be given for $(A \vee \neg A)$, in cases when neither a proof of A nor one of $\neg A$ exists. Therefore, one supposes that the proofs required for the truth of the above-displayed clauses are intuitionistic. Unfortunately, we intuitionists often have relatively little experience with specifically intuitionistic proofs, compared with, say, the collective experience of mathematicians with proofs in higher algebra. The newcomer to intuitionism has, presumably, even less relevant experience. It remains to be explained, therefore, how Heyting's so-called explanations of meaning are to be of any use to the neophyte. This objection is especially worth pressing if it is thought that intuitionists speak their own private argot, unknown to outsiders.

In these quarters, it is also assumed that a sentence is true when and only when there is to hand a proof of it of the sort Heyting imagined. We are told that this is what it means for a sentence to be to be true *intuitionistically*. But that cannot be so. When mathematical justification via mathematical proof is the issue, and we push hard enough, insisting upon proof after proof, either in intuitionism or in classical mathematics, we arrive sooner or later at axioms. I do not mean that we accept axioms entirely without justification. For the familiar axioms, we have justifications aplenty. We have justifications for axioms both before and after the fact, both *a priori* and *a posteriori*. But these justifications are not proofs. Most of the time, we accept axioms as true without proof, but with justification. It would seem, therefore, that we offend the Heyting-inspired account of truth just described whenever we tell our students that the axioms of intuitionistic set theory are true but they have no proofs and may never have them. It is difficult to imagine how one would go about proving that the union of every set of sets is a set.

8 A Realizability Theory of Constructions

In Heyting's definitions of the proof relation, or at least in the foregoing renditions of them, there lurks a confusion between forms of words or thoughts that provoke in mathematicians conviction—that is, proofs—and a wider class of objects, abstract constructions. Decent mathematical proofs house the springs and mechanisms of their own acceptation. By contrast, an abstract construction is something a ken of whose existence and efficacy may well require a separate proof, as exemplified for Kleene realizability above. A construction is that battery of mathematical facts in virtue of which an operation has its power. In intuitionistic mathematics, an operation cannot be a licentious "anything goes" mathematically, a bunch of outputs thrown up at random from a bunch of inputs. Some manner of uniformity, some sort of mathematical control is demanded. Constructions provide that control.

That being said, I plan in the remainder of this essay to chart the consequences of allowing constructions to fall outside the relevant ground structure, even to be indefinable over that structure. This suggestion of mine stands in clear opposition to what has been a ruling assumption in these regions, an assumption according to which constructions are simple—both ontologically and epistemically—and, when it comes to constructing truths about natural numbers, relatively lower-order. Until further notice, I plan to reason conventionally in the metatheory. Experts know that this sort of reasoning—generally invalid—is allowable within a suitable Boolean extension of my basal intuitionistic universe. Under such harsh, classical conditions, the intuitionist can survive for a time, perhaps just long enough to show a theory consistent.

Let $\mathfrak{N}$ be the standard model for full classical third-order arithmetic. Let $\mathfrak{C}$ be the collection of all finite binary sequences drawn from the first-order part of the domain of $\mathfrak{N}$, ordered in the standard tree-like way. Let $\mathfrak{M}$ be an elementary Henkin extension of $\mathfrak{N}$ that encodes as nonstandard integers all the infinite branches in infinite subtrees of $\mathfrak{C}$, and (of course) continues to satisfy full, true third-order arithmetic. König's Lemma shows that every such subtree has at least one infinite branch. I will define a relation of recursive realizability $\Vdash$ on which quantifiers within strictly arithmetic formulae range over members of $\mathfrak{N}$ exclusively, while the Turing indices of $\mathfrak{M}$ provide realizability witnesses, i.e., constructions.

Lowercase Roman letters k, l, m, and n stand for standard natural numbers, members of the first-order part of $|\mathfrak{N}|$, the domain of $\mathfrak{N}$. Uppercase Roman letters E, F, and so on stand for constructions, that is, members of the first-order reduct of the domain of $\mathfrak{M}$. A, B and C are realizability subsets of $\mathfrak{N}$ or realizability collections of those subsets. Here, a realizability subset is a set of pairs $\langle E, n\rangle$ in the domain $|\mathfrak{M}|$ of $\mathfrak{M}$ in which E is an element of the first-order domain of $\mathfrak{M}$ and n belongs to that of $\mathfrak{N}$. A realizability collection of such subsets is a set of pairs $\langle E, A\rangle$ from $|\mathfrak{M}|$ in which E comes from $\mathfrak{M}$ and A is a realizability set, as just described.

Closed atomic formulae over $\mathfrak{N}$ with parameters from that structure get interpreted as follows.

For ϕ first-order and atomic, $E \Vdash \phi$ iff $\mathfrak{N} \vDash \phi$.
For A a realizability set and n from $\mathfrak{N}$, $E \Vdash n \in A$ iff $\langle E, n\rangle \in A$.
For A a realizability collection of realizability sets and B a realizability set, $E \Vdash B \in A$ iff $\langle E, B\rangle \in A$.

Otherwise, the realizability is very much as in Kreisel-Troelstra realizability for second-order Heyting arithmetic (Troelstra, 1973), respecting differences between elements (or constructions) from $\mathfrak{M}$ and elements of $\mathfrak{N}$'s domain. $E.F$ stands for the result of applying index E to input F in $\mathfrak{M}$, if it is defined. $\lambda E.(E)_0$ and $\lambda E.(E)_1$ are primitive recursive (again, in $\mathfrak{M}$) unpairing functions relative to an $\mathfrak{M}$-primitive recursive pairing $\lambda E \lambda F.\langle E, F\rangle$. As usual, formulae realized are closed formulae with parameters from $\mathfrak{N}$.

Here is the general definition of the $\Vdash$ relation for compound formulae:

$E \Vdash \phi \wedge \psi$ iff $E_0 \Vdash \phi$ and $E_1 \Vdash \psi$,
$E \Vdash \phi \vee \psi$ iff, when $E_0 = 0$, $E_1 \Vdash \phi$ and, when $E_0 = 1$, $E_1 \Vdash \psi$,
$E \Vdash \phi \rightarrow \psi$ iff, whenever $F \Vdash \phi$, $E.F$ is defined and $E.F \Vdash \psi$,
$E \Vdash \neg\phi$ iff, for all F, $F \nVdash \phi$,
$E \Vdash \forall x\phi$ iff, for all n from the first-order part of $|\ \mathfrak{N}\ |$, $E.n$ is defined and $E.n \Vdash \phi[x/n]$,
$E \Vdash \exists x\phi$ iff E_0 is in the first-order part of $|\ \mathfrak{N}\ |$ and $E_1 \Vdash \phi[x/E_0]$,
For a second-order variable X^2, $E \Vdash \forall X^2\phi$ iff, for all realizability sets A, $E \Vdash \phi[X^2/A]$,
For a second-order variable X^2, $E \Vdash \exists X^2\phi$ iff, for some realizability set A, $E \Vdash \phi[X^2/A]$, and
Analogous conditions govern quantification binding third-order variables X^3.

Let $\mathbf{HA}^3$ be full impredicative third-order arithmetic in intuitionistic logic (Troelstra and van Dalen, 1988, 164). For ϕ a formula of the language of $\mathbf{HA}^3$, take $\forall\phi$ to be a universal closure of ϕ. When closed formula ϕ has a realizability witness as just defined, in other words, when $\exists E.\ E \Vdash \phi$, we say that $\mathfrak{K}$ (for Kleene) satisfies ϕ, in symbols $\mathfrak{K} \vDash \phi$.

Lemma 1: [Soundness] If $\mathbf{HA}^3 \vdash \phi$, then $\mathfrak{K} \vDash \forall\phi$.

Proof $\mathfrak{M}$ is a model of classical third-order arithmetic. *A fortiori* its internal Turing indices possess all properties requisite to the usual proof of such a soundness theorem (Troelstra, 1973). ∎

In $\mathfrak{K}$, one can carry out the reasoning earlier described to puncture the supposition that there can be more than one intuitionistic negation, since the Uniformity Principle **UP** holds there.

Theorem 1: $\mathfrak{K} \vDash$ **UP**.

Proof Throughout this proof, the variable X is assumed to be second-order. Let $E \Vdash \forall X\exists x.\ \phi$. Then, for all realizability sets A of $\mathfrak{K}$,

$$E \Vdash \exists x.\ \phi(A).$$

Hence, E_0 is a standard natural number, and, for all appropriate A,

$$E_1 \Vdash \phi(A, E_0).$$

So,

$$E_1 \Vdash \forall X.\, \phi(E_0),$$

and

$$E \Vdash \exists x \forall X.\, \phi \quad \blacksquare$$

But more is true. Ordinary Kleene number realizability (Kleene, 1945) over the standard model $\mathfrak{N}$ verifies intuitionistic Church's Thesis **CT**. Ordinary realizability has this property because its realizability witnesses, the constructions that underlie operations, are themselves elements of the ground interpretational structure, namely $\mathfrak{N}$. In the present, higher-order realizability, constructions are not generally members of that ground structure, and Church's Thesis does not hold in $\mathfrak{R}$.

Theorem 2: $\mathfrak{R} \nvDash$ **CT**.

Proof Let E be a member of the first-order domain of $\mathfrak{M}$ that encodes a noncomputable branch of an infinite subtree of $\mathfrak{C}$. Let F be a Turing index in $\mathfrak{M}$ for the function that, on input $G \in\mid \mathfrak{M} \mid$, outputs the Gth sequence in the E-encoding. Here, $\Lambda G.\Theta(G)$ denotes a code for the function $\lambda G.\Theta(G)$.) Then, in $\mathfrak{R}$,

$$\Lambda G.\langle F.G, G\rangle \Vdash \forall x \exists y.\, \langle x, y\rangle \in \widehat{F},$$

where $\widehat{F}$ is the realizability set

$$\{\langle F(n), \langle n, F(n)\rangle\rangle : n \in\mid \mathfrak{N} \mid\}.$$

However, there is, in $\mathfrak{N}$, no Turing index computing the function $\widehat{F}$, since the chosen branch is noncomputable. ■

The Fan Theorem, **FT**, is a central pillar in the edifice of Brouwerian analysis. In effect, it is an intuitionistic compactness result for $[0, 1]$. **FT** is the statement

> Assume that every branch in a decidable subtree T of $\mathfrak{C}$ is finite. Then T is bounded uniformly: there is a natural number n such that every branch in T is of length less than n.

Again, $\mathfrak{C}$ is the collection of all finite binary sequences drawn from the first-order part of the domain of $\mathfrak{N}$, ordered in the usual fashion. Assuming Church's Thesis, one can use an argument of Kleene (Troelstra and van Dalen, 1988, 217ff) to show that, under ordinary realizability, **FT** is false. Church's Thesis fails in $\mathfrak{R}$, and the way is clear for the Fan Theorem to hold there.

Theorem 3: $\mathfrak{R} \vDash$ **FT**.

Proof Assume that E is a member of the first-order domain of $\mathfrak{M}$ and that

$$E \Vdash (T \text{ is a decidable subtree of } \mathfrak{C} \wedge \text{ every branch of } T \text{ is finite}).$$

Let $\widetilde{T}$ be

$$\{n : \exists F.\, F \Vdash n \in T\}.$$

A calculation with realizability witnesses confirms that, in $\mathfrak{N}$, $\widetilde{T}$ is a finite subtree of $\mathfrak{C}$. It follows that there is a natural number m that bounds every branch of $\widetilde{T}$ uniformly. Therefore, there is an index G in $\mathfrak{M}$ such that

$$G.E \ = \ m.$$

Consequently, the $\mathfrak{M}$-pair $\langle G.E, \Lambda H.0\rangle$ witnesses

$$\exists x \forall F(F \text{ is a branch in } \mathfrak{C} \to (F \restriction x) \notin T),$$

where $F \restriction x$ is the initial segment of F of length x. ■

Ordinary Kleene realizability does not validate Brouwerian analysis. The extent to which intuitionists can embrace a higher-order version of realizability, such as that over $\mathfrak{K}$, may be a measure of the extent to which intuitionists can legitimately embrace a leading, correct idea of Brouwer and Heyting: that the mathematical principles capturing facts of intuitionism can be deduced, with the aid of set theory, from a theory of constructions. As is obvious from the definition of $\mathfrak{K}$-realizability, the needed fundamental properties of those constructions are neither immediately apparent nor unshakeably certain. Even so, the constructions that are realizability witnesses over $\mathfrak{K}$ may show us why such intuitionistic principles as **UP** and **FT** are true and, thereby, why **TND** must be invalid.

References

Heyting, A. (1956) *Intuitionism: An Introduction*, Studies in Logic and the Foundations of Mathematics, Amsterdam: North-Holland Publishing Co., VIII+133.

Kleene, S.C. (1945) On the Interpretation of Intuitionistic Number Theory, *The Journal of Symbolic Logic*, **10**, 109–124.

McCarty, C. (2008) Intuitionism and Logical Syntax, *Philosophia Mathematica*, Series III, **16**(1), Special Issue on Legacies of Logical Positivism, N. Tennant (ed.), 56–77.

Oliveira e Silva, T. (2009) Goldbach Conjecture Verification, *http://www.ieeta.pt/~tos/goldbach.html*, April 2009.

Rogers, H., Jr. (1987) *Theory of Recursive Functions and Effective Computability*, Second Edition, Cambridge: MIT Press, xxi+482.

Troelstra, A. (1973) *Metamathematical Investigation of Intuitionistic Arithmetic and Analysis*, Springer Lecture Notes in Mathematics, Volume 344, Berlin: Springer, xvii+486.

Troelstra, A. (1980) Intuitionistic Extensions of the Reals, *Nieuw Archief Voor Wiskunde*, Volume XXVIII, 63–113.

Troelstra, A. and van Dalen, D. (1988) *Constructivism in Mathematics: An Introduction*. Volume I, Amsterdam: North-Holland, xx+342+XIV.

Euclidean Arithmetic: The Finitary Theory of Finite Sets

J.P. Mayberry

1 The Sorites Fallacy

There is a central fallacy that underlies all our thinking about the foundations of arithmetic. It is the conviction that the mere *description* of the natural numbers as the "successors of zero" (i.e., as what you get by starting at 0 and iterating the operation $x \mapsto x + 1$) suffices, *on its own*, to characterise the order and arithmetical properties of those numbers absolutely. This is what leads us to suppose that the dots of ellipsis in

$$0, 1, 2, 3, 4, 5, \cdots$$

or in

$$0, 1, 2, \cdots, n$$

are somehow *explanatory*.

The conventional conception of the natural numbers is based on this fallacy—I shall call it the *sorites fallacy*. It is incorporated in the widespread idea that the notion of *finite iteration* is simply "given" to us.[1] In fact, a major task—*the* major task—for the foundations of arithmetic is to provide a precise, mathematical analysis of the notion of finite iteration.

It might seem extravagant to suggest that what seems to be our simplest and most familiar mathematical concept—the concept of natural number—is based on an out-and-out fallacy. But the two most acute minds that have considered the problem of characterising the natural numbers arrived at essentially just this conclusion over one hundred and twenty years ago. I refer, of course, to Frege and Dedekind who,

[1] Dedekind discusses this fallacy in (new 1971).

G. Sommaruga (ed.), *Foundational Theories of Classical and Constructive Mathematics*, The Western Ontario Series in Philosophy of Science 76,
DOI 10.1007/978-94-007-0431-2_12,

each in his own way, arrived at the diagnosis I have just given, and who, moreover, recommended essentially the same solution to it.[2]

Dedekind's axiomatic characterisation of the natural number system up to isomorphism is, in a sense, a vindication of the conventional conception of natural number, for it is based on a rigorous, set-theoretical *definition* of "finite iteration" that does not rest on the Sorites fallacy. Dedekind thus provides us with a proper *proof* that the system of natural numbers is "absolute" in the sense simply assumed by the naive, conventional conception of natural number. But his proof requires a very strong set theoretical assumption, namely, the existence of a transfinite set with a power set; and this assumption immediately gives rise to Cantor's apparently intractable continuum problem.

However, the familiar alternatives to Dedekind's infinitary theory—the various "finitary" or "constructive" approaches to natural number arithmetic—simply incorporate the naive conception of "finite iteration" into the foundations of arithmetic by taking proof by induction and definition by recursion as fundamental truths not requiring justification. For proof by induction rests on the idea of iterating inferences and definition by recursion rests on the idea of iterating calculations.

My intention is to sketch another way of looking at these questions, one that does not rest on the sorites fallacy. I shall follow Dedekind in using set-theoretical concepts and methods to give an account of theoretical arithmetic, and, in particular, of finite iteration; but I shall avoid his infinitistic assumptions.

2 The Ancient Concept of Number

It is sometimes said that the notion of natural number is the oldest and most basic mathematical notion that we possess. This, however, is simply not the case. The natural numbers, along with the rational and real numbers, were invented in the seventeenth century, as "abstracted ratios of concrete quantities",[3] and, as such, were quite different from numbers as traditionally conceived.

In classical antiquity a number (*arithmos*) was defined to be a *finite plurality composed of units*, a unit being whatever counts as one thing in the finite plurality under consideration. A number in this ancient sense is always a number *of*—of horses, of sheep, of numbers, $\cdots$. In a number of horses the units are horses, in a number of sheep, sheep, and in a number of numbers, numbers.

[2] From a mathematical point of view, Dedekind's analysis is the more profound, recognising, as it does, the central importance of definition by recursion. See section §126 of Dedekind (1893). Both, however, agree on the correct way to define "finite iteration". See point (6) of Dedekind (new 1971).

[3] Newton's definition of "number" in Newton (1728) runs as follows:

> By a *Number* we understand not so much a Multitude of Unities, as the abstracted Ratio of any Quantity to another Quantity of the same Kind, which we take for Unity.

Clearly numbers in this sense are more akin to what we call "sets" than to what we call "numbers", and the units of a number correspond to the members of a set. In fact, because Euclid, and indeed all mathematicians before Cantor, accepted the principle that *the whole is greater than the part*,[4] the "numbers (*arithmoi*)" of tradition really correspond to our "Dedekind finite sets".

Thus ancient arithmetic was a form of set theory. I am going to describe an up-to-date version of this theory, which I shall call *Euclidean arithmetic*, that can serve as an alternative to conventional *natural number arithmetic*.

3 Euclidean Arithmetic

In Euclidean arithmetic it is the notion of *finite set*, rather than the notion of *natural number*, that is taken as fundamental.[5] But we also need to acknowledge as fundamental the *global operations* (or *global functions*) and *global relations* that apply to sets.

Thus, for example, we have the global operation of *power set* ($S \mapsto \mathrm{P}(S)$) which assigns to each set S, the set $\mathrm{P}(S)$ of all subsets of S. The *boolean union* ($S, T \mapsto S \cup T$) is a binary global operation which assigns to each pair S, T of sets their union $S \cup T$. Similarly we must employ the binary global relations of *membership* ($\in$), *inclusion* ($\subseteq$), and *identity* ($=$).

These functions and relations are "global" in the sense that they are defined *everywhere*, at all finite sets and their members.[6] In this regard they are to be strongly distinguished from *local functions* and *local relations* which are defined, in the conventional way, to be sets of ordered pairs, which in Euclidean arithmetic means *finite* sets of ordered pairs, so that, for example, local functions have finite domains and codomains.

All the basic operations of set theory are global functions since they are everywhere defined. Their values are always sets.

Pair set: *Let x and y be any well-defined objects. Then x and y taken together are finite in multitude and therefore form a set, the* (**unordered**) **pair** *of x and y. Thus we can define the global function*

$$x, y \mapsto \{x, y\} \text{ (the set whose only members are } x \text{ and } y)$$

Power set: *Let S be any set. Then the subsets of S, taken together, are finite in multitude and therefore form a set, the* **power set** *of S. Thus we can define the global function*

$$S \mapsto \mathrm{P}(S) \text{ (the set of all subsets of } S)$$

[4] This is Common Notion 5 in Book I of the *Elements*.

[5] I shall allow for sets some or all of whose members are *individuals*, which are not sets. Sets and individuals together comprise the *objects*—the subject matter—of arithmetic.

[6] If a is an individual it behaves like the empty set, $\emptyset$, in set-theoretical contexts: thus $\emptyset \subseteq a$, $\mathrm{P}(a) = \{\emptyset\}$, and $a \cup x = x$, $a \subseteq x$, etc., for any x.

Subset selection (Comprehension): *Let S be any set and Φ be any* definite property[7] *applying, truly or falsely to members of S. Then the elements $x \in S$ for which $\Phi(x)$, taken together, are finite in multitude and therefore form a set. Thus we can define the global function*

$$S, \Phi \mapsto \{x \in S : \Phi(x)\} \text{ (the set of all } x \in S \text{ such that } \Phi(x))$$

Sum set (Union): *Let S be any set. Then the members of members of S, taken together, are finite in multitude and therefore form a set, the* **union** *or* **sum set** *of S. Thus we can define the global function*

$$S \mapsto \bigcup S \text{ (the set of all members of members of } S)$$

Replacement: *Let S be any set and σ be any one-place global function. Then the plurality obtained by replacing each element $x \in S$ by its value, $\sigma(x)$, under σ is finite in multitude and therefore forms a set, the* **image** *of S under σ. Thus we can define the global function*

$$S, \sigma \mapsto \{\sigma(x) : x \in S\} \text{ (the set of all } \sigma(x) \text{ such that } x \in S)$$

These operations are the familiar set formation operations of the Zermelo-Fraenkel system of set theory. Each incorporates a *finiteness principle*, a presupposition that certain pluralities are finite. In the context of Euclidean arithmetic they are obviously well-defined, and the finiteness assumptions they incorporate are self-evidently true, since Euclidean arithmetic is the theory of *finite* sets.[8]

All the familiar concepts of general set theory—ordered pair, (local) function, (local) relation, cartesian product, linear ordering, etc.—can be defined from the basic operations in the usual way.

It is technically useful to use Kuratowski's definition of linear ordering in accordance with which a linear ordering is identified with the set of its non-empty initial segments.[9] This has the advantage that if L is a Kuratowski linear ordering of a set S, then $L \cong_c S$, i.e., L and S are *equivalent in cardinality* in that there is a bijection $f : L \to S$.

We can use the notion of linear ordering to characterise the last of our basic set-theoretical operations. Define a *membership chain* for a set S to be a linear ordering whose first term (if S is a non-empty set) is a member of S and each succeeding term (if any) is a member of the immediately preceding term. We count the empty linear ordering, $[\,]$ ($=_{\text{def}}$ $\emptyset$) as a membership chain for the empty set, $\emptyset$, or for an individual, a.

---

[7] A *property* is a one place global relation. What it means for a property to be "definite" will be explained later, after *Brouwer's Principle* has been laid down.

[8] The axioms of Zermelo-Fraenkel corresponding to these global operations give necessary and sufficient conditions for membership in the sets obtained by applying them to arbitrary arguments. I take those conditions to be implicit in my descriptions of the operations.

[9] Thus given a three element set $\{a, b, c\}$, the linear ordering $[b, a, c]$ of its elements in which b comes first followed by a and c in that order is defined by $[b, a, c,] = \{\{b\}\{b, a\}\{b, a, c\}\}$.

Membership fan: *Let S be any set. Then the membership chains for S, taken together, are finite in multitude and therefore form a set, the* **membership fan** *for S. Thus we can define the global function*

$$S \mapsto \epsilon(S) \ \ \textit{(the set of all membership chains for } S\textit{)}$$

Like the more familiar set-theoretical operations considered above, the membership fan incorporates a finiteness principle.

From the membership fan we can define the transitive closure of a set:

$$\mathrm{TCL}(S) = \bigcup\bigcup \epsilon(S)$$

Since the transitive closure of a set consists of all those sets and individuals that are involved, directly or indirectly, in determining what that set *is*, sets are *well-founded* in the sense that if $S \in \mathrm{TCL}(T)$ then $T \notin \mathrm{TCL}(S)$, so that if S must be employed in determining what T is, then it cannot be the case that T must be employed in determining what S is. This is the principle of **Foundation**, which, together with the familiar principle of **Extensionality** characterises sets as *purely extensional pluralities.*[10]

We cannot describe the finiteness of sets in Euclidean arithmetic in the conventional way using "natural numbers"; but we can adopt

***Euclid's Principle*:** *The whole is greater than the part, that is to say, every set is of strictly larger cardinality than any of its proper subsets.*

This entails that all sets are finite. But the theory I am presenting here under the name of Euclidean arithmetic is intended to be a *finitary* theory of finite sets. Since the theory deals with all possible finite sets together with all their members, an infinity of objects fall within its scope. Thus if the theory is to be *finitary* it must conform to

***Brouwer's Principle*:** *A quantifier is subject to ordinary, classical logic only if its domain of quantification is finite, that is to say, is a set. If its domain of quantification is infinite, then it is subject to constructive laws.*

Brouwer's Principle has a profound effect on the development of the theory. For by virtue of that principle we can not guarantee that

$$\mathrm{Q}(a) \textit{ or not } \mathrm{Q}(a)$$

for assertions $\mathrm{Q}(a)$ that require for their expression, directly or indirectly, global quantification over all sets and individuals. It follows that such conditions $\mathrm{Q}(a)$ cannot be used to form properties (one-place global relations) Φ to use as arguments

[10] Both Foundation and Extensionality are essential here. See Maberry (2000, sections 4.11 and 8.6).

in the comprehension operator. This is in conformity with Zermelo's requirement that the property Φ used in the definition of $\{x \in S : \Phi(x)\}$ should be *definite* (see Zermelo, 1908; Zermelo, 1930).

This restriction on the argument Φ in Comprehension strongly restricts the versions of proof by induction and definition by recursion that we can justify, *even when the induction or recursion is applied to a finite linear ordering*

$$L = [\text{First}_L, \cdots, x, \text{Next}_L(x), \cdots, \text{Last}_L]$$

We can establish *local* versions of induction and recursion along these orderings as I shall explain in Section 4.

Note that *local* (bounded) quantifiers can actually be *defined* using the Comprehension operator even in the absence of genuine quantifiers, eg.

$$(\forall x \in S)\Phi(x) \Leftrightarrow \{x \in S : \Phi(x)\} = S$$

These quantifiers satisfy conventional (i.e., classical) laws. Moreover, definite properties are obviously closed under classical truth functions, so any expression we can write down using symbols for the basic global operations and relations, together with bounded quantifiers (as just defined) and classical propositional connectives, will define a "definite" property in Zermelo's sense to which Comprehension can be applied.

In fact, if we formalise Euclidean arithmetic as a free variable theory in the obvious way, with Comprehension and Replacement operators, but without unbounded quantifiers, then the global functions may be regarded as corresponding to λ-terms $\lambda \mathbf{x} t[\mathbf{x}/\mathbf{a}]$, where t is a term of the formalised theory, and global relations to λ-terms $\lambda \mathbf{x} A[\mathbf{x}/\mathbf{a}]$, where A is a formula. There is a theoretical difficulty here, however: since I am calling the uniqueness of natural number systems into question, I must call into question the uniqueness of conventional systems of formal syntax as well.

4 Induction and Recursion

In Euclidean arithmetic it is possible to establish induction along a finite linear ordering, but only with respect to definite properties.

Induction Along Linear Orderings: *Let L be a linear ordering, and Φ be a definite property. Then from the premises*

(i) $\Phi(0_L)$.
(ii) $\Phi(x) \Rightarrow \Phi(\text{Next}_L(x))$, *for all* $x \in \text{Field}(L)$ *except* $x = \text{Last}_L$.

we may conclude that $\Phi(x)$ for all $x \in \text{Field}(L)$.

The proof here is along customary lines, except we must use the fact that Φ is definite in order to form $\{x \in \text{Field}(L) : \Phi(x)\}$.

We can use this version of induction to establish definition by recursion with respect to a *local* function.

Local Recursion Along Linear Orderings: *Let L be a linear ordering, S a non-empty set, $a \in S$, and $g : S \to S$ be a local function. Then there is exactly one function $f : \text{Field}(L) \to S$ such that*

(i) $f(0_L) = a$.
(ii) $f(\text{Next}_L(x)) = g(f(x))$, *for all* $x \in \text{Field}(L)$ *except* $x = \text{Last}_L$.

One might be tempted to say that this simply acknowledges the possibility of starting at a and iterating the function g as many times as there are terms in the linear ordering $L = [0_L, \cdots k_L]$:

$$a\,(= f(0_L)),\; g(a)\,(= f(1_L)), \cdots, \overbrace{g(g(\cdots(g(a)\cdots))}^{k_L}\,(= f(k_L))$$

But that would be a mistake, because the seemingly similar principle of global recursion does *not* hold in Euclidean arithmetic.

> *Let L be a linear ordering and γ be a global function. Then for any a there is exactly one function f defined on* Field(L) *such that*
>
> (i) $f(0_L) = a$.
> (ii) $f(\text{Next}_L(x)) = \gamma(f(x))$, *for all* $x \in \text{Field}(L)$ *except* $x = \text{Last}_L$.

This too seems simply to acknowledge the possibility of starting at a and iterating the global function γ as many times as there are terms in the linear ordering $L = [0_L, \cdots k_L]$:

$$a\,(= f(0_L)),\; \gamma(a)\,(= f(1_L)), \cdots, \overbrace{\gamma(\gamma(\cdots(\gamma(a)\cdots))}^{k_L}\,(= f(k_L))$$

In both these cases we can, indeed, iterate these functions as often as we like, but *that* is irrelevant to the justification of recursion. To suppose otherwise is to fall into the sorites fallacy.[11]

The fact that "iteration" works for a local function $g : S \to S$ but fails for a global function γ arises from the fact that in the local case, the values of the function f being defined are confined to the set S which is given *in advance*, whereas in the global case we cannot, in general, confine the anticipated values of the function f to a set fixed in advance.[12]

If we try to adapt the purely set-theoretical proof justifying local recursion to the global case we find that, at a crucial point where the local proof requires us to define certain initial segments of L by comprehension, the global argument would require

[11] See section 8.4 of Mayberry (2000).

[12] It is possible, however, to justify *limited* recursion in Euclidean arithmetic. See Mayberry (2000, sections 9.2 and 10.3).

comprehension with respect to a property defined using unbounded existential quantification (a Σ_1 property), which is therefore not definite in Zermelo's sense, and consequently is not allowed in Euclidean arithmetic. Thus it is Brouwer's Principle which prevents us from establishing a general principle of recursion with respect to global functions.

5 Arithmetical Functions and Relations

There are two different senses in which Euclidean arithmetic can be said to "contain" a version of natural number arithmetic. The first is via the theory of *arithmetical* functions and relations, which I am about to describe. The second is via the theory of *natural number systems* (called "simply infinite systems" in Dedekind's monograph (1893)), which I shall describe later (Section 6).

A global function, φ, is said to be *arithmetical* if the cardinality of its value depends only on the cardinalities of its arguments. Thus, for example, a binary global function φ is arithmetical if, for all sets S, S_1, T, and T_1

$$S \cong_c S_1 \text{ and } T \cong_c T_1 \Rightarrow \varphi(S,T) \cong_c \varphi(S_1,T_1)$$

Note that, on formalisation, this definition is Π_1.

Similarly, a relation, Φ is said to be *arithmetical* if its truth value depends only on the cardinalities of its arguments. Thus a global binary relation Φ is arithmetical if, for all sets S, S_1, T, and T_1

$$S \cong_c S_1 \text{ and } T \cong_c T_1 \Rightarrow [\Phi(S,T) \Leftrightarrow \Phi(S_1,T_1)]$$

There are global functions and relations arithmetical in this sense corresponding to the familiar basic functions and relations of conventional natural number arithmetic:

(i) **Power of two:**

$$x \mapsto P(x)$$

(This corresponds to the natural number function $x \mapsto 2^x$.)

(ii) **Successor:**

$$x \mapsto x \cup \{x\}$$

(This corresponds to the natural number function $x \mapsto x + 1$.)

(iii) **Addition (disjoint union):**

$$x, y \mapsto x + y = (x \times \{\emptyset\}) \cup (y \times \{\{\emptyset\}\})$$

(iv) **Multiplication (Cartesian product):**

$$x, y \mapsto x \times y \text{ (the set of all ordered pairs } (u, v) \text{ where } u \in x \text{ and } v \in y)$$

(v) **Exponentiation:**

$$x, y \mapsto x^y = \{f \in \mathrm{P}(y \times x) : f \text{ is a function from } y \text{ to } x\}$$

(v) **Cardinal equivalence and ordering:**

$$x \cong_c y \text{ and } x <_c y$$

Set-theoretical analogues of bounded sums and bounded products can also be defined.

We can think of these arithmetical functions as *defining* their "natural number" correlates. This means that the "natural number" functions of Euclidean arithmetic are constructive in the strongest conceivable sense. For, informally, a one place "natural number" function φ_{number} is well-defined only if we can exhibit a set-theoretical operation, φ_{set}, which, when applied to a set S of size n, yields a set $\varphi_{\text{set}}(S)$ of size $\varphi_{\text{number}}(n)$.

What we cannot do in Euclidean arithmetic is to lay down definitions whose legitimacy rests on fantasies of "arbitrarily long" finite computation. We cannot, for example, simply write down recursion equations and, construing them as instructions for calculation, regard them as defining a global function. To accept such "definitions" as efficacious, for such a reason, is obviously to fall victim to the sorites fallacy.[13]

6 Natural Number Systems

The naive idea of a natural number system (called a simply infinite system by Dedekind (1893)) is that of an infinite sequence generated from an *initial element* a, by a (global) *successor function* σ:

$$a,\ \sigma(a),\ \sigma(\sigma(a)),\ \sigma(\sigma(\sigma(a))), \sigma(\sigma(\sigma(\sigma(a)))),\ \cdots$$

For technical reasons, however, it is necessary in Euclidean arithmetic to identify a natural number system with the corresponding infinite sequence of linear orderings that are the proper initial segments of its sequence of terms

$$[\,](=\emptyset), [a],\ [a, \sigma(a)],\ [a, \sigma(a), \sigma(\sigma(a))],\ [a, \sigma(a), \sigma(\sigma(a)), \sigma(\sigma(\sigma(a)))] \cdots$$

Of course the presence of the dots of ellipsis, with their appeal to the idea of "iterating" the successor function, means that this is obviously not a legitimate definition. In fact, it is the very idea of iteration itself that we need to characterise in a mathematically precise way. Indeed, the notion of "infinite sequence" used here is ambiguous, as we shall see.

[13] It can be shown that the arithmetical functions and relations of Euclidean arithmetic suffice to provide an interpretation of $I\Delta_0 + \exp$. This was established in Homolka (1983).

Definition: *Let σ be a unary global function, a an arbitrary object, and L a linear ordering. We say that L* ***is generated by*** σ ***from*** *a if $L = \emptyset$ or*

(a) $\text{First}_L = a$
(b) $\text{Next}_L(x) = \sigma(x)$, *for all* $x \in \text{Field}(L)$ *except* $x = \text{Last}_L$.

This definition is of fundamental importance because it characterises the notion of a finite sequence of distinct terms "generated" by a global function σ in a completely "static", set-theoretical way, without appealing, in any way whatsoever, to the naive notion that we can iterate (repeatedly apply) the function σ starting from an argument a.

Now we can define the notion of a natural number system:

Definition: *Let σ be a unary global function and a an arbitrary object. We say that* σ ***generates a natural number system from*** (***the initial term***) *a if for all non-empty linear orderings L*

$$\sigma \text{ generates } L \text{ from } a \Rightarrow \sigma(\text{Last}_L) \notin \text{Field}(L)$$

If expressed in formalised language, this definition would be Π_1 since it requires an initially placed global universal quantifier. Accordingly the proposition that σ generates a natural number system from the initial term a is not definite in Zermelo's sense.

We can think of a natural number system as the species (proper class), $\mathcal{N}_{\sigma,a}$, of linear orderings generated by its successor function σ from its initial term a.[14] The *numbers* of $\mathcal{N}_{\sigma,a}$ are the linear orderings which compose it; its *terms* are the terms of the linear orderings that are its numbers.

If L is a number of $\mathcal{N}_{\sigma,a}$, then it can be thought of both as a cardinal number and as an ordinal number. In either role it is a particular example of the kind of thing that it numbers. For L (considered as a cardinal number) is a set of the cardinality designated by L and a linear ordering of the length (order type) designated by L (considered as an ordinal number). This is a consequence of adopting Kuratowski's definition of a linear ordering.

Here are some examples of natural number systems:

(i) **The *von Neumann* natural number system** $\mathcal{VN}$:

a. $\sigma(x) = x \cup \{x\}$.
b. $a = \emptyset$.

(ii) **The *Zermelo* natural number system** $\mathcal{Z}$:

a. $\sigma(x) = \{x\}$.
b. $a = \emptyset$.

(iii) **The *Cumulative Hierarchy* natural number system** $\mathcal{CH}$:

[14] We can speak of the species $\mathcal{N}_{\sigma,a}$, of linear orderings generated by its successor function σ from its initial term a, even when it does not form a natural number system.

a. $\sigma(x) = x \cup \mathrm{P}(x)$.
b. $a = \emptyset$.

(iv) **The *Infimum*[15] of the natural number systems $\mathcal{N}$ and $\mathcal{M}$** $\mathrm{Inf}(\mathcal{N}, \mathcal{M})$:
Let $\mathcal{N}$ and $\mathcal{M}$ be natural number systems. Define a global function τ by

$$\tau((x, y)) = (\sigma_{\mathcal{N}}(x), \sigma_{\mathcal{M}}(y))$$

Then τ generates a natural number system, $\mathrm{Inf}(\mathcal{N}, \mathcal{M})$ *(the* ***infimum*** *of $\mathcal{N}$ and $\mathcal{M}$), from the initial term $(a_{\mathcal{N}}, a_{\mathcal{M}})$.*

(v) **The *Ackermann* natural number system $\mathcal{ACK}$**:

a. $\sigma(x) = \mathrm{ACK}(x)$, where $\mathrm{ACK}(x)$ is the pure set whose Ackermann code is the next number after the Ackermann code of x.[16]
b. $a = \emptyset$.

The four natural number systems defined outright in examples (i), (ii), (iii), and (v) are by no means equivalent. Indeed, in general, natural number systems can differ both in their lengths and in their closure properties. But I must supply constructive meanings for "length" and "closure" before I can provide examples.

Suppose we are given a species $\mathcal{S}$ of sets. Then a *measure for $\mathcal{S}$ with respect to the natural number system $\mathcal{N}$* is a unary global function, μ, such that, whenever a set T lies in $\mathcal{S}$, $\mu(T)$ lies in $\mathcal{N}$ and $T \cong_c \mu(T)$. In these circumstances we may say that $\mathcal{N}$ is a *scale* for $\mathcal{S}$ with *measure* μ. This definition becomes Π_1 on formalisation.

In accordance with these definitions we may say that the natural number system $\mathcal{N}$ is *shorter than, or equal to* the natural number system $\mathcal{M}$ (in symbols: $\mathcal{N} \preceq \mathcal{M}$) if it is possible to exhibit a measure μ with respect to $\mathcal{M}$ for the species of all linear orderings, L, that lie in $\mathcal{N}$.

When I say "it is possible to exhibit a measure μ" I mean that it is posssible, using the basic set-theoretical operations of pair set, power set, union, comprehension, replacement, and membership fan *explicitly to define* a global function μ with the necessary properties. This means that *the global operations of Euclidean arithmetic are to be identified with operations that can be explicitly defined from the standard set-theoretical operations of that theory.*[17]

In a similar way we need to give a constructive meaning to the assertion that a natural number system is *closed* under a given arithmetical function φ.

Suppose we are given an arithmetical function φ (of two arguments, say) and a natural number system $\mathcal{N}$. Then we may say that a binary global function ψ *represents φ in $\mathcal{N}$* if, whenever L_1 and L_2 are linear orderings lying in $\mathcal{N}$, $\psi(L_1, L_2)$ also lies in $\mathcal{N}$ and

[15] This system is what is generated when we attempt to specify an isomorphism from $\mathcal{N}$ to $\mathcal{M}$ in the obvious, intuitive way. We don't in general get an isomorphism but a new natural number system with the properties suggested by its name.

[16] A purely set-theoretical definition of ACK is given in section 10.6 of Mayberry (2000).

[17] If we add a new global function φ to our list of initial global functions, the notion of global function must be changed accordingly. See Mayberry (2000), section 10.7.

$$\psi(L_1, L_2) \cong_c \varphi(L_1, L_2)$$

(This definition becomes Π_1 when formalised.)

Now we can say that a natural number system $\mathcal{N}$ is *closed* under an arithmetical global function φ if we can exhibit a global function ψ that represents φ in $\mathcal{N}$.

Armed with these definitions we can establish the following result[18]

Theorem 1:

(i) $\mathcal{CH} \preceq \mathcal{VN}$
(ii) $\mathcal{CH} \preceq \mathcal{Z}$
(iii) $\mathrm{Inf}(\mathcal{N}, \mathcal{M}) \preceq \mathcal{N}$ *and* $\mathrm{Inf}(\mathcal{N}, \mathcal{M}) \preceq \mathcal{M}$.
(iv) $\mathcal{N}_1 \preceq \mathcal{N}$ *and* $\mathcal{N}_1 \preceq \mathcal{M}$ *imply that* $\mathcal{N}_1 \preceq \mathrm{Inf}(\mathcal{N}, \mathcal{M})$.
(v) $\mathcal{ACK}$ *is closed under addition, multiplication, and exponentiation.*

From a classical, infinitary point of view, all the global functions employed in Euclidean arithmetic correspond to formal terms t in the natural formal language, $\mathcal{L}$, for set theory in which the set-theoretical operations of pair set, power set, union, comprehension, replacement, and membership fan are incorporated directly into the primitive vocabulary of the formal language.

Thus a global function φ of two arguments corresponds to the formal $\mathcal{L}$-term $\mathbf{t}_\varphi$ with free variables a and b in that, by definition $\varphi(S, T) = \mathbf{t}_\varphi[S/a, T/b]$: classically we are dealing with functions of the form

$$\lambda x_1 \cdots x_n \mathbf{t}[x_1/a_1, \cdots, x_n/a_n]$$

for $\mathcal{L}$-terms $\mathbf{t}$, with free variables from among the distinct free variables $a_1, \cdots, a_n$.

Now we can ask, from a classical, infinitary standpoint, whether, given a natural number system $\mathcal{N}$, it is possible to find an $\mathcal{L}$-term, $\mathbf{t}_+$ with free variables a and b such that $\lambda xy\mathbf{t}_+[x/a, y/b]$ defines addition in $\mathcal{N}$ *when the $\mathcal{L}$-terms are given their standard interpretations in the hereditarily finite sets*, i.e., whether

$$\mathbf{V}_\omega \models \lambda xy\mathbf{t}_+[x/a, y/b] \textbf{ defines addition in } \mathcal{N}$$

where $\mathbf{V}_\omega$ is the standard, infinitary model of the hereditarily finite pure sets which assigns the global operators in $\mathcal{L}$ their canonical values.

In the cases in which $\mathcal{N}$ is the von Neumann natural number system, $\mathcal{VN}$, the Zermelo system, $\mathcal{Z}$, or the Cumulative Hierarchy system, $\mathcal{CH}$, the answer is *no*. For a straightforward *classical* induction over the *classical* (transfinite) set of $\mathcal{L}$-terms shows that no term $\mathbf{t}_+$ exists which defines addition in $\mathcal{N}$.

This constitutes a classical, infinitary, motivation for positing the following assumption:

[18] The proofs of these claims are straightforward and are to be found in Mayberry (2000), chapter 10.

Postulate I: *For any binary global function φ, it is impossible that, for all linear orderings L_1 and L_2 lying in $\mathcal{N}$*

$$\varphi(L_1, L_2) \text{ lies in } \mathcal{N} \textit{ and } \varphi(L_1, L_2) \cong_c (L_1 +_c L_2)$$

where $\mathcal{N}$ is $\mathcal{VN}$, $\mathcal{Z}$, or $\mathcal{CH}$.

From a conventional, infinitary standpoint, we can consistently add Postulate I to the axioms of a first order formalisation of Euclidean arithmetic.[19] If a contradiction resulted we would have an inconsistency in a theory that is logically equivalent to a weak version of natural number arithmetic ($I\Delta_0$ + exp)—a highly unlikely, but not *absolutely* inconceivable, event. But from the finitary standpoint incorporated in Euclidean arithmetic, *the argument to establish this consistency result is not valid.*

Accordingly, I shall call a posit of this kind a *classically witnessed postulate* and treat it as if it were an axiom, like those which characterise the basic set-theoretical operations. But, of course, it can't be an axiom proper, because it is not self-evident.

A similar, but more difficult argument[20] shows us that there is no $\mathcal{L}$-term **t**, with one free variable a such that

$$\mathbf{V}_\omega \models \lambda x\mathbf{t}[x/a] \textbf{ is a measure for } \mathcal{Z} \textbf{ in } \mathcal{VN}$$

which, as in the previous case, provides a classical, infinitary justification for laying down another classically witnessed postulate:

Postulate II: *For any unary global function φ, it is impossible that, for all linear orderings L lying in $\mathcal{Z}$*

$$\varphi(L) \text{ lies in } \mathcal{VN} \wedge \varphi(L) \cong_c L$$

which can be expressed symbolically by

$$\mathcal{Z} \not\preceq \mathcal{VN}$$

But an essentially similar argument leads to the classically witnessed postulate

Postulate III: *For any unary global function φ, it is impossible that, for all linear orderings L lying in $\mathcal{VN}$*

$$\varphi(L) \text{ lies in } \mathcal{Z} \wedge \varphi(L) \cong_c L$$

or, in brief

$$\mathcal{VN} \not\preceq \mathcal{Z}$$

and, of course, many other such classically witnessed postulates could, in principle, be laid down.

Thus in Euclidean arithmetic, augmented by the classically witnessed Postulates I, II, and III, not only are there non-isomorphic natural number systems, *there are pairs of natural number systems that are incomparable in length.*

---

[19] On formalisation, this becomes a Π_1 condition on the free function variable φ, where we must include a rule of substitution for such variables.

[20] This was discovered by Richard Pettigrew (2008) who modified a model-theoretic argument of S. Popham (1984). A version of Popham's argument is given in section 11.2 of Mayberry (2000).

This seems at odds with our intuition, and indeed it is. Our intuition of "the" natural numbers is *extensional*, so that we imagine "them" as arranged in an unending progression extending to "infinity"

$$0, 1, 2, 3, 4, 5, \cdots \text{ (those ubiquitous dots of ellipsis!)}$$

In Euclidean arithmetic natural number systems are *intensional* in character, and the relative lengths of two such systems are determined by the detailed logical and set-theoretical relations among their initial objects and, especially, among their respective successor functions. In Euclidean arithmetic there is more than one way of "passing to infinity".

In fact, there is a bewildering variety of natural number systems with differing lengths and closure properties.[21] In the next section I shall describe a straightforward way to illustrate this.

7 Binary Expansions

A *binary numeral* is just a finite sequence (L, f) of the binary digits 0 $(=\emptyset)$ and 1 $(=\{\emptyset\})$ (so that $L = [0_L, \cdots K_L]$ is a linear ordering and $f : \{0_L, \cdots K_L\} \to \{0, 1\}$ is a suitable function). Given a natural number system $\mathcal{N}$ define its *binary expansion,* $\mathcal{N}[2]$, to be the natural number system whose terms are the binary numerals (L, f) whose underlying linear ordering L lies in $\mathcal{N}$.

The *numbers* of $\mathcal{N}[2]$ are thus linear orderings $[\mathbf{0}, \cdots, \mathbf{m}]$ whose terms, $\mathbf{0}, \cdots, \mathbf{m}$, are binary digits whose lengths lie in $\mathcal{N}$, and which are arranged in their natural order.

The essential facts about these binary expansions are contained in the following proposition:

Theorem 2[22]: *Let $\mathcal{N}$ be a natural number system.*

(i) $\mathcal{N} \preceq \mathcal{N}[2]$
(ii) $\mathcal{N}[2]$ *is closed under addition.*
(iii) *A necessary and sufficient condition for $\mathcal{N}$ to be closed under addition is that $\mathcal{N}[2]$ be closed under multiplication.*
(iv) *A necessary and sufficient condition for $\mathcal{N}$ to be closed under multiplication is that $\mathcal{N}[2]$ be closed under* $\mathrm{Lexp}_2^{\mathbf{1}}$ $(= \lambda x, y(x^{\log_2(y)}))$.
(v) *A necessary and sufficient condition for $\mathcal{N}$ to be closed under* $\mathrm{Lexp}_2^{\mathbf{n}}$ *is that $\mathcal{N}[2]$ be closed under* $\mathrm{Lexp}_2^{\mathbf{n+1}}$ $(= \lambda x, y(x^{\mathrm{Lexp}_2^{\mathbf{n}}(\log_2(y), \log_2(y))}))$.

It is important to realise that, in the (implied) definition of $\mathrm{Lexp}_2^{\mathbf{n}}$, the expression $\mathbf{n}$ is *not* a variable ranging over "natural numbers". The "recursion" in the definition is

---

[21] Many such examples are given in Pettigrew (2008).

[22] See section 10.4 of Mayberry (2000). It is straightforward to define a set-theoretical analog of $\log_2$.

really a recipe for laying down as many *particular* definitions of these higher "logarithmic exponentials" as one wishes (or as one can). We cannot even "internalize" this recursion to finite linear orderings $L = [0_L, \cdots, k_L]$.

The method of binary expansion will not lead to closure under exponentiation, however.

Theorem 3[23]: *Let $\mathcal{N}$ be a natural number system. Then the following are equivalent*:

(i) *$\mathcal{N}[2]$ is closed under exponentiation.*
(ii) *$\mathcal{N}$ measures $\mathcal{N}[2]$.*
(iii) *$\mathcal{N}$ is closed under exponentiation.*

Thus if we start with a natural number system, $\mathcal{N}$, not closed under addition (e.g., if $\mathcal{N}$ is $\mathcal{VN}, \mathcal{Z}$ or $\mathcal{CH}$ and we adopt Postulate I) then successive binary expansions produce longer and longer natural number systems

$$\mathcal{N} \prec \mathcal{N}[2] \prec (\mathcal{N}[2])[2] \prec ((\mathcal{N}[2])[2])[2] \prec \cdots$$

closed under stronger and stronger arithmetical functions

$$\lambda x(x+1),\ +,\ \times,\ \lambda xy(x^{\log_2(y)}),\ \lambda xy(x^{(\log_2(y)^{log_2(\log_2(y))})})\ \cdots$$

though all these "logarithmic exponentials" grow more slowly than the exponential function. If we ignore the first two natural number systems $\mathcal{N}$ and $\mathcal{N}[2]$, the hierarchy of *natural number systems* starting from $(\mathcal{N}[2])[2]$, corresponds to the $\Omega_{\mathbf{n}}$ hierarchy of *theories* of weak arithmetic described in Hajek and Pudlak (1993).

It is worth pointing out that, from the standpoint of Euclidean arithmetic, *none of these natural number systems is "nonstandard"*.

8 Conclusions

Euclidean arithmetic is easily formalisable from the informal treatment I have given. The straight-forward classical, first order formalisation of the basic theory (without the classically witnessed postulates) yields a theory that is intertranslatable with the weak arithmetic $I\Delta_0$ + exp (Homolka, 1983). It can also be formalised as a predicative, second order theory,[24] or in intuitionistic first and second order predicative variants. Of course only the intuitionist versions conform to Brouwer's Principle.[25]

---

[23] Mayberry (2000), 10.4.15.

[24] By "predicative" I mean that, comprehension for global relations holds only for formulas free of *both* first or second order global quantifiers.

[25] In these formalised theories we must include a *singleton selector*, ι, such that $\iota(S) = x$ if S is the singleton $\{x\}$ and $\iota(S) = S$ otherwise.

I have not expounded it as a formal theory here, however, because I have been dealing with foundational questions and it is obvious (or should be) that *no formal theory can serve as a foundation for arithmetic*. This is because the basic principles underlying formal syntax are essentially the same as those underlying arithmetic itself, and conventional treatments of formal syntax rest upon the Sorites fallacy in exactly the same way that conventional treatments of natural number arithmetic do.[26]

Of course in Dedekind's infinitary exposition of natural number arithmetic we have a unique (up to isomorphism) natural number system which can be used to "count out" all (and only) finite sets.

But in Euclidean arithmetic this is emphatically not the case. For in that theory no natural number system is "long enough" to count out all finite sets.[27] Euclidean arithmetic thus separates the notion of counting from the general notion of finiteness which is incorporated in Euclid's Principle: The whole is greater than the part.

If we introduce appropriate classically witnessed postulates, we can introduce the notion of a *long linear ordering*, which is a linear ordering (necessarily finite in Euclidean arithmetic) that contains an injective image of an entire natural number system as a proper initial segment.[28]

Using long linear orderings we can develop a version of the Infinitesimal Calculus in which all "real numbers" are to be found among any suitable finite set of rationals. Here the classical notion of limit is replaced by the distinction between large and small whole numbers. But a discussion of these ideas would take me beyond the subject of this paper.

References

Dedekind, R. (1893) *Was sind und was sollen die Zahlen?* Braunschweig: Vieweg, Translated by Wooster W. Berman as *The Nature and Meaning of Numbers* in *Essays in the Theory of Number*, New York: Dover, 1963.

Dedekind, R. (new 1971) Letter to Kefferstein. Translated by H. Wang and S. Bauer-Mengelberg in J. van Heijenoort, ed. (1971) *From Frege to Gödel. A Source Book in Mathematical Logic, 1879–1931*, in J. van Heijenoort, Cambridge, MA: Harvard University Press, 1971.

Frege, G. (1974) *The Foundations of Arithmetic*, Translated by J.L. Austin, Oxford: Basil Blackwell.

Hajek, P. and Pudlak, P. (1993) *Metamathematics of First-Order Arithmetic*, in the series *Perspectives in Mathematical Logic*, Berlin: Springer.

Homolka, V. (1983) *A System of Finite Set Theory Equivalent to Elementary Arithmetic*, Ph.D Thesis, University of Bristol.

---

[26] In fact, the logical and philosophical difficulties that arise in the treatment of infinite totalities in Euclidean arithmetic also arise in infinitary set theory in the guise of difficulties surrounding what is sometimes called *Cantor's Absolute* (See Mayberry, 2000, section 3.5).

[27] This can be formulated as a classically witnessed postulate.

[28] Here we exploit the fact that it is possible to have sub*species* of a finite set that are not sub*sets*. Vopenka calls these species *semi-sets* in Vopenka (1979).

Mayberry, J.P. (2000) *The Foundations of Mathematics in the Theory of Sets*, Cambridge: Cambridge University Press.
Newton, I. (1728) *Universal Arithmetic: or a Treatise of Arithmetical Composition and Resolution*, London. Reprinted in *The Mathematical Works of Isaac Newton* Volume II (ed. by Whiteside, D. T.), New York: Johnson Reprint Corporation, 1966.
Pettigrew, R. (2008) *The Theories of Natural Number Systems and Infinitesimal Analysis in Euclidean Arithmetic*, Ph.D. Thesis, University of Bristol.
Popham, S. (1984) *Some Results in Finitary Set Theory*, Ph.D. Thesis, University of Bristol.
Vopenka, P. (1979) *Mathematics in the Alternative Set Theory*, Leipzig: Teubner Verlagsgesellschaft.
Zermelo, E. (1908) Untersuchungen über die Grundlagen der Mengenlehre, *Mathematische Annalen* **65**, 261–281.
Zermelo, E. (1909) Sur les ensembles finis et le principe de l'induction complète, *Acta Mathematica* **32**, 185–193.
Zermelo, E. (1930) Über Grenzzahlen und Mengenbereiche. Neue Untersuchungen über die Grundlagen der Mengenlehre, *Fundamenta Mathematicae* **14**, 29–47.

Intentionality, Intuition, and Proof in Mathematics

Richard Tieszen

Mathematicians, in the ordinary course of their work, are directed in their thinking toward objects, structures, or states-of-affairs in their various domains of research but they are typically not directed toward the cognitive acts and processes in which the awareness of their research domain is constituted. Being directed toward the normal *objects* of their research—natural numbers, sets, functions, spaces, groups, and so on, and properties of or relations between such objects—is quite different from being directed toward the *consciousness* of such objects. If we turn to the cognitive acts and processes of mathematicians, however, then we immediately notice a basic feature that is characteristic of many forms of consciousness: intentionality.

The concept of intentionality can already be found in medieval philosophy. It has been at the center of the philosophy of mind for at least forty years now in the work of John Searle, Daniel Dennett, Hilary Putnam, Jerry Fodor, Donald Davidson, W.V. Quine, Roderick Chisholm, and many others, although it was of course introduced even earlier in modern philosophy by Franz Brentano. It figured into a lot of Edmund Husserl's thinking about logic and mathematics (see, e.g, Husserl, 1900, 1901, 1929). In the nineteen twenties and early nineteen thirties it made its way into some of the thinking in the foundations of mathematics in the work of Oskar Becker (1927), Arend Heyting (1931), Hermann Weyl (1926), and Felix Kaufmann (1930). Since that time it has hardly been investigated at all in connection with mathematical consciousness and mathematical thinking. The idea of the intentionality of consciousness has thus been around for a long time but, depending on which way the winds of reductionism are blowing in the philosophy of mind, it has at times been pushed into the margins. Although I do not have space to defend the view here, I would argue that the last century saw the development of one mistaken reductionistic view after another (e.g., behaviorism, identity theory, eliminative materialism, computational functionalism) about the nature of human consciousness. In this paper I want to bring the idea of the intentionality of consciousness back into the center of focus in considerations in the foundations of mathematics, especially in connection with the line of thinking that was started in the work of Husserl, Becker, Weyl, and Heyting. I think that a fruitful and philosophically rich view of

G. Sommaruga (ed.), *Foundational Theories of Classical and Constructive Mathematics*, The Western Ontario Series in Philosophy of Science 76,
DOI 10.1007/978-94-007-0431-2_13,

the foundations of mathematics emerges if we begin with the idea that mathematical consciousness exhibits intentionality. My argument for this claim in this paper will have to be truncated. It would be especially useful to provide many examples of the kinds of proofs I describe below but, due to space limitations, that will have to wait for another occasion.

1 Intentionality

'Intentionality' refers to the 'aboutness' or 'directedness' of consciousness. It is not to be confused with the more specific idea of 'intention', as when I say that "I intend to go to Paris". Many kinds of conscious states exhibit intentionality. Knowing, believing, remembering, intending, desiring, and willing, for example, are all conscious states that are object-directed. In mathematics in particular our thinking seems to be directed toward various kinds of objects and states of affairs, e.g., natural numbers, real numbers, sets, functions, groups, spaces, and properties of and relations concerning such objects.

One could try to deny that our thinking in mathematics exhibits intentionality. The denial could take several forms: (1) Thinking in general is not really 'about' anything. This claim, however, is quite dubious. If we allow that there are beliefs at all, for example, then how could believing not be about anything? Similarly, if we allow that there is knowledge at all, then how could knowing not be about anything? The denial could take other, less radical forms. One might argue that (2) some thinking is object-directed, such as in natural science, but some is not, such as in mathematics. The early Wittgenstein, Carnap, and the logical positivists held such a view. Mathematical statements, as mere tautologies, are said to be without content or object. They are true, for example, on the basis of linguistic conventions. Another possibility is that (3) mathematical thinking is object-directed only when applied. On any of these views we have to explain away the appearance and the natural sense of the directedness of thinking in mathematics, even in the case of pure mathematics. The idea would be that the object-directedness of mathematical thinking is somehow only an appearance and not the reality. Is it really the case, however, that when I believe that $2 + 2 = 4$ my belief is not about anything, or not about anything in particular, and that the appearance of its being about something is sheer illusion?

I am unhappy with views that deny the intentionality of mathematical consciousness. It would take a long time to lay out all of the arguments against such views and I propose that I not do so here. In any case, as I said above, I think a fruitful and philosophically rich view of the foundations of mathematics emerges if we start with the idea that mathematical consciousness does exhibit intentionality. We will then be led to look at intuition in a new way, and to look at the concept of proof in a new way.

Suppose we view mathematical sentences as expressions of the contents (or 'intentions') of our acts of mathematical thinking, as in "Mathematician (a human subject) M thinks that 7 + 5 = 12", "M believes that there is no largest prime number", "M thinks that every even integer > 2 is the sum of two primes", or "M believes that a supercompact cardinal exists". I will say that the 'content' or intention is expressed by the sentences following the word 'that' in these examples. We are directed in our thinking toward certain objects or states of affairs by way of these contents or intentions. Based on the phenomenological view that human subjects are meaning-bestowing beings, I will call these meaning-intentions. Human beings bestow meaning. We are interpreters. On the theory of intentionality I would want to adopt we can say that directedness is due to the meaning-intentions of our acts of thinking, not to the existence of the objects. I might be directed toward objects in my thinking, and my thinking and behavior might be determined by this directedness even if the object or state of affairs toward which I am directed does not exist. This is what happens in misperceptions, illusions, and hallucinations. Generally, it is not a condition for the possibility of mere object-directedness in any particular case that there first be an object that affects me in some way, but it is a condition for the possibility of object-directedness that there be a meaning-intention.

Intentionality in mathematics is ubiquitous. It is even present in a particular way in strictly formalistic views of the foundations of mathematics, as we will see below. Intuition in mathematics, however, is not ubiquitous.

2 Intuition as Fulfillment of Meaning-Intention

The term 'intuition' has been taken to have many different meanings. Here is a partial list of meanings that have been attached to the term: the intuitive is the informal, or the non-inferential, or the non-rigorous, the visual, the holistic, the incomplete, the convincing in spite of lack of proof, a feeling, the mysterious or inexplicable, the first thing that comes to mind, a guess, that which is easy to follow. In this paper I would like to set all of these uses of the term 'intuition' aside. I will provide a particular characterization of intuition that could be compared with the various other characterizations but I will not enter into the comparisons here. In the foundations of mathematics in particular, one can find distinct conceptions of intuition in the work of Hilbert, Brouwer, and Gödel. I cannot go into these in this paper but I will make a few comments about them below.

On the conception of intuition I want to develop in this paper intuition is to be understood in terms of the intentionality of human consciousness. In order for me to say that the objects or states of affairs toward which I am directed do exist I need to have *evidence* for their existence. *Knowledge* of objects or of states of affairs requires evidence. In the language of intentionality, we can say that the meaning-intention by virtue of which we are directed toward objects or states of affairs should be *fulfilled* or *fulfillable*. Fulfillment can in principle admit of types and degrees but I will not go into these details at the moment. I will say that the ('dynamic') fulfillment of an intention consists of a sequence of mental acts in (internal) time in which we come to see that something is true. 'Internal time' refers to the temporality of cognitive processes, as distinct from the (external) temporality of objects in space and time. We either come to see, experience, or grasp an *object* or we come to see, experience, or grasp a *state of affairs*. The fulfillment of an intention is 'static' when the object is immediately present to consciousness. In this case, no sequence of acts in time is required in order to experience the object or state of affairs.

Thus, we distinguish empty intentions from fulfilled or fulfillable intentions. We can speak of meaning-intention and meaning-fulfillment. We proceed from mere belief or thought to knowledge when our meaning-intentions are fulfilled or fulfillable. If, for example, mathematician M thinks that $7 + 5 = 12$ then we can say that M knows that $7 + 5 = 12$ if the meaning-intention expressed by "$7 + 5 = 12$" is fulfilled for M.

We now define *intuition* in terms of fulfillment of intentions (see also Tieszen, 1989, 2005). An *intuition* just consists of a sequence of *mental* acts in *(internal) time* in which we come to see something or to see that something is true. It has to be a sequence of *mental* acts or it is not an intuition. Something that was formerly only intended but not seen or experienced, not known, is now seen or known. This much will be true of intuition in the case of either ordinary perception or mathematical intuition. On the view I am describing, there will in fact be various analogies between ordinary perception (perceptual intuition) and mathematical intuition due to similarities of the structure of intention and fulfillment across ordinary perception and mathematical thinking. I will say more about some of these analogies as we proceed.

3 A General Conception of Proofs as Fulfillments of Mathematical Meaning-Intentions

Proceeding to the concept of mathematical proof, we can say that a *proof* is a fulfillment of a mathematical intention. A proof, that is, is a sequence of mental acts in (internal) time in which we come to see that something in mathematics is true. It has to be a sequence of *mental* acts or it is not a proof. Something that was formerly only intended but not seen or experienced, not known, is now seen or known when we have a proof. Proofs provide evidence. They take us from mere thought or belief to knowledge. Thus, on the view I am outlining, a proof just is an intuition. Since proofs are proofs of sentences (or propositions), we might say that the kind of intuition embodied in proof is *intuition that*. The idea of *intuition of*, where this is understood as intuition of objects that are taken to have certain properties or to stand in certain relations, will be mentioned in §11 below.

Several additional points about this general conception of proof can be made immediately:

(i) A proof is a fulfillment of a mathematical intention, but it is not necessarily the case that every fulfillment of a mathematical intention is a proof. One might argue, for example, that solving an equation is an example of fulfilling a mathematical intention but that it is not a proof. In this case, we would be construing the notion of proof as narrower than the notion of fulfillment of a mathematical intention. A proof is a type of fulfillment.

(ii) The concept of proof, i.e., intuition or fulfillment, is subject to a process/product distinction. Sometimes we use it to refer to a process, and sometimes to the product of a process. The context usually allows us to determine how it is being used.

(iii) An analogy: In the case of either perceptual or mathematical intuition an intuition is a filling of an empty meaning-intention. This means that the state of affairs toward which we are directed by the meaning-intention is now better understood. We have evidence for it. We have a better grasp of it. We are, in this sense, filling in indeterminacies involved in the mere meaning-intention. In the case of mathematics, this means we are filling in indeterminacies in the proposition to be proved, thus making it more determinate, clearer, exact, convincing, believable.

(iv) Another analogy: When we are trying to acquire knowledge either in ordinary perception or in mathematics there is goal-directedness. We are trying to see whether we can reach, realize, or find the state of affairs intended in our act. In the case of mathematics, the proposition to be proved (the empty intention to be fulfilled) is, as it were, a goal. It is a problem to be solved. Goal-directedness is always involved in conscious knowledge-seeking. Machines, by way of contrast, do not have goals, except insofar as they are designed by us to fulfill certain goals we have. There is no conscious knowledge-seeking in their case. Machines do not have intrinsic intentionality. I will return to this issue below in the section on purely formal proofs.

Some of the central issues in the foundations of mathematics are concerned with the differences between constructive and classical non-constructive proofs. In terms of the conception of proof I am discussing in this paper, we might ask whether everything that is called a 'proof' of a proposition in mathematical practice should count as a type of intuition or fulfillment of a mathematical meaning-intention. I already mentioned above that there could be degrees and types of evidence provided by intuition (= proof), but it should also be noted that there is a historical precedent for identifying only constructive proofs with fulfillments of mathematical meaning-intentions. The identification was first made in the work of Oskar Becker and Arend Heyting in the late nineteen twenties and early nineteen thirties in the context of intuitionism. The idea was that proofs in the sense of intuitionistic constructions are just forms of intuition or fulfillment of mathematical intentions. Hence, both the notion of proof and the notion of intuition in this context are intuitionistic. In the middle of the foundational debate in the nineteen twenties Becker (1927) and Weyl (1921) (see also Tieszen (2000b), and Mancosu and Ryckman (2002)) were tempted by Brouwer's position that not everything that is called a 'proof' in mathematics really is a proof. Objections were of course made to non-constructive 'proofs' on a number of grounds.

In §§4–10 below I will continue to discuss some features associated with the general conception of proofs as fufillments of mathematical meaning-intentions. In

§11 I will come back to the matter of constructive proof. On the way to §11 I will argue that there are meaning-intentions in mathematics that are provably not fulfillable (§6). Between this kind of case, where there can be no intuition of the states of affairs the intentions are putatively about, and the purely constructive understanding of fulfillability or intuitability, discussed in §11, lies much that is of interest concerning the foundations of classical non-constructive mathematics. I do not go very far into this area in this paper, apart from making some fairly innocuous observations in §§4–10 about the general conception of proofs as fulfillments of mathematical meaning-intentions. I am, however, very interested in the question whether we can generalize, or abstract, or idealize from a constructive basis in such a way as to preserve something of the idea that proofs in other parts of mathematics are fulfillments of mathematical intentions.

4 Proofs and Purely Formal Proofs

I would like to say something right away about purely formal proofs. Should we count purely formal proofs as a type of intuition in our sense, i.e., as fulfillments of meaning-intentions? By 'formal' one might mean that one uses a language in mathematics that differs from ordinary language in picking out the forms or structures of phenomena, so that one works with formulas in mathematical thinking and reasoning instead of working with ordinary non-mathematical language. It has been noted on many occasions that mathematics as we know it would simply not be possible without the creation of distinctively symbolic or formal notation, that there are better and worse forms of such notation, and so on. In mathematical practice, a meaning is often attached to such symbols or formulas as they are introduced, or at least they are taken to have a meaning, even if it is only indicated quite informally and roughly. In this sense of the word 'formal', formal proofs are just a standard part of mathematical practice and no special problems are posed by this feature alone to the effect that formal proofs could not count as intuitive fulfillments of mathematical intentions.

By way of contrast, 'formal' might refer to the kind of formalism that has been associated with Hilbert's program and that is nowadays also part of the precise mathematical characterization of computation. This is the sense of 'formal' in which one abstracts completely from the meaning or reference of the symbols and from the origins of the symbols and mechanically manipulates the resulting meaningless concrete sign configurations in accordance with finite sets of rules solely on the basis of their external (outer) form as it is given to us in sense experience. Hilbert's conception of intuition in the foundations of mathematics is that we have immediate sensory intuition of concrete, finite sign configurations as tokens (see, e.g., Hilbert, 1926).

If we are really serious about the idea that formal or mechanical proofs involve only the mechanical manipulation of meaningless syntax then what is the sentence (string of signs) *about* that is to be proved? If mathematics is just syntax of language then, as Carnap was very concerned to point out in the *Logical Syntax of Language* (Carnap, 1934, §§1–2), it is not about anything. To put the point into clearer relief, if I have a rule for manipulating strings of signs that tells me that I can 'derive' the string *xyxyxyz* from the strings *xy* and *yz*, then there is no doubt that I can apply such rules, or program a computer to apply such rules, without knowing what *xyxyxyz* is about (if anything). We cannot, as I said, consider such a 'proof' to be the fulfillment of a meaning-intention expressed by *xyxyxyz* because no meaning-intention is expressed by *xyxyxyz*.

In a similar vein, consider a proof of a sentence in a purely formalized system such as Peano arithmetic. By telling you that it is Peano arithmetic I am already indicating an interpretation, but please just focus on the syntax manipulation. Keep in mind the difference between Peano arithmetic and number theory as it is practiced by working mathematicians. Should the purely formal 'proof' in Peano arithmetic count as an intuition in the sense that it is a fulfillment of a mathematical meaning-intention? If there is no meaning-intention associated with the sentence to be 'proved' in the first place then there is certainly no possibility of meaning-fulfillment. It would follow that the 'proof' in this case should not count as a mathematical intuition.

One might argue that the uninterpreted signs do take on a kind of meaning in this purely formal context: their 'meaning' is determined solely on the basis of the rules of the game, so to speak, just as the meaning of chess pieces is determined by the rules of that game. Thus, we should distinguish what we might call the 'games meaning' from the original meaning-intention (if there was one). The 'games meaning' changes if the rules of the game change. In particular, purely formal proofs are always *relative* to a given set of rules, to a given formal system, but the concept of proof as fulfillment of mathematical meaning-intentions is not one that is always relative to some formal system or other. There is a widespread feeling that 'games' are detached from reality, and this is reflected in the easy-going relativism associated with some types of formalism.

If we pay attention to the intentionality of consciousness then we see that in the case of the move to purely formal proofs there is a shift in the directedness of consciousness. When operating in a purely formal system in this sense we are directed toward concrete, finite sign configurations and their properties and relations. This is directedness toward concrete syntax, which is present to us in sense perception. As noted above, this is Hilbert's conception of intuition. In the language I have used above, we can be directed toward the outer form of signs and such objects can also be immediately present to consciousness, i.e., we can intuit them.

Can we say that there is a sense in which our understanding of the sign string *xyxyxyz* is filled out by the 'proof' of it here? Even if we do not know what any of it is about or what any of it means, what we come to understand more fully is the system of sign configurations itself, i.e., what kinds of sign configurations will be involved in obtaining other sign configurations on the basis of *these* rules. However, this is the *only* thing we come to understand more fully. One can become more or less skilled at this kind of thing. There is, for humans, still a kind of goal-directedness in this purely formal context. The 'proofs' in this case, however, do not involve understanding or knowledge of the objects or states of affairs the signs were about prior to abstracting away from the meaning of the signs. Think of all the deep and interesting results and open problems in number theory, real analysis, set theory, group theory, and other areas of mathematics. What do we come to learn about natural numbers, real numbers, functions, sets, groups, and so on, on the basis of manipulating meaningless symbols in precise formal systems? Nothing, or very little, unless the formal systems are understood in relation to intended interpretations.

In making these remarks I do not mean to disparage the methods and results of proof theory. There is a lot of very interesting and important work in proof theory. In addition to learning many things through the analysis of the ordinal strength of formal theories and other techniques, we also obtain a kind of exactness and clarity with proof- theoretic results. Formal systems can help to simplify work with abstract concepts and ease reasoning about such concepts in some ways. They can certainly aid in communication and, hence, in obtaining objectivity in mathematics. It is just that pure formalism or mechanism about proof cannot by itself give us the whole picture, or a completely accurate picture, of how proof works in human understanding or human consciousness.

Brouwer's worry about formalism is closely related to the point here. In the terminology I have used, Brouwer is worried about the shift away from the directedness involved in ordinary mathematical thinking, in which we are directed toward mathematical objects such as numbers, functions, sets, and so on, to directedness toward *language* in which we are no longer thinking about mathematical objects properly speaking but instead about concrete signs in formal languages, along with their properties. For Brouwer this is already several steps removed from mathematics and thus from mathematical intuition in its original sense (Brouwer, 1907, pp. 94–95).

Gödel makes some comments in a similar vein about purely formal proofs in an interesting paper from the nineteen thirties in which he explores some of the philosophical consequences of his incompleteness theorems (Gödel 193?). Gödel says that Hilbert's belief in the decidability of every clearly posed mathematical problem is not shaken by the proof of the incompleteness theorems. The incompleteness theorems only show that something was lost in translating the concept of proof as "that which provides evidence" into a purely formalistic (and hence, relative) concept. Gödel concludes that it is not possible to formalize mathematical *evidence* even in the domain of number theory. The incompleteness theorems show that Hilbert's concept of intuition, in which intuition is of concrete finite sign configurations, will not suffice for the foundations of mathematics. On the view of proof I am putting forward, by way of contrast, mathematical evidence is provided when our mathematical meaning-intentions are fulfilled or are at least partially fulfillable. Gödel says that the problem of finding a mechanical procedure for deciding every sentence of a class for certain classes of mathematical sentences, however, *is* absolutely unsolvable. Another way to put this, according to Gödel, is to say that it is not possible to mechanize mathematical reasoning completely.

In discussions with Hao Wang (1996, p. 169) about Husserl's philosophy Gödel said that

> One fundamental discovery of introspection marks the true beginning of psychology. This discovery is that the basic form of consciousness distinguishes between an intentional object and our being pointed (directed) toward it in some way (willing, feeling, cognizing). There are various kinds of intentional object. There is nothing analogous in physics. This discovery marks the first division of phenomena between the psychological and the physical.

Gödel (1953/59, p. 342, footnote 20) says that the concept of proof, in its original contentual meaning, is an 'abstract concept'. This is the concept according to which a proof is not "a sequence of expressions satisfying certain formal conditions, but a sequence of thoughts convincing a sound mind." Similar comments about proof are made in various places in Gödel's writings. If we combine these remarks of Gödel on intentionality, proof and evidence with the Husserlian view that it is intuition that provides evidence, then proofs in the 'abstract' sense can be seen as the fulfillments of mathematical intentions. In other words, proofs are just expressions of mathematical intuitions. Mathematical intentions without proofs are empty, but 'proofs' without mathematical intentions, such as purely formal proofs, are blind. Open problems in mathematics can, in accordance with what we have said, be viewed as expressions of intentions that can be either fulfilled or frustrated. We can see a sentence such as $2^{\aleph_0} = \aleph_1$ as an expression of an intention that we expect to be either fulfilled or frustrated instead of viewing it as no more than an uninterpreted string of signs such that either $2^{\aleph_0} = \aleph_1$ or $\neg\, 2^{\aleph_0} = \aleph_1$ can be derived from other uninterpreted sign configurations in a mechanical way on the basis of uninterpreted sets of 'rules'. Gödel of course argues that this is a meaningful, clearly stated mathematical problem.

5 Proofs, Practice, and Axioms

In practice mathematicians do not always work in strict formal systems, or at least they do not work exclusively in strict formal systems. I would like to take existing mathematical practice seriously. Mathematicians also do not always work in axiom systems, or at least they do not work exclusively in axiom systems. Axioms and axiom systems, like formalizations, often emerge out of and against the background of mathematical practice.

In mathematical practice it seems that one can very well have a sense of what sentences in many areas of mathematical practice are *about*, even when the mathematics is not constructive. That is, intentionality (or meaning-intention) in mathematical thinking can be present even if constructivity is absent. Let us agree with Carnap that purely formal proofs are not about anything at all. We have also mentioned intuitionistic mathematics, which must be based entirely on building up constructions. We have, however, many mathematical theories and parts of mathematical practice that are not intuitionistic and are not purely formalistic. Mathematical practice in different regions or domains of mathematics has a *history* and we can often see how the sentences in these regions have their origins in this history. The meaning-intentions associated with mathematical sentences are acquired through this history, through the sedimentation of concepts and results, false starts and readjustments, applications, connections with other theories, and so on. We do not just arrive at them arbitrarily. The sentences in such cases are not just fabricated from nothing. They are motivated in certain ways. They are formulated against a historical background of intentional acts and processes in the ongoing development of mathematics. Underlying them is a sedimentation of sense (meaning), a 'meaning history', and it is on the basis of this meaning sedimentation that we at least have a sense of what they are *about*.

The history of which I am speaking here, by the way, does not have to be viewed as merely contingent or empirical history. We need not adopt a historicism about it, as though each era has its own mathematics that is relative to and accessible only to that era. On the contrary, in mathematics there is a remarkable invariance of meaning and of results across historical periods, places, times, and persons. We find a kind of universality or invariance that is distinct from the changing empirical particularities of places, times and persons.

The *proofs* we find in mathematical practice also seem to have some general characteristics of fulfillments of meaning-intentions that we noted above: empty meaning-intention, filling when a proof is provided (so that there is greater understanding, greater clarity and determinacy, etc.), goal-directedness and realization, and so on. The analogies we noted continue to hold to some extent.

Now if we turn to axiomatic presentations of mathematics it appears that the intuitive evidence provided by a proof can only be as good as the intuitive evidence associated with any axioms or assumptions on which the proof is based. In particular, if there are axioms or assumptions that are not themselves intuitively evident then what should we say about sentences proved on the basis of such axioms or assumptions? Should the proof in this case count as a mathematical intuition in the sense that it is a fulfillment of a meaning-intention? In intuitionism this should presumably not create a major problem, or at least not the kind of problem we would have in non-constructive mathematics, because no axioms should be allowed that are not founded on intuitionistic constructions in the first place. The axioms would have their origins in constructions, i.e., certain structures and intuitive processes of the mind. Brouwer was not especially interested in axiomatizing intuitionistic mathematics but for anyone who is interested in axiomatization in this setting it would be necessary to choose as axioms only those sentences that express meaning-intentions for which meaning-fulfillments are fulfillments for finite, limited beings who must carry out in time (and possibly space) the cognitive processes in which they can make the individual objects of their intentions present.

The intuitionistic view of the origins of axioms and proofs is not as complicated as the case where we are interested in the more general features of proof as intuition. Accounting for the more general features of the idea of proof as intuition, in the sense of fulfillment of meaning-intentions, would require giving up some of the elements that are involved in standard conceptions of constructive or intuitionistic proof. A central element that would clearly be omitted is the demand that there should always be a method or procedure for finding the objects one is thinking about.

There would, in effect, be an *abstraction* from this condition or an *idealization*. The non-constructive or classical mathematician does not ask for methods that will make mathematical objects such as transfinite sets completely available to consciousness as individuals. The domains of quantification do not need to be constructible.

In considering general features of proof as fulfillment of intention that go beyond the constructivists' conception of mathematics there are some well-known issues that would need to be considered. These include issues about the use of the principle of the excluded middle in proofs, or the use of double negation elimination in indirect 'existence' proofs; issues about proofs or propositions involving impredicativity; issues about proofs involving the potential or actual infinite (transfinite); and issues about proofs involving the axiom of choice. I think it is of great interest to analyze these issues in connection with the idea of proofs as fulfillments of mathematical meaning-intentions but, as I indicated above, I will not go into these matters here. What I will do, however, is to consider the limit case in which there definitely (provably) cannot be fulfillments of mathematical meaning-intentions, whether the concepts of fulfillment and intuition are construed constructively or not.

6 Frustrated Meaning-Intentions

Let us now consider the case where there can be no intuition that the proposition is true, much less intuition of the intended object(s) the proposition is about. The meaning-intention expressed by the naïve comprehension axiom in set theory, $(\exists x)(\forall y)(y \in x \leftrightarrow Py)$, can be viewed as an example of such a case. It gives the impression, one might say, of being true, or at least it gave some mathematicians this impression. They might even have thought that, in some sense, it was intuitively supported.

I think that on the usual interpretation of this formula (which takes the domain of quantification to be unrestricted) we must count the sentence as an expression of a meaning-intention. It is not a meaningless sentence. It is not just gibberish, and it is not, except perhaps subject to subsequent stipulation (convention), ungrammatical. If it were gibberish or ungrammatical then no one would have been able to derive a contradiction from it in the first place. One cannot derive contradictions from nonsensical or ungrammatical sentences. Generally, one cannot have nonsensical or ungrammatical sentences in proofs, i.e., in expressions of fulfillments. There cannot, however, be an intuition that $(\exists x)(\forall y)(y \in x \leftrightarrow Py)$, in the sense of the fulfillment of an intention. Not only can there not be a proof of this sentence but it is an intention that leads to a contradiction. I will call such intentions *frustrated intentions*. (We have these in mathematics and logic and, analogously, in everyday life.) Thus, we can speak of fulfilled intentions but also of frustrated intentions. Frustration, in effect, negation, is itself based in intuition. A consequence of the general view of proof as intuition is not only that there can be no intuition that $(\exists x)(\forall y)(y \in x \leftrightarrow Py)$ but, in fact, there is an intuition that $\neg(\exists x)(\forall y)(y \in x \leftrightarrow Py)$, for it is provable that $(\exists x)(\forall y)(y \in x \leftrightarrow Py)$ leads to a contradiction: 1. Assume that $(\exists x)(\forall y)(y \in x$

$\leftrightarrow$ Py). 2. $(\exists x)(\forall y)(y \in x \leftrightarrow y \notin y)$, from 1 by substitution. 3. $(\forall y)(y \in r \leftrightarrow y \notin y)$, from 2 by substitution. 4. $r \in r \leftrightarrow r \notin r$, from 3 by substitution.

Note that I am saying that an expression such as $(\exists x)(\forall y)(y \in x \leftrightarrow Py)$ can very well have a meaning apart from intuition (fulfillment). In fact, it must if we are to see that it leads to a contradiction and that we must therefore somehow qualify it. It can have a meaning apart from proof but also apart from disproof. It is only subsequent to the discovery that it leads to a contradiction that the motivation might set in to regard such an expression as ungrammatical, as one would, for example, in a theory of types. We might then set up a language in which such a formula cannot be expressed in order to avoid the contradiction, even if the language introduces other complications or infelicities relative to mathematical practice.

Of course we do not always have such definitive cases of frustration. Sometimes, however, we do. That is, we have *provably* empty intentions. In such cases we can say that the meaning-intention is empty *a priori* (or permanently).

Before the contradiction was discovered one could have worked with this meaning-intention, had one's thinking directed by it, used it, and so on, albeit with perhaps some concern about whether it was really evident or not. It could have determined a person's thinking and behavior whether there were or were not suspicions about the evidence for it. If it was ever taken to have been supported by 'intuition' in some sense or other then such an 'intuition' is subsequently corrected later by intuition. (I will argue in §10 that, on the conception of intuition I am describing, intuitions in mathematics can be corrected by subsequent intuitions in mathematics.) Hence, I would like to say that there can be directedness by way of such a meaning-intention but that there cannot be a reliable intuition that it is true. On the contrary. This does not mean, however, that the mere directedness necessarily disappears. Someone who does not know about the contradiction will probably find it easy to use the naïve comprehension principle in reasoning about sets.

7 Multiple Proofs for the Same Meaning-Intention

Let us consider a few more features of the general concept of proofs as fulfillments of mathematical meaning-intentions. When there is more than one proof of a proposition then we should say that there are multiple but different intuitions or fulfillments of the same meaning-intention. There are different ways to fulfill one and the same meaning-intention. This also has an analogy in ordinary perception. If I believe or think that my bicycle is still parked at the library across campus then there is more than one way in which I can try to fulfill this intention. In the case of mathematics, I might devise different proofs or it might happen that different people provide different proofs for the same proposition. What we have here are different intuitions that lead to the same result. In building up different intuitions of the same result it appears that we generally increase our evidence that the result is true. Different proofs lead to it, or different mathematicians from different places, times, cultures, and so on, are all converging on the same result. This is a sign of the objectivity of the result, a sign that the intuitions are intuitions of something objective. Indeed, anything that in the end should count as a proof (intuition) should possess this intersubjective dimension. We have group intentionality in mathematics and, against this background, the possibility of fulfillment of mathematical intentions for mathematicians in the group.

Different proofs of the same proposition can of course be simpler or more complex, use more or less familiar concepts and previous results, can be more or less perspicuous, more or less elegant, and so on. All of these things should be true of our intuitions. There has been a lot of oversimplification in the literature on mathematical intuition.

8 Proofs That Exceed Meaning-Intention, and Mismatches Between Proofs and Meaning-Intentions

It sometimes happens that proofs provide more information than is needed to obtain the proposition under consideration. Our intuitions exceed the meaning-intention in a given case, i.e., the fulfillment is richer than the meaning-intention at issue. This means that the mathematical experience is fuller than the meaning-intention that is fulfilled by the experience. We could draw a stronger conclusion from the intuition. This also has an analogy with some meaning-intentions involved in ordinary perceptual experience. We do not draw all the conclusions we could draw from our experience. The experience is fuller or richer than the meaning-intention we take it to fulfill.

Sometimes the proof (intuition) actually supports a different proposition (intention) from the one that we take to follow. In this case adjustments have to be made in order for the meaning-intention to fit the fulfillment, such as adding to or subtracting from either the proof or the proposition to be proved. This phenomenon is perhaps less likely to occur in more polished parts of mathematical presentation than it is in the rougher work that eventually leads to such presentation.

9 Internal and External Proofs for Meaning-Intentions

Another interesting situation arises when there are different fulfillments of a meaning-intention but at least one of these proofs appears to use concepts or results that lie outside the 'scope' of the expressed meaning-intention. Consider, for example, a case in which we seem to be directed by an intention only to natural numbers and properties of or relations between natural numbers. We might have one proof of a sentence expressing such a meaning-intention that refers only to natural numbers and their properties and relations and another proof of the sentence that refers at certain points to concepts or results from topology. I will refer to the former kind of proof as an *internal* proof and to the latter kind as an *external* proof. An external proof is 'external' only in the sense that some of the contents of the proof lie outside the given contents of the problem, not in the sense that all of them do. Distinctions like this one have been drawn elsewhere in the literature on the foundations of mathematics (see, e.g., the division between pure and impure proofs in the unpublished work of Andrew Arana). On the basis of the account I am describing in this paper, the distinction concerns the relation of meaning-intentions to meaning-fulfillments in mathematics. We can speak of intuitions that are internal to meaning- intentions and intuitions that are external to meaning-intentions.

The first thing I want to note about this distinction is that it reflects the sense we have in mathematical practice that there are different areas or 'regions' of mathematics, that mathematical sentences are not, as has been held by some philosophers (such as logicists or even some formalists), topic-neutral. What lies behind this sense of different regions of mathematics is just the kind of meaning theory we have been discussing. Mathematical sentences have meanings, and we should pay attention to the different meaning-intentions expressed by these sentences, not ignore them or attempt to impose some kind of grand eliminative reductionist scheme on mathematical practice that levels or eliminates differences in meaning-intention.

We can think of each meaning-intention as having a 'horizon'. Corresponding to our distinction between internal and external proofs is the distinction between the internal and external horizon of a meaning-intention. Our thinking is always directed by meaning-intentions and what is in the internal horizon of a meaning-intention is the range of possible experiences we could have that would be determined by and would be consistent with the given meaning-intention. This is closely related to the idea that intensions determine extensions and that intensions determine particular ranges of possibilities. An internal proof, i.e., an internal fulfillment of a meaning-intention, would be a fulfillment that does not appeal to anything outside of the internal horizon of the given meaning-intention in supplying evidence for the state of affairs the intention is about. An external fulfillment, on the other hand, uses concepts, results, or methods that are not in the internal horizon of the given meaning-intention in order to supply evidence for the state of affairs the intention is about. It brings the problem or intention into relation with other areas of mathematics that are not referred to explicitly or directly by the meaning- intention. There are clear examples of both kinds of proofs in mathematical practice. Andrew Wiles' proof of Fermat's conjecture, for example, would count as an external proof. One might try to imagine, by way of contrast, what it would be like to have an internal proof of this proposition.

In the theory of founded levels of intentionality of the type that we find in mathematics there are interesting questions about the relationship of internal to external horizons of meaning-intentions, and the relationship of internal to external fulfillments of meaning-intentions. At the moment, I will simply note that as a matter of fact there is in mathematical practice a cross-fertilization of intuitions from different areas of mathematics. There is holism or contextualism of a sort. Why shouldn't intuitions from one area of mathematics fertilize another area of mathematics?

One might wonder whether there is some reason why we should value one of these kinds of proofs (internal or external) more for some reason. This is, I think, an interesting question but I will not discuss it here.

Proofs in axiomatic systems are automatically internal to the expressed meaning-intentions if no concepts, methods or results outside of what is specified in advance in the primitive terms, definitions, axioms, and rules of inference of the axiom system are used in the proofs of theorems. There cannot be the kind of crossover between fields that one finds in mathematical practice where the axiomatic method is not in play or is not used directly. Axiomatization is distinct from formalization. There are or have been axiom systems that are not fully formalized and others that are formalized in a strict sense. Throughout most of its history Euclidean geometry, for example, was not what we nowadays think of as formalized axiomatic system. In axiomatic systems there is, in effect, great regimentation, ordering, and organization of intuitions. Later fulfillments of intentions build on earlier fulfillments of intentions in a systematic, organized way. Earlier intuitions are used in the later intuitions. This kind of very rigorous build-up and ordering of fulfillments of intentions is not present in unaxiomatized mathematical practice.

10 Mistaken Proofs (Intuitions)

There is a long history of mistaken proofs in mathematics. This can be described, in terms of our account, as a history of mistaken intuitions or fulfillments.

There is a possibility that a proof will need to be corrected (or refined, improved) in light of subsequent mathematical experience or evidence, just as we correct sensory perceptions in light of subsequent sensory perceptions. We can only correct intuitions on the basis of subsequent intuitions, and this will be true in mathematics and in ordinary sense experience. Just as there is a history of mistaken proofs and mistaken axioms in mathematics, so there is a history of mistaken human sense perceptions. Intuition, experience, is foundational in this sense, but it is not foundational in the sense that it is infallible. Notice, however, the qualification: we said that we cannot rule out the possibility that a mathematical proof will need to be corrected in light of subsequent *mathematical* experience or evidence. We did not say that a mathematical proof might need to be corrected in light of subsequent *sensory* experience. We do not identify mathematical experience with sensory experience or hold that only the latter exists. To take the intentionality of consciousness seriously is to recognize different types of experience, types of experience and evidence correlated with types of directedness. Experience is not confined only to sensory experience, as empiricism and various forms of reductionism would have us believe. A broadening of the notion of experience is required if we are to do justice to the intentionality of consciousness.

When we have a proof of a proposition in mathematics, as opposed to verifying a hypothesis in natural science or even in everyday sensory experience, it has often been held by philosophers that we have established a result that is a priori, at least in the sense that no sensory experience could overturn our result. Could further mathematical experience overturn our result? Yes, but it is not sensory experience. If this is correct then we can keep the idea that mathematical proofs are a priori in the sense that they are not revisable on the basis of sense experience, but we can allow that they are in principle revisable on the basis of mathematical experience, of distinctively mathematical evidence.

11 Constructive Proof

As mentioned earlier, the idea of identifying proofs with fulfillments of mathematical meaning-intentions was first put forward in the work of Oskar Becker and Arend Heyting in the late nineteen twenties and early nineteen thirties in the context of the debate in the foundations of mathematics at that time. The debate itself had its background in the discovery of paradoxes in set theory, such as the contradiction generated by the naïve comprehension principle discussed in §6 above. One can think of Brouwer's response to the crisis in foundations as attempting to insure security by limiting mathematics to what could be founded on not just any old conception of intuition (or fulfillment) but on a reliable conception of intuition. (Hilbert was also doing this, although his 'linguistic' conception of reliable intuition was of course different from Brouwer's and was criticized by Brouwer.) For Brouwer, a reliable conception of intuition would be one that recognized, among other things, that the human subject has a finite and limited form of consciousness. Intuitions are mental processes carried out through internal time in which the mathematical (not 'linguistic') objects one thinks about must be made present to consciousness. There must be a process in which the subject can engage in order to find the mathematical objects she thinks about if mathematics is to be reliable. Human subjects are incapable of generating and hence of grasping completed infinite objects or structures. Such objects or structures would lie outside of or would transcend all possible experience for human subjects (see, e.g., Brouwer, 1912, 1952). All we can do is to carry out some steps in a process and then, at best, defer in our reasoning to a rule or determination that specifies what is in the horizon of possible experience associated with our meaning-intention. Intuition cannot take us outside of possible experience. Brouwer's remedy was to hold that there are no non-experienceable objects or truths in mathematics. For Brouwer this also meant that there could be no non-mental objects or truths in mathematics (see also Tieszen, forthcoming).

It was in this context that Oskar Becker, who was an assistant to Edmund Husserl in the nineteen twenties, first identified proofs with fulfilled or fulfillable mathematical meaning-intentions, and he did so on the basis of ideas in the work of both Husserl and Brouwer (see Tieszen, 1992, 2000a). Becker corresponded with Heyting about his views (see van Atten, 2005), and the idea of proof as fulfillment of intentions was then also used by Heyting to interpret the intuitionistic logical constants. (For relevant views of Brouwer on logic see Brouwer (1951a, b, 1955.) In this role it played an important part in the formulation of what is now called the BHK interpretation of the intuitionistic logical constants. Another variant of this interpretation was formulated by Kolmogorov (1932) at around the same time, in terms of problems (intentions) and solutions (fulfillments). ("BHK" stands for "Brouwer-Heyting-Kolmogorov".) In recent times, Per Martin-Löf (1984, 1987) has again used the language of intention and fulfillment in some presentations of his work on intuitionistic type theory. The Curry-Howard propositions-as-types idea can also be described in this language.

Let us consider briefly how the BHK interpretation of the logical constants is formulated if we use the language of intention and fulfillment:

(i) Fulfillment of the intention expressed by $A \wedge B$ consists of the fulfillment of A and the fulfillment of B
(ii) Fulfillment of the intention expressed by $A \vee B$ consists of the fulfillment of A or the fulfillment of B
(iii) Fulfillment of the intention expressed by $A \rightarrow B$ consists of a cognitive process (construction) that transforms any fulfillment of A into a fulfillment of B
(iv) Fulfillment of the intention expressed by $\neg A$ consists of the frustration of the intention expressed by A; that is, of the fulfillment of $A \rightarrow \perp$, where $\perp$ is a contradiction
(v) Fulfillment of the intention expressed by $(\exists x)Ax$ consists of the fulfillment of an intention directed toward an object c in a domain D and fulfillment of Ac
(vi) Fulfillment of the intention expressed by $(\forall x)Ax$ consists of a cognitive process (construction) that assigns to every element c of a domain D a fulfillment of Ac.

Given the manner in which we characterized fulfillment and intuition, we can rewrite this in terms of *intuition*:

(i) Intuition that $A \wedge B$ consists of intuition that A and intuition that B
(ii) Intuition that $A \vee B$ consists of intuition that A or intuition that B
(iii) Intuition that $A \rightarrow B$ consists of a cognitive process (construction) that transforms any intuition that A into intuition that B
(iv) Intuition that $\neg A$ consists of intuition that $A \rightarrow \perp$, where $\perp$ is a contradiction
(v) Intuition that $(\exists x)Ax$ consists of intuition of an object c in a domain D and intuition that Ac
(vi) Intuition that $(\forall x)Ax$ consists of a cognitive process (construction) that assigns to every element c of a domain D an intuition that Ac.

The BHK interpretation, it has been noted, does not by itself force an intuitionistic interpretation of logic (see Troelstra and van Dalen, 1988, Volume 1, pp. 9, 33). It is the way that one understands the notion of construction, fulfillment, or intuition that is crucial.

The fulfillment-intention scheme, as I have suggested in earlier sections of this paper, might have some application outside of a purely constructive context, although it will then involve certain idealizations, abstractions, and other modifications. This would not have to mean that nothing at all is left of the notion of fulfillment of meaning-intention or of intuition.

The intuitionistic conception of proof, as opposed to other constructivist and nonconstructivist conceptions, however, has been especially relevant to the philosophy of mind and the study of mathematical consciousness because, historically speaking, it is in intuitionism more than anywhere else in the foundations of mathematics that features of human consciousness and human mental processes are not only highlighted and investigated but are also held to be essential. Issues about human consciousness, subjectivity, and intersubjectivity are of concern right from the start.

12 Conclusion

I think that enough has been said above to support the following conclusions. Intentionality in mathematics is ubiquitous. How could it not be, as long as mathematics involves meaning, thinking, believing, imagining, and knowing? Intuition in mathematics, whether one understands fulfillment in a purely constructive manner or not, is not ubiquitous. According to the view I have discussed here, however, intuition is also far from being non-existent in mathematics.[1]

References

Becker, O. (1927) Mathematische Existenz, *Jahrbuch für Philosophie und Phänomenologische Forschung* **8**, 439–809.

Benacerraf, P. and Putnam, H., eds. (1983) *Philosophy of Mathematics: Selected Readings*, 2nd ed., Cambridge: Cambridge University Press.

Brouwer, L.E.J. (1907) *Over de grondslagen der wiskunde*, Ph.D. thesis, Universiteit van Amsterdam, English translation On the Foundations of Mathematics, in Brouwer (1975), 13–101, citations are to the pagination in the English translation.

Brouwer, L.E.J. (1912) *Intuïtionisme en Formalisme*, Amsterdam: Clausen. English translation Intuitionism and Formalism, in Benacerraf and Putnam, eds. (1983), 77–89.

Brouwer, L.E.J. (1951a) Changes in the Relation Between Classical Logic and Mathematics, typescript of a lecture, published as Appendix 9 of van Stigt (1990), 453–458.

Brouwer, L.E.J. (1951b) Notes for a Lecture, typescript of a lecture, published as Appendix 8 of van Stigt (1990), 447–452.

Brouwer, L.E.J. (1952) Historical Background, Principles and Methods of Intuitionism, *South African Journal of Science* **49**, 139–146.

Brouwer, L.E.J. (1955) The Effect of Intuitionism on Classical Algebra of Logic, *Proceedings of the Royal Irish Academy* **57**, 113–116.

Brouwer, L.E.J. (1975) *Collected Works*, Vol. I. ed. by A. Heyting, Amsterdam: North Holland.

Carnap, R. (1934) *Logische Syntax der Sprache*, Berlin: Springer, English translation as *The Logical Syntax of Language*, New York: Harcourt Brace, 1937.

Feferman, S. et al., eds. (1995) *Kurt Gödel: Collected Works*, Vol. III. Oxford: Oxford University Press.

[1] Versions of this paper were presented at the Archives Poincaré/CNRS/Université Nancy 2, at REHSEIS/CNRS/Université Paris 7, in the UC Berkeley Logic Colloquium, and the CSLI Cognitive Science Lunch at Stanford University. I thank members of those audiences for comments, and especially Gerhard Heinzmann for making possible an extended visit in Nancy. Work on the paper was partially supported by a National Endowment for the Humanities (NEH) Fellowship, which I hereby gratefully acknowledge.

Gödel, K. (193?) Undecidable Diophantine Propositions, in Feferman, S. et al., eds., (1995), pp. 164–175.

Gödel, K. (1953/59) III and V. Is Mathematics Syntax of Language? in Feferman, S. et al., eds. (1995), pp. 334–363.

Heyting, A. (1931) Die intuitionistische Grundlegung der Mathematik, *Erkenntnis* **2**, 106–115, English translation in Benacerraf, P. and Putnam, H., eds. (1983), pp. 52–61.

Hilbert, D. (1926) Über das Unendliche, *Mathematische Annalen* **95**, 161–190, translated as On the Infinite in van Heijenoort, J., ed. (1967), pp. 367–392.

Husserl, E. (1900) *Logische Untersuchungen. Erster Theil: Prolegomena zur reinen Logik*, Halle: M. Niemeyer. 2nd edition 1913.

Husserl, E. (1901) *Logische Untersuchungen. Zweiter Theil: Untersuchungen zur Phänomenologie und Theorie der Erkenntnis*, Halle: M. Niemeyer. 2nd edition in two parts 1913/22.

Husserl, E. (1929) Formale und transzendentale Logik. Versuch einer Kritik der logischen Vernunft, *Jahrbuch für Philosophie und phänomenologische Forschung* **10**, v–viii and 1–298. Published simultaneously as Separatum.

Husserl, E. (1969) *Formal and Transcendental Logic*, translated by D. Cairns, The Hague: Nijhoff.

Husserl, E. (1973) *Logical Investigations*, Vols. I, II. Translation of the 2nd edition. London: Routledge and Kegan Paul.

Kaufmann, F. (1930) *Das Unendliche in der Mathematik und seine Ausschaltung*, Wien: Deuticke. Translated as *The Infinite in Mathematics*, Dordrecht: Reidel, 1978.

Kolmogorov, A. (1932) Zur Deutung der Intuitionistischen Logik, *Mathematische Zeitschrift* **35**, 58–65, English translation in Mancosu, P. ed. (1998), 328–334.

Mancosu, P., ed. (1998) *From Brouwer to Hilbert*, Oxford: Oxford University Press.

Mancosu, P. and Ryckman, T. (2002) Mathematics and Phenomenology: The Correspondence Between O. Becker and H. Weyl, *Philosophia Mathematica* **10**, 130–202.

Martin-Löf, P. (1984) *Intuitionistic Type Theory*, Napoli: Bibliopolis.

Martin-Lof, P. (1987) Truth of a Proposition, Evidence of a Judgment, Validity of a Proof, *Synthese* **73**, 407–420.

Tieszen, R. (1989) *Mathematical Intuition: Phenomenology and Mathematical Knowledge*, Dordrecht: Kluwer.

Tieszen, R. (1992) What is a Proof?, in Detlefsen, M., ed. *Proof, Logic and Formalization*, London: Routledge, 57–76, reprinted with revisions as Proofs and Fulfilled Mathematical Intentions in Tieszen (2005).

Tieszen, R. (2000a) Intuitionism, Meaning Theory and Cognition, *History and Philosophy of Logic* **21** (2000), 179–194, reprinted with minor revisions in Tieszen (2005).

Tieszen, R. (2000b) The Philosophical Background of Weyl's Mathematical Constructivism, *Philosophia Mathematica* **3**, 274–301, reprinted with minor revisions in Tieszen (2005).

Tieszen, R. (2005) *Phenomenology, Logic, and the Philosophy of Mathematics*, Cambridge: Cambridge University Press.

Tieszen, R. (2008) The Intersection of Intuitionism (Brouwer) and Phenomenology (Husserl), in van Atten, M. Boldini, P. Bourdeau, M. and Heinzmann, G., eds. *100 Years of Intuitionism*, Birkhäuser, pp. 78–95.

Troelstra, A., and van Dalen, D. (1988) *Constructivism in Mathematics*, Vols. I, II. Amsterdam: North-Holland.

van Atten, M. (2005) The Correspondence Between Oskar Becker and Arend Heyting, in Peckhaus, V., ed. *Oskar Becker und die Philosophie der Mathematik*, München: Wilhelm Fink, pp. 119–142.

van Heijenoort, J., ed. (1967) *From Frege to Gödel*. Cambridge, MA: Harvard University Press.

van Stigt, W. (1990) *Brouwer's Intuitionism*, Amsterdam: North-Holland.

Wang, H. (1996) *A Logical Journey: From Gödel to Philosophy*, Cambridge, MA: MIT Press.

Weyl, H. (1921) Über die neue Grundlagenkrise der Mathematik, *Mathematische Zeitschrift* **10**, 39–79, reprinted in Weyl (1968), II, 143–180, English translation in Mancosu (1998), 86–118.

Weyl, H. (1926) *Philosophie der Mathematik und Naturwissenschaft*, München: Leibniz Verlag, *Philosophy of Mathematics and Natural Science*, 1949, is an expanded English version.

Weyl, H. (1949) *Philosophy of Mathematics and Natural Science*, Princeton: Princeton University Press.

Foundations *for* Computable Topology

Paul Taylor

"In these days the angel of topology and the devil of abstract algebra fight for the soul of every individual discipline of mathematics." (Hermann Weyl.)

"Point-set topology is a disease from which the human race will soon recover." (Henri Poincaré)

"Logic is the hygiene that the mathematician practises to keep his ideas healthy and strong." (Hermann Weyl)

"Mathematics is the queen of the sciences and number theory is the queen of mathematics." (Carl Friedrich Gauss)

"Such is the advantage of a well constructed language that its simplified notation often becomes the source of profound theories." (Pierre-Simon Laplace)

Foundations should be designed for the needs of mathematics and not *vice versa*. We propose a technique for doing this using the correspondence between category theory and logic and is potentially applicable to several mathematical disciplines.

This method is applied to devising a new paradigm for general topology, called Abstract Stone Duality. We express the duality between algebra and geometry as an abstract monadic adjunction that we turn into a new type theory. To this we add an equation that is satisfied by the Sierpiński space, which plays a key role as the classifier for both open and closed subspaces.

In the resulting theory there is a duality between open and closed concepts. This captures many basic properties of compact and closed subspaces, even without adding an explicitly infinitary axiom. It offers dual results that link general topology to recursion theory.

The extensions and applications of ASD elsewhere that this paper surveys include a purely recursive theory of elementary real analysis in which, unlike in previous approaches, the real closed interval [0, 1] is compact.

G. Sommaruga (ed.), *Foundational Theories of Classical and Constructive Mathematics*, The Western Ontario Series in Philosophy of Science 76,
DOI 10.1007/978-94-007-0431-2_14,

1 Foundations *for* Mathematics

1.1. Mis-quoting both Gauss and Eric Bell, Logic is the Queen and Servant of Mathematics.

Ever since Logic became a mathematical discipline in the nineteenth century, foundational discussions (whether using sets, types or categories) have been based on four premises:

(a) logicians know best, and mathematicians should be grateful for what they are given;
(b) it is up to mathematicians to glue the continuum back together from the dust that logicians have provided; whilst
(c) they have convinced each other that foundations ought to allow them to make *arbitrary* abstractions of mathematical processes; and
(d) anyone who tinkers with the foundations risks bringing the whole edifice of science crashing down.

1.2. The first two of these attitudes flew in the face of the tradition of the preceding two millennia, in which, for example, Euclid proved lots of *rigorous theorems* about lines, circles and other figures *as things in themselves*. Frege and Hilbert believed the third, and were admonished for it by Russell and Gödel. We now have a compromise situation in which the admissible forms of abstraction are in equilibrium with the extent to which the typical mathematician understands them. The result of this is that we debate the "truth" of the axiom of choice, *etc*, without any point of reference in the real world against which to assess it.

So it is impossible to axiomatise *arbitrary* abstraction, and nor is the *status quo* clearly defined.

There is, however, a natural *baseline* for foundations, namely *computation* (also called *recursion*), all forms of which are essentially equivalent by the Church–Turing thesis. Mathematics in the time of Gauss and before was computational. As we shall see in this paper, this also agrees with many of the intuitions of modern general topology. However, we also need to consider logically stronger systems, in particular in order to study the *behaviour* of computation and topology, for example whether a program terminates or an open subset covers a compact one.

1.3. What do we mean by "foundations" in mathematics anyway?

The basics are *counting* and *measurement*. Over the millennia, mathematicians have found clever ways of deriving results about these things by inventing things whose originally fictitious nature is still remembered in their names: *irrational*, *imaginary* and *ideal* numbers.

For example, the front-line issue in mathematics during the sixteenth century was how to solve cubic equations. Scipione del Ferro and Niccolò Tartaglia had found a method for the case where there is only one (real) root. Rafaele Bombelli showed how to extend this to the case where there are three real roots, but his *intermediate* calculations involved square roots of -1, which had not been needed in the original case (Fauvel, 1987 §8.A).

The foundational question here is not so much how to "define" $\sqrt{-1}$ as to show that, if we introduce it in the middle of a calculation but then eliminate it, we will not obtain a contradiction. More precisely, whatever we do get by this method could have been found using real numbers alone. In logical jargon, complex numbers are a *conservative extension*.

So what is their value? Just try doing Fourier analysis without them, and see what a mess you get into!

An *axiom* was originally a statement that is obvious and needs no justification, but the meaning of this word has changed. It is still a *starting point*, but one that is carefully chosen to facilitate the development of a particular body of abstract theory. The purpose of *foundations* should be reinterpreted in the same way, on a larger scale.

1.4. As the quotation from Laplace says, language and notation can be powerful driving forces behind a mathematical theory, whilst a bad notation (such as using the letters CDILMVX to write numbers) obstructs even the simplest task. As the subject has got more powerful and sophisticated, so the design of language and notation has become a professional discipline in itself. This discipline is, or at least ought to be, called Logic. It has now enjoyed well over a century of its own mathematical development and applications.

However, this does not justify the prescriptive role that Logic*ists* and their successors have claimed. The "orthodox" foundations of mathematics were based on the ideas from quite a short part of the history of Logic, and even in this period were not the most important things what were going on in Mathematics. Moreover, the *conceptual* structure of the modern subject, in particular Emmy Noether's Modern Algebra, developed after the foundations that were allegedly designed for it. At the very least, these foundations should be reviewed in the light of other developments over the last century and the technological needs of the present one.

1.5. The philosophical thesis of this paper is that we can make Logic the *servant* of (a particular discipline in) Mathematics. Starting from what we see as the principal theorems of the Mathematics that we want to do, we apply certain results that link categorical with symbolic Logic in order to create a new language for our chosen Mathematical discipline. The formal method that we propose is set out in the next two sections.

In motivating this, we shall use the metaphor that a logician is like an *engineer* or an *architect* who has been commissioned to design a new gadget or building for a *client*, who in turn has *customers*. However, we stress that our proposal is not just philosophy, but exploits extensive technical development in a number of mathematical disciplines.

Being a mathematician, I only discuss this thesis very briefly and make my points perhaps more tersely than philosophers and historians would do. So I leave it them to explore them more fully. I feel compelled to say some of these things because I have found it impossible to do a substantial piece of research on the reformulation of general topology within the "orthodox" framework that my colleagues, referees, conferences, journals, universities and funding agencies say that I should use.

1.6. The instruments that we shall apply to designing a foundation *for* (some part of) mathematics are category theory and modern symbolic logic, by which we mean proof theory and type theory. Very many excellent Mathematicians "work in an elementary topos" or "in Martin-Löf type theory", but if we were to do this we would once again be bowing to (other) Logicians as our masters. So, in order to employ these disciplines as servants of mathematics, we make our own *selection* from the menu of techniques that they offer.

Category theory grew out of (algebraic) topology and algebraic geometry, with no foundational pretensions in the beginning (Mac Lane, 1988; McLarty, 1990). It is now used throughout mathematics, primarily because of the notion of *universal property* or *adjunction*. Mathematicians in other fields who have a feeling for the unity of the subject often express the headline theorems of their speciality in this form. Category theory seems to be very good at distilling decades' worth of experience in one discipline into adjunctions and a small number of other abstract but widely applicable concepts and transferring it to other subjects. These now include the theoretical parts of computer science and physics.

Thus the client's brief to the architect of the new foundations will be provided as universal properties.

On the other hand, symbols are the coinage of the everyday business of mathematics, so they have to be the way in which any new foundations will be *used*. However, any *pre-existing* symbolic language acquiesces to a lot of foundational prejudgements without giving them proper scrutiny. We aim to replace these old assumptions with our own principles. These are, in the first instance, formulated in terms of category theory, because it is *agnostic*.

We shall demonstrate the *inter-relationship* between categories and symbolic logic in the next section. It is this that makes them together a powerful method for devising new language and notation for mathematics.

1.7. We will apply the proposed methodology to the case for which the client is general topology, whose customers include geometric topologists, analysts and theoretical computer scientists. The extended version of this paper provides a survey of the whole of this research programme (ASD) as it currently stands.

The new theory axiomatises continuity *directly*, without any recourse whatever to set theory or its usual alternatives. It begins from ideas concerning the duality of algebra and topology that were introduced by Marshall Stone (§§4, 5 & 7). These are put in an abstract form (hence the name ASD) using a monadic adjunction (§6), along with an algebraic equation that characterises the way in which the Sierpiński space uniquely classifies open subspaces (§8). Surprisingly much of the basic theory of open, closed, compact and Hausdorff spaces and subspaces can be recovered in this setting, and the resulting theory is computable, at least in principle.

However, when we try to use this theory in practice in discrete mathematics and computation, we find that we need *something* to play the role of "sets". For this, we use those spaces that come with $\exists$ and $=$, which we call *overt discrete* (Taylor, 2005). They can be made to behave like traditional set theory if we add the non-computable hypothesis that all spaces have "underlying sets" (Taylor, 2004a).

The infinitary joins in general topology appear explicitly at a surprisingly late stage in our approach, completing the re-axiomatisation of locally compact spaces (Taylor, 2006a) and providing the basis for recursive elementary real analysis (Taylor, 2009a, 2010a). To go beyond that, however, we need to re-think the underlying categorical framework, which is work in progress (Taylor, 2009b).

1.8. The most prominent mathematical achievement of the ASD programme so far is the account it gives of recursive elementary real analysis, *including the Heine–Borel theorem*. This states that $[0, 1] \subset \mathbb{R}$ is compact in the sense that every cover by a family of open subspaces contains a finite sub-family that also covers (Taylor, 2009a).

This is the practical reason why no computable foundations of analysis have been developed before that are appropriate for mathematics. Suppose we take ordinary set theory and analysis on the one hand together with computability theory on the other, and put them together in the obvious way by defining $\mathbb{R}$ as the *set of computably representable real numbers*. Then we find that this important result *fails*. Roughly speaking, this is because we may *list* the computable real numbers as $a_0, a_1, a_2, \ldots$ according to the codes that represent them, and cover each of them by intervals of length $\frac{1}{4}, \frac{1}{8}, \frac{1}{16}, \ldots$, no finite subset of which would suffice. See Bridges and Richman (1987, §3.4) for a very clear account of this phenomenon.

The real interval $[0, 1]$ *is* compact in ASD, on the other hand, so this is more in line with both the traditions and applications of topology.

1.9. The applications of Abstract Stone Duality to elementary real analysis in (Taylor, 2010a) also demonstrate the utility of the new language. That paper provides an entirely different introduction to ASD that is aimed at a general mathematical audience—it assumes only first year analysis and a facility with formal systems, completely avoiding the underlying category theory. It is complementary to the account in this paper, neither of them assuming familiarity with the other.

Although we advocate computable foundations, we shall say very little about practical implementation here. This is a very interesting topic in itself, particularly in relation to elementary analysis, so it will be studied more extensively in future work (Taylor, 2006b, 2010b; Bauer, 2008).

2 Category Theory and Type Theory

We now sketch the technical background to our general method for the development of foundations *for* mathematics.

2.1. Gerhard Gentzen (1935), the father of proof theory, set out the way in which we reason with the logical connectives ($\wedge$, $\vee$, $\Rightarrow$) and quantifiers ($\forall$ and $\exists$).

Consider, for example, how we use and prove implications ($X \Rightarrow Y$). If we know both X and $X \Rightarrow Y$ then we also know Y; this is traditionally called *modus ponens*. More generally, we may have *proofs* ∇_1 and ∇_2 of X and $X \Rightarrow Y$ respectively, from a set Γ of *assumptions*.

Putting these together gives a proof of Y, as illustrated on the left:

$$\dfrac{\dfrac{\nabla_1}{\Gamma \vdash X} \qquad \dfrac{\nabla_2}{\Gamma \vdash X \Rightarrow Y}}{\Gamma \vdash Y}\,(\Rightarrow E) \qquad\qquad \dfrac{\dfrac{\nabla}{\Gamma, X \vdash \quad Y}}{\Gamma \quad \vdash X \Rightarrow Y}\,(\Rightarrow I)$$

Conversely, how can we prove the implication $X \Rightarrow Y$? *By definition*, it is by giving a proof $\triangledown$ of Y that may *assume* X, as on the right. We need to maintain a list Γ of such assumptions in order to keep track of the extra ones that we add and discharge in proofs of implications; this list Γ is called the *context*.

These two patterns are called respectively the *elimination* and *introduction* rules for $\Rightarrow$ (hence the labels $\Rightarrow E$ and $\Rightarrow I$), because of the disappearance or appearance of this symbol in the conclusion. There are similar rules for the other connectives and quantifiers.

2.2. Under Haskell Curry's analogy between propositions and types (Howard, 1980; Seldin, 2002), type connectives ($\times$, $+$, $\rightarrow$) and quantifiers (Π and Σ) were later expressed in a similar way. But now, in place of the bald assertion *that* $X \Rightarrow Y$, we have a *function* $X \rightarrow Y$ that takes values of one type to those of the other. This function needs a name. Bertrand Russell and Alfred North Whitehead (1910–1913), *20 originally called it $\hat{x}(\phi x)$, but the hat evolved into a lambda:

$$\dfrac{\Gamma \vdash a : X \qquad \Gamma \vdash f : X \rightarrow Y}{\Gamma \vdash fa : Y}\,(\rightarrow E) \qquad\qquad \dfrac{\Gamma, x : X \vdash \quad p \quad : Y}{\Gamma \quad \vdash \lambda x.\, p : X \rightarrow Y}\,(\rightarrow I)$$

Since types come with terms, these too need rules for their manipulation. The *beta* rule arises from first introducing and then eliminating the symbol, and the *eta* rule from the converse,

$$\dfrac{\Gamma \vdash a : X \qquad \Gamma, x : X \vdash p : Y}{\Gamma \vdash (\lambda x.\, p)a \;=\; [a/x]^* p : Y}\,(\rightarrow\beta) \qquad\qquad \dfrac{\Gamma \vdash f : X \rightarrow Y}{\Gamma \vdash f \;=\; \lambda x.\, fx : Y}\,(\rightarrow\eta)$$

where $[a/x]^* p$ indicates substitution (§2.7). Now the context Γ must include both assumptions and typed variables or *parameters*.

2.3. If we have such a system of rules for the connectives *that we use*—instead of encoding them in a much more general and powerful all-encompassing foundational monolith—then we have some chance of *computing* in our system. Of course, mathematicians always did that, before logicians came along: the difference is that the system is now much bigger (§1.4), and so requires a more formal relationship amongst its parts, and so amongst the specialists in very diverse disciplines (§3.6).

In the simply typed λ-calculus, for example, we merely have to keep applying the β-rule, and then we are guaranteed to get to the normal form . . . eventually. In order to prove termination of this process we use induction on the complexity of the types that are involved. For uniqueness, we need to know that, whenever we have a choice of *redexes* (reducible expressions), the two results can be brought back together by further reductions (Church & Rosser 1936). Both of these results depend in delicate ways on the formulation of the system.

Whilst it may be rather naïve to rely on this "guarantee" in practice, it turns out to be very profitable to hand the formal system over to the experts in manipulating symbols. They do not need to understand the original meaning of the notation, but may use their own experience of similar systems to make the computation more efficient. Even so, it is the mathematical system—not the way in which some machine happens to work—that determines the fundamental rules. This has been carried out very successfully in functional programming, which is now almost as fast as machine-oriented approaches (Hudak, 1989; Hudak et al., 2007).

There is, in fact, much profit to be made from collaborating with computational specialists only *via* a formal language. They may be able to transform it into something entirely different, that works qualitatively faster than the methods that the mathematician originally had in mind. One example of this that is indirectly related to the ideas in this paper (§6.4) is the continuation-passing method of compiling programs, because of its similarity to the way in which microprocessors handle subroutines (Appel, 1992; Thielecke, 1997).

Another example arises in the solution of equations involving real numbers. Interval-halving methods are commonly used to prove theorems in constructive analysis (Taylor, 2010a, §1), but if we were to compute with them *verbatim* we would only get *one new bit* per iteration. Newton's algorithm, by contrast, *doubles* the number of bits each time. But the *verbatim* reading is a mis-communication—we only *halve* the interval for the purpose of explanation. A practical implementation may choose the division in some better place, based on other information from the problem. In fact, this is what is done in logic programming with constraints (Cleary, 1987). Putting the ideas together, we can turn one algorithm into the other (Taylor, 2006b; Bauer, 2008).

Altogether, we can think mathematically in one way, but obtain computational results in another.

2.4. Bill Lawvere Lawvere (1969) recognised the situation in type theory (§2.2) as a special case of that studied by categorists. The functors $\mathsf{F} : \mathcal{S} \to \mathcal{A}$ and $\mathsf{U} : \mathcal{A} \to \mathcal{S}$ between categories $\mathcal{A}$ and $\mathcal{S}$ are said to be *adjoint* ($\mathsf{F} \dashv \mathsf{U}$) if there is a bijection (called *transposition*) between morphisms

$$\frac{\mathsf{F}X \longrightarrow A \qquad \text{in } \mathcal{A}}{\overline{X \longrightarrow \mathsf{U}A \qquad \text{in } \mathcal{S}}}$$

that is *natural* in X and A, which means that it respects pre- and post- composition with morphisms in the two categories. The transposes of the identities on $\mathsf{F}X$ and $\mathsf{U}A$ are called the *unit* and *counit* of the adjunction, respectively,

$$\eta_X : X \to \mathsf{U}(\mathsf{F}X) \qquad \text{and} \qquad \varepsilon_A : \mathsf{F}(\mathsf{U}A) \to A,$$

and these satisfy equations known as the *triangle laws*:

$$\mathsf{U}\varepsilon_A \cdot \eta_{\mathsf{U}A} \;=\; \mathrm{id}_{\mathsf{U}A} \qquad \text{and} \qquad \varepsilon_{\mathsf{F}X} \cdot \mathsf{F}\eta_X \;=\; \mathrm{id}_{\mathsf{F}X}.$$

The use of the letter η for different things in category theory and type theory is unfortunate, but this historical accident is well established in both subjects.

2.5. For example, the correspondence

$$\frac{\Gamma,\ x : X \vdash \quad p : \quad Y}{\Gamma \quad \vdash \lambda x.\, p : X \to Y} \qquad \text{or} \qquad \frac{\Gamma \times X \xrightarrow{\ p\ } Y}{\Gamma \xrightarrow[\tilde{p}]{} Y^X}$$

is bijective on the left by the symbolic rules (§2.2), and on the right because of the adjunction $(-)\times X \dashv (-)^X$. However, purely categorical textbooks cover many pages with diagrams expressing arguments that could be written much more briefly by adopting the notation $\lambda x.\, p$ instead of $\tilde{p}$ (*cf.* §§1.4 & 6.1).

This analogy can be applied systematically to all of the logical connectives:

(a) The adjoint transposition corresponds to the symbolic rules like λ-abstraction, case analysis, recursive definitions and pairing that typically bind variables or hypotheses, changing the context.
(b) The "algebraic" operation ev, π_0 or π_1 that leaves the context unchanged is provided by the co-unit ε of the adjunction in the case of $\to$ and $\times$, but the unit η gives the inclusions ν_0 and ν_1 for $+$.
(c) The triangle laws provide the β- and η-rules.
(d) One of the naturality conditions for the adjunction shows how the construction works for morphisms. The other says that we may substitute within $\langle -, - \rangle$ and λ, but in the case of $\vee$ and $\exists$, it gives the so-called *commuting conversions* or *continuation rules*.
(e) Substitution under $\vee$ and $\exists$ must be stated categorically by means of additional conditions (§2.9).

The reason why some of the connectives ($\Rightarrow$, $\wedge$, $\to$, $\times$) work one way round and the others ($\vee$, $\exists$, $+$) differently is that they are respectively right and left adjoints to something that is simpler. This is described more fully in (Taylor, 1999, §7.2).

2.6. In order to formalise this relationship between adjunctions and systems of introduction, elimination, β- and η-rules, we need a *fluent* translation between the diagrammatic and symbolic languages. In fact, we just require a way of turning a type theory that has all of the structural rules (identity, cut, weakening, contraction and exchange, as they are called in proof theory) into a category that has finite products. Then we can ask whether this category obeys other universal properties and read the corresponding symbolic rules off from them.

The following method also works for *dependent types, i.e.* possibly containing parameters, but we shall avoid them in this paper. There are, conversely, ways of defining a language from a *pre-existing* category (Taylor, 1999, §7.6), but these are more complicated and will not be needed here. Systems such as linear logic that do not obey all of the structural rules correspond to other categorical structures. These structures might, for example, be *tensor* products ($\otimes$), something that categorists understood long before they did predicate logic (Mac Lane, 1963). It is therefore often the *syntax* that requires the most innovation, for example using Jean-Yves Girard's proof nets (Girard, 1987).

A type theory $\mathcal{L}$ of the kind that we shall consider has recursive rules to define

(a) *types* (X);
(b) *contexts* (Γ), which are lists of typed variables (*parameters*) and equations between terms, understood as *hypotheses*;
(c) *terms* ($\Gamma \vdash a : X$) of each type X, whose free variables belong to the given context Γ; and
(d) *equations* ($\Gamma \vdash a = b : X$) between these terms.

2.7. We have already used the *substitution* operation $[a/x]^* c$, whose definition must be presented carefully in order to avoid *variable capture* by λ-abstraction. An operation, called $\hat{x}^* c$ in (Taylor, 1999, §1.1) but not given a name elsewhere, *weakens* c by pretending that it involves the actually absent variable x. These satisfy the *extended substitution lemma*,

$$[a/x]^*[b/y]^*c \;=\; [[a/x]^*b/y]^*[a/x]^*c \qquad\qquad [x/y]^*\hat{x}^*[y/x]^*\hat{y}^*c \;=\; c$$

$$[a/x]^*\hat{x}^*c \;=\; c \qquad\qquad [a/x]^*\hat{y}^*c \;=\; \hat{y}^*[a/x]^*c \qquad\qquad \hat{x}^*\hat{y}^*c \;=\; \hat{y}^*\hat{x}^*c,$$

where x and y are distinct variables with $x, y \notin \mathsf{FV}(a)$ and $y \notin \mathsf{FV}(b)$.

2.8. Like so many other operations in mathematics, substitution is defined by its *action* on other objects. So, when we abstract the operation from its action, it has an *associative* law of composition, which brings it into the realm of category theory.

The *objects* of the category are, abstractly, the *contexts* Γ of the type theory. More concretely, Γ is represented by the set of terms that are definable using free variables chosen from Γ. The substitution and weakening operations define *morphisms*

$$\begin{array}{ccccc} [a/x]^*c & \longleftarrow\!\!\dashv & c \qquad \hat{x}^*d & \longleftarrow\!\!\dashv & d \\ \Gamma & \xrightarrow{\;[a/x]\;} & [\Gamma,\ x : X] & \xrightarrow{\;\hat{x}\;} & \Gamma. \end{array}$$

It is convenient to draw the arrows in such a way that substitution looks like *inverse image*, which is why we write it with a star. We also draw weakening with a special arrowhead $\longrightarrow$ and call it a *display map*. The five equations of the extended substitution lemma say that certain diagrams formed from these arrows commute, *i.e.* segments of paths of arrows that match one pattern may be replaced by another. For example, the composite of the two arrows above is the identity on Γ.

Altogether, we have defined, not the category itself, but an *elementary sketch*. That is, whilst we have spelt out *all* of the objects, we have only given a *generating set* of arrows. The category $\mathsf{Cn}_{\mathcal{L}}^{\times}$ itself consists of all paths formed from these arrows, subject to the equations of the given symbolic theory $\mathcal{L}$ and those of the extended substitution lemma. There is a normalisation theorem that says that any morphism $\Gamma \to \Delta$ is essentially an assignment to each variable in Δ of a term of the same type, but formed using variables from the context Γ. In keeping with the traditional names of categories ("of objects X and morphisms Y"), We therefore call this the *category of contexts and substitutions*. See Taylor (1999, §4.3) for the details of this construction in the simply typed case and Chapter VIII there for its generalisation to dependent types.

The category $\mathsf{Cn}^{\times}_{\mathcal{L}}$ has the universal property that interpretations of the theory $\mathcal{L}$ in another category $\mathcal{T}$ correspond bijectively (up to unique isomorphism) to functors $\mathsf{Cn}^{\times}_{\mathcal{L}} \to \mathcal{T}$ that preserve the relevant structure. This means that it is the *classifying category* for models of this structure.

Type theories are presented in a recursive way, whilst categories have associative composition, but anyone who has written any functional programs knows that associativity and recursive definitions do not readily mix. Andrew Pitts defined the classifying category in this way in (Pitts, 2001, §6); it was because even showing that the identity is well defined as a morphism was so complicated in this approach that I was led to the construction *via* a sketch. It demarcates the things that are naturally described in a recursive and symbolic way from those that are best done associatively and diagrammatically.

2.9. We said in §1.6 that there are reasons for using diagrammatic methods for some aspects of the presentation of mathematics but symbolic ones for others. Many of the comparisons and contrasts between them are illustrated by Lawvere's notion of *hyperdoctrine* (Lawvere, 1969). This is technically much more complicated than anything that we shall need in this paper, but that fact is one of the methodological points that I would like to make.

A hyperdoctrine represents the quantifiers Π and Σ as the right and left adjoints (*cf.* §2.5) to the substitution functors $f^* : \mathcal{A}[Y] \to \mathcal{A}[X]$ that are defined by a map $f : X \to Y$. However, induced and restricted modules along a *ring* homomorphism also provide adjunctions like these. The difference is that, in symbolic logic, we may *substitute* for parameters *within* the scope of the quantifiers. This is not part of the correspondence in §2.5, so category theory must express it by another axiom, known as the *Beck–Chevalley condition*, which is not true for general ring homomorphisms. Another property that is peculiar to the existential quantifier is the equation

$$\sigma \ \wedge \ \exists x.\, \phi x \iff \exists x.\ \sigma \wedge \phi x,$$

which Lawvere called the *Frobenius law* because of analogies with linear algebra.

So category theory is

(a) *agnostic* in treating both modules and quantifiers in a general way;
(b) more *precise* in identifying the substitution operations in the former that go unremarked in symbolic logic; and
(c) more *efficient* than giving pages of proof rules like those in §§2.1–2 in order to specify which structures we want to include in our system.

On the other hand, it is

(d) *clumsy* and very difficult for the student to learn if we are actually just interested in using the quantifiers to express other mathematics.

2.10. Another criticism of hyperdoctrines is that, in requiring this structure for *all* maps $f : X \to Y$, they have a hidden bias towards set theory. This remark applies also to type theories that allow quantification over types with arbitrary parameters. In order to apply Lawvere's ideas to topology or domain theory (for example, to model second-order quantification over type variables), we must *restrict* the class of maps that admit quantification. In fact, it is more accurate to say that quantification is adjoint to *weakening* rather than to substitution (Taylor, 1999, Chapter IX).

Set theory will play a role in this paper, but not as *the standard* or a *competing logical* theory. Instead, we see it as an *analogous mathematical* one, providing a second example of Stone duality (§4.5). We aim to study topology, which is about *continuous* things, so by "set theory" here we shall mean any treatment of *discrete* objects, which may be in an elementary topos, using type theory or in some axiomatisation of the $\in$ relation.

The key difference that will emerge is that in the logic of discrete things we may use quantifiers ranging over *any* object, whilst in topology only *some* spaces may play this role. An object X that may serve as the range of a universal quantifier is called *compact*. More generally, a map that admits universal quantification with parameters *à la* Lawvere is called *proper*, and we think of the map as a continuous family of compact spaces. Similarly, a map with existential quantification is *open*, but we need to introduce a new word, *overt*, for the corresponding objects.

2.11. Nevertheless, Lawvere's insight is the key to our method, *when we turn it around*. Having seen many systems of rules and adjunctions for different *logical* connectives, we may recognise the general pattern and apply it to *mathematics*.

Many theorems in mathematics (our client's brief) may be stated as adjunctions, to which we apply Lawvere's correspondence in reverse: we invent *new* symbols whose introduction, elimination, abstraction, beta and eta rules exactly capture the adjunctions. To illustrate this process, we discuss the duality between topology and algebra in §§4–5, and translate the adjunction into a new symbolic language in §6. We shall go on to apply the same technique to some other ideas in topology in the extended version of this paper. These include definition by description and Dedekind completeness, which are the manifestations for $\mathbb{N}$ and $\mathbb{R}$ of our abstract notion of sobriety.

2.12. The new symbols are not meant to be merely abbreviations for a collection of lemmas that happen to be used frequently in the old structure, but to stand on their own feet as principles of reasoning in the new one. As with complex numbers (§1.3), we want to be able to compute with them without having to worry about what they mean (§2.3). We therefore need to ensure that they are accompanied by a *complete* set of rules of inference and show that they are conservative over simpler systems.

One good indicator of sufficiency is the ability to prove a *normalisation theorem*. Typically, this says that say that any term that makes arbitrary use of the introduction and elimination rules is provably equivalent to one that uses

(a) the new elimination rule only for free variables of the new type; and
(b) the new introduction rule only on the outside of the term.

The equivalence between the categorical and symbolic constructions is deduced from theorems of this kind, as is conservativity. This will be a key step in the general method that we now propose.

3 Method and Critique

We observed in §1.6 that category theory is adept at capturing the significant ideas in a mathematical discipline in the form of universal properties, whilst axioms may be expressed in symbolic logic as systems of introduction and elimination rules. Then in §2 we demonstrated the formal connection between these.

3.1. On this basis, we propose a methodology for devising a new foundational system *for* a subject. This is summed up by the diagram

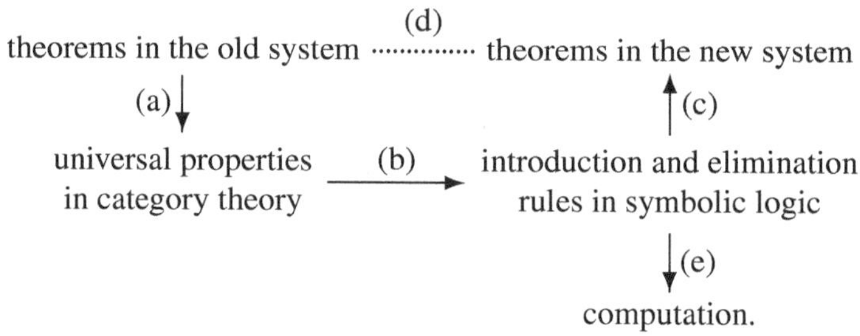

In this, the task of the journeyman mathematician—the deduction of theorems from axioms—is to fill in the arrow (c) on the right. Hopefully, this will make the rectangle (a–b–c–d) "commute" in the sense that we recover the old theorems from the new axioms.

However, this job only makes sense in a professional framework in which axioms have *already* been chosen to capture the intuitions and applications of the discipline. This choice is represented by the arrow (a) on the left.

The constructions that we sketched in §§2.8f provide the horizontal arrow (b), whilst the computational interpretation (e) depends on the details of the resulting calculus (§2.3).

We shall show how this programme works for the example of general topology, with some hints about how to apply the ideas to other subjects.

3.2. Arrow (a) in some sense derives the axioms from the theorems.

Euclid and Bourbaki gave us a style of presenting mathematics that begins from axioms which (so far as their text is concerned) come out of the blue, and deduces theorems from them as if the results were obvious and inevitable. Abel described Gauss as "like a fox, who effaces his tracks in the sand with his tail". Consequently, lesser mathematicians think that it is enough to *state* a bunch of axioms in order to justify the *relevance* of their results, whilst students learn nothing about the ways in which their masters *discovered* their theorems.

In following our methodology, we cannot "write down all the axioms first"—because *finding* the right axioms is the *goal* of the investigation. We shall want to be able to deduce the old theorems from the new axioms, but this is an *experimental test* of the axioms that we propose. So before we have discovered the *final version* of the axiom system, we must conduct experiments in *ad hoc* systems first, *cf.* §5.

Along the way, like any engineer, we shall assemble several *prototypes*. This means that a certain amount of *repetition* of tests in needed. Of course this happens all the time in mathematics too (Lakatos, 1963)—it is just the textbooks that falsify history by saying otherwise. At the end of the day, we want to state our chosen axioms and deduce the important theorems from them. But that will indeed be the *end* of the process, which will then no longer be an active piece of *foundational* research.

For a modern study of a wide range of techniques for discovering the principles of several mathematical disciplines, see Corfield (2003).

3.3. Counterexamples play a very important part in the empirical development of mathematics. Unfortunately, they are sometimes given such a degree of prominence that they can stifle subsequent progress.

Typically, the argument behind a counterexample may prove

$$D, \quad E, \quad F \quad \vdash \quad \neg G,$$

entirely rigorously. Then the reader is expected to agree that,

therefore, E is false.

But why? D, F and G had exactly the same status in the proof, so why pick on E?

This is an argument *by authority*, not a valid piece of mathematics: the author has simply used sleight-of-hand to make us accept his prejudices. These become embedded in the literature and treated as if they were *theorems*, on which whole theories are built. Having forgotten that the counterexample depended on cultural assumptions, people subsequently resist the introduction of a new paradigm.

This abuse of counterexamples is very common, but the one that has caused the greatest intellectual paralysis is surely Georg Cantor's, that no set is in bijection with its powerset. On this he built his theory of cardinality, which is far too coarse to be of any use in mainstream mathematics. He said of the discovery that $\mathbb{R}^2 \cong \mathbb{R}$, "I see it but I don't believe it", but this did not deter him from jumping into the abyss.

Because of Cantor's theorem, a set X can only satisfy $X \cong X^X$ in the trivial cases $X \equiv \emptyset$ or $\mathbf{1}$. Apparently, the untyped λ-calculus is therefore inconsistent. Dana Scott eventually saw through this and went on to construct *topological lattices* with this property (Scott, 1993). As these can be embedded in presheaf toposes, where we may treat them as intuitionistic sets, we see that Cantor's supposedly fundamental theorem of set theory actually relies on excluded middle.

The message that we ought to take from a counterexample is no more than that you can't do it *like that*. Maybe someone someday will think of a *different* way. They will identify underlying assumptions A, B and C that had previously been overlooked, and achieve all of D, E, F and G on some *other* basis, using A', B' and C' instead.

In particular, just as in the use of formal methods for security (putting a strong lock on a weak door), it is the very rigour of a fragment of argument that can be most misleading about the bigger picture. In our architectural metaphor, counterexamples are reasons for choosing one blueprint over another, whilst theorems hold the building up once we have chosen which plan to carry out. A tower block may later turn out to be a mistake for sociological reasons that have nothing to do with how accurately its plans were drawn.

It would, perhaps, be useful to make a clear distinction between the principal track of a *theory* (from axioms to theorems) and its *peri-theory*. Peri-theory is the discussion *around* the theory, consisting of the main examples, the prototypes and the counterexamples that *led us to the choice* of statements of the axioms and the theorems. Other examples of peri-theory are converses that lead back from the main theorem to (the necessity of) the chosen axioms, and more generally the lists of equivalent alternative formulations of the axioms that some textbooks like to give.

This entire paper, apart from §§6 & 8, concerns the peri-theory of topology.

3.4. Unfortunately, it is not as straightforward as we have suggested to work from a client's brief: this cannot be taken literally, and has to be negotiated.

If the client is allowed to dictate what is *fundamental*, the new foundations will be no more flexible than the old ones. It takes an outsider's perspective to distinguish the key elements of an enterprise from the accumulation of detail. Arrow (a) does not therefore just copy the theorems and chapter headings of the old theory.

In approaching a mathematical discipline, category theory often focuses on ideas that its specialists previously regarded as trivial. For example, the Sierpiński space Σ and subobject classifier Ω (§7) are merely two-element sets in classical set theory and topology. Also, whilst mathematicians have acknowledged that the theorems that *create* important structures are adjoints, an adjunction is a relationship between two parties. In this, the other partner typically does something that appears to be mere bureaucracy, namely to "forget" the same structure.

3.5. Another issue on which the clients' preconceptions need to be challenged is the identification of their most valuable products. It is up to their *customers* to do this. For example, set theorists put much of their effort into considering infinite cardinals, but the demand from the market is for more prosaic things, such as quotients of equivalence relations. These are axiomatised explicitly in categorical logic.

On the other hand, led in some cases by an over-emphasised counterexample, specialists sometimes develop variant forms of a subject that *omit* properties which the customers may regard as essential. An example of this is compactness of the real interval $[0, 1] \subset \mathbb{R}$ (§1.8), which is absent from Bishop's theory (Bishop & Bridges 1985).

3.6. Negotiation of this kind is they way in which we answer the final objection to changing foundations, namely that the prevailing set-up is crucial for the whole of science (§1.1(d)).

Science and mathematics are *systems* whose *component* disciplines interact across an *interface*. The higher-level components depend only weakly on the details of those below, but do have requirements for certain features. This is a commonplace in modern technology. It is also well known in the wider picture of science: whilst chemistry depends on physics, it only uses four sub-atomic particles, then organic chemistry for the most part only uses four elements, and genetics only four bases and twenty amino acids.

Physics very likely relies on compactness of the interval, for example to ensure that local patches of solutions to a differential equation fit together into a global one, instead of forming a singular cover (§1.8). On the other hand, I would be very sceptical if you told me that some property of black holes depends on excluded middle. Have you developed a constructive theory as an experimental control, and found observational evidence to distinguish it from the classical one? This is, after all, what the empirical method says that you should do.

However, the question of whether some application in science actually depends on a particular foundational principle is never an easy one. Settling the necessity of *each* particular invocation of excluded middle, Heine–Borel or a large cardinal in the chain of theorems about functional analysis or other subjects may require a decade's research. Unfortunately, the commonest way of handling this is by "megaphone philosophy" that, if it contains any mathematics at all, consists largely of the abuse of counterexamples.

3.7. When we replace a component of a system, it needs to be backwards-compatible in its *function*, not necessarily its *implementation* or *extent*. After all, an architect who has been commissioned to replace a building will only do so with an exact copy if it is to be a museum.

In the case of general topology, the methods of construction in this paper are such that the new building must either be smaller than the old one (consisting of just locally compact spaces) or substantially bigger. This is entirely consistent with the historical development of the subject, which grew from figures in Euclidean three-dimensional space, to $\mathbb{R}^{\mathbf{n}}$, to projective and non-Euclidean geometry, to real manifolds, to spaces of functions in analysis, and to domains for the denotational semantics of programming languages. The intuition of continuity has been captured in numerous quite different ways, making use of metrics, uniformity and convergence of sequences, as well as systems of points and open neighbourhoods.

For comparison, there can be little doubt that groups and fields have the right axioms for the intuitions that they seek to capture. This is indicated by the fact that we use the axioms directly for computation. Algebra textbooks are also able to *classify* finite fields, and they make a serious attack on the similar (albeit intractible) problem for groups.

The situation in general topology is much less clear. It resembles a medieval "world" chart that more or less correctly depicts the Mediterranean, but which is decorated with mythological creatures around the outside. The various approaches to continuity accurately capture real manifolds, just as the old cartographers recorded their own familiar territory, but we cannot be confident in using the outer parts of the chart.

In both cases we have very reasonable approximations locally: the error lies in trying to extend the *co-ordinate system* too far. Flat charts are useful, but long ago Eratosthenes had known, not only that the Earth is spherical (from lunar eclipses), but even how big it is. A flat system of co-ordinates imposes an *artificial boundary* on the world that it can describe.

Similarly, it has been known since the 1960s that points and open sets provide the wrong co-ordinate system for topology. Sheaves in algebraic geometry were based on open sets and not points, whilst algebraic topologists sought more "convenient" categories (Brown, 1964; Steenrod, 1967), *i.e.* those that admit general spaces of functions. However, it is by no means clear what topologies these function-spaces should carry, especially if we want to invesigate the properties of $\mathbb{N}^{\mathbb{N}^{\mathbb{N}}}$ and $\mathbb{R}^{\mathbb{R}^{\mathbb{R}}}$.

There is, therefore, nothing special about the boundaries of the category whose objects the textbooks call "topological spaces". These books treat non-Hausdorff spaces with derision, and make little attempt to explore the full extent of even the world that is measured out by their own co-ordinate system. This investigation only began when the analogy with the $\exists\wedge$-fragment of logic was recognised, *cf.* (Vickers, 1988). We therefore *un*define the terms "space" and "topological space", leaving them open to new definitions. The textbook spaces will be re-branded as *Bourbaki spaces* (Bourbaki, 1966) to strip them of their authority.

3.8. Even when we have found the right co-ordinate system, it may still not be appropriate to *describe* it in the same way for both foundations and applications. Any engineer knows that the user manual for a gadget should not be written in the same way as its technical specification. So this paper discusses the *foundations* of ASD, whilst (Taylor, 2009a, 2010a) provide separate introductions that are suitable for its *applications* to elementary real analysis.

One reason for this is that the technical ideas may later be redeployed to make something else with an entirely different use. In our case, we shall find that there is a new underlying abstract structure (§6) that may have other applications besides topology, and deserves to be studied in its own right. It includes certain definitions that, in the presence of the specifically topological structure, can be characterised in other ways (*cf.* §7.7). This is in line with the usual experience of applying category theory to mathematical disciplines, namely that phenomena with various different traditional formulations turn out to be examples of a common abstract theme.

Secondly, the demands from users are often the driving forces behind advances in technology. However, as we have seen for computing hardware and software, the biggest improvements do not result from adding bells and whistles, but from re-thinking and strengthening the fundamental principles. In ASD, the main challenge for the future is how to extend the boundaries beyond local compactness, in such a way that Banach spaces are given "the right" topology, whatever that means. In order to do this, it is the initial formulation of the underlying abstract structure in §6, rather than its adaptation to topology, that needs to be replaced.

3.9. Finally, there is an issue for which architecture is entirely the wrong metaphor and the departure of Columbus from the Mediterranean is much more appropriate. *Los Reyes Catolicos* certainly did not promote free intellectual exploration in their domestic and colonial policy, but they did at least *fund* a "blue sea" project.

Nowadays, one is asked to give advance notice of all of the theorems that one intends to prove. Such planning is appropriate when building a house, but it can be done *if and only if* there are *no* original ideas. A mathematician with a plan for a theorem wants to carry it out straight away, and the only pieces of equipment that are needed are a clear head and a clear blackboard. Although we don't put our *lives* at stake as Columbus did when we embark on scientific experiments or attempt to prove mathematical theorems, if there is no *intellectual* risk of failure in a proposed piece of research, then it is redundant, and probably not worthy of funding.

We like to think that the *finished product* of mathematics is the most precise of any branch of science or engineering. The corollary of this is that the *vision* of a mathematical project in advance of its detailed plan is necessarily *much more vague* than in any other discipline.

And things may not go according to plan even if we do succeed, because there may be a new continent to discover. For a powerful account that has a far wider relevance than to physics, see Part IV of Smolin (2006), which demonstrates a familiarity with the real experience of those who do revolutionary science that is absent from Kuhn (1962).

It is time to apply this method to some mathematics.

4 Stone Duality

Now we begin to describe the ideas from traditional general topology that we shall take as the "client's brief" for our new theory of that subject. We sketch how this programme is executed in §§6 & 8 and in the extended version of this paper. These ideas are presented in the form of universal properties, providing arrow (a) of the diagram in §3.1. Following arrow (b) there, we shall then develop a corresponding syntax. As far as possible, we do this in a general way that might be adapted to other subjects.

4.1. The principal mathematical insight that we shall use is the *duality between algebra and geometry.*

Figures like circles and parabolae are defined by polynomials such as $x^2+y^2-a^2$ and x^2-2by, and *superposition* of the figures corresponds to *multiplication* of their polynomials. Polynomials over a field that is algebraically closed may, like natural numbers, be expressed *uniquely* as products of irreducible or *prime* factors, which therefore capture the irreducible geometrical figures in a purely algebraic way.

By no means all commutative rings admit unique factorisation into primes, for example

$$3\cdot 2 = 6 = (1+\sqrt{-5})\cdot(1-\sqrt{-5}) \quad \text{in} \quad \mathbb{Z}[\sqrt{-5}].$$

However, Ernst Kummer recovered the situation by introducing *ideal numbers*. The definition of an ideal that is now found in textbooks, as a subset of the ring with certain properties, is due to Richard Dedekind, who began the use of set theory to create mathematical objects.

Leopold Kronecker, on the other hand, preferred to treat equations themselves as names for their own roots. This is perhaps the point at which the classical and constructive traditions in the development of mathematics diverged.

The notions of ideal and prime may be transferred from rings to lattices using the analogy between their operations. The systematic study of these ideas was undertaken by Marshall Stone (1937, 1938). Since geometry and algebra each have their own natural generalities, they need not coincide exactly: it may be necessary to make extensions, restrictions or other adjustments on both sides to achieve agreement.

4.2. Putting Stone's programme in categorical language, let $\mathcal{A}$ be some category of "algebras" and $\mathcal{S}$ one of "spaces", the exact nature of which we leave open. Then by a *Stone duality* we mean an adjunction

$$\mathcal{A}^{\mathrm{op}} \underset{T}{\overset{P}{\rightleftarrows}} \mathcal{S}, \quad (P \dashv T)$$

in which TX is the algebra (maybe of open subspaces) associated with a geometrical object X, and PA is the space of primes of an algebra A.

Because of the contravariance, the unit and counit of the adjunction (§2.4) go in the same direction in the categories $\mathcal{S}$ and $\mathcal{A}$,

$$\eta_X : X \to P(TX) \qquad \text{and} \qquad \iota_A : A \to T(PA).$$

These say how each point defines a prime, and how the algebra is represented by its space of primes.

We call X a *sober space* if η_X is an isomorphism, *i.e.* every prime corresponds to a unique point, so a sober space is one that has exactly the points that are required by algebra. Conversely, A is a *spatial algebra* if it has enough primes to provide a faithful representation, making ι_A invertible. The situation is a *Stone equivalence* if all spaces are sober and all algebras spatial, in which case the functors P and T are equivalences.

4.3. Stone dualities often arise from a *schizophrenic object* Σ, *i.e.* one that in some sense belongs to both categories, where

$$TX \equiv \mathcal{S}(X, \Sigma) \qquad \text{and} \qquad PA \equiv \mathcal{A}(X, \Sigma).$$

The abstract description of Stone duality that we have given comes, essentially, from the book (Johnstone, 1982), which also provides an excellent and wide-ranging survey of concrete examples. In those that directly interest us, Σ is at least a distributive lattice, whose elements we sometimes regard as "truth values" (§7).

If you would like to apply the methods of this paper in another discipline, you will of course want to see what it can do for topology first. However, on your second reading you should stop *here*. You need to identify the object Σ in your subject. It is likely to be the most important algebraic or geometrical structure, albeit one that the undiscerning may have dismissed as trivial (§3.4). Do not be misled by our discussion of subobject classifiers in §7, or by the fact that all but two of the examples in (Johnstone, 1982, §VI 4) are based on a two-element set—the other two use the line $\mathbb{R}$ and the circle $\mathbb{R}/\mathbb{Z}$. The natural numbers $\mathbb{N}$, integers $\mathbb{Z}$, complex numbers $\mathbb{C}$, the closed interval $[0, 1] \subset \mathbb{R}$ and the ring $K[x]$ of polynomials in one variable over a ground field play this key role in other mathematical settings.

The most important property of Σ that you should check is that all of the other desirable objects are sober with respect to it; this fact may already be a headline theorem in your subject. Then, ignoring the rest of §7, it may be worth looking for an algebraic property like that in §7.7; unlike sobriety, this may not have been recognised before.

4.4. The traditional definition of topological spaces gives rise to a Stone duality

$$\mathcal{A} \equiv \mathbf{Frm} \equiv \mathbf{Loc}^{\mathsf{op}} \qquad \mathcal{A}^{\mathsf{op}} \underset{T}{\overset{P}{\rightleftarrows}} \mathcal{S} \quad (\top) \qquad \mathcal{S} \equiv \mathbf{BbkSp}$$

in which **BbkSp** is the category of *Bourbaki spaces* (the things that the textbooks call *topological spaces*, *cf.* §3.7) and continuous functions between them.

The algebra TX associated with a Bourbaki space is simply its lattice of open subspaces. Discarding the points, what kind of algebra is it? It has finite meets and arbitrary joins, where binary meets distribute over arbitrary joins. A *frame* is exactly an algebra for these operations, and we write **Frm** for the category of frames and homomorphisms.

The functor $\mathbf{Frm} \to \mathbf{Set}$ that forgets this algebraic structure has a left adjoint, which provides the *free frame* $\mathsf{F}N$ on a set N. The elements of the free frame may be characterised in two ways, either as *upper families* $\mathcal{U}$ of finite subsets of N, *i.e.* if $T \supset S \in \mathcal{U}$ then $T \in \mathcal{U}$ too (Johnstone, 1982, II 1.2), or as *Scott open families* of arbitrary subsets of N.

How do we recover the points or primes of a frame A? They are continuous maps $\mathsf{1} \to PA$, which correspond under the adjunction to frame homomorphisms $A \to T\mathsf{1}$. Since $T\mathsf{1} \cong \mathcal{P}(\mathsf{1})$, these are subsets $F \subset A$ whose membership predicate is a frame homomorphism, so $\top \in F$ and if $\alpha, \beta \in F$ then $\alpha \wedge \beta \in F$, whilst if $\bigvee \alpha_i \in F$ then $\alpha_i \in F$ for some i.

4.5. In our investigation of topology, we shall often find it helpful to consider the analogy with discrete objects ("sets" in the sense of §2.10). Classically, there is a Stone equivalence $\mathbf{CABA}^{\mathsf{op}} \simeq \mathbf{Set}$, where **CABA** is the category of complete atomic Boolean algebras and homomorphisms for $\bigvee$ and $\bigwedge$, for which TX is the powerset of the set X and PA is the set of atoms (minimal non-$\bot$ elements) of A. This equivalence was proved by Adolf Lindenbaum and Alfred Tarski (1935).

4.6. We want to design a new building to replace the old one functionally and not extensionally (§3.7), so it is really the *underlying intuition* of Stone duality that we shall use. The concrete examples can be a little distracting, especially with regard to the notion of *sobriety*. We defined this above in a way that depends on the choice of adjunction, and so may vary from one application to another.

Classically, an object is sober with respect to the concrete Stone duality between frames and Bourbaki spaces (§4.4) iff every irreducible closed subspace is the closure of a unique point. In particular, every Hausdorff space is sober in this sense (Johnstone, 1982, II 1.6). On the other hand, it turns out that the Hausdorff spaces $\mathbb{N}$ and $\mathbb{R}$ are sober with respect to our *abstract* Stone duality iff they admit definition by description (Taylor, 2002a, §9) and by Dedekind cuts (Taylor, 2009a, §14) respectively.

There is no *theorem* to say that the new definition of sobriety is *coextensive* with the old one, simply because there is no common genus of which these are species.

There is nothing to be gained by trying to find such a generality, as it would be artificial.

4.7. One way to achieve Stone *equivalence* is just to replace $\mathcal{S}$ by $\mathcal{A}^{\mathrm{op}}$. In the cases where algebras are frames or commutative rings, the new "algebraic" spaces are called *locales* or *affine varieties*, respectively. A *continuous function* f between the locales that correspond to frames A and B is by definition a frame homomorphism $f^* : B \to A$ in the opposite direction (§4.4). However, since any such f^* preserves (finite meets and) all joins, it has a right adjoint, and these are written $f^* \dashv f_*$.

Stone duality is now vacuous as an extensional *theorem*, instead serving as a vehicle for the *intuitions* of general topology and algebraic geometry, which may be used to reconstruct these disciplines in a new, algebraic, form (Hartshorne, 1977; Johnstone, 1982).

Locale theory mitigates a lot of the evils of point-set topology. For example, it is largely free of the axiom of choice, without which it is next to impossible to prove anything about Bourbaki spaces. In particular, Peter Johnstone proved Tychonov's theorem (that a product of compact spaces is compact) without it (Johnstone, 1982, Thm. III 1.7). Also, the closed interval is Dedekind complete and compact (§1.8) in the localic reals over any elementary topos, whereas if we use the point-based definitions in the internal language of a sheaf topos then the object of Cauchy reals is typically smaller than the Dedekind one, so the Heine–Borel theorem fails (Fourman & Hyland, 1979).

4.8. Whilst locale theory makes significant advances over traditional *topology*, it still relies on the old *foundations* of the category of sets, as of course do the other concrete examples in this section. Our goal is to build new foundations *for* topology.

Nevertheless, locales provide a useful test-bed for many of the ideas that we shall consider. We shall in particular need to know about *sublocales* $i : S \rightarrowtail X$, by which we mean coequalisers of frames. These have $i_* \cdot i^* = \mathsf{id}$, and are therefore captured by the other composite, $j \equiv i^* \cdot i_*$, which satisfies $\mathsf{id} \leqslant j = j \cdot j$ and preserves $\wedge$, but usually not any kind of joins. An endofunction j of a frame with these properties is called a *nucleus* and always arises from a sublocale in this way (Johnstone, 1982, II 2.2). In particular, an element $a \in A$ of a frame gives rise to the nuclei $a \Rightarrow (-)$ and $a \vee (-)$, which encode the corresponding open and closed sublocales, respectively. We shall encounter localic nuclei in §§5.10, 6.6 and 7.5.

5 Always Topologize

5.1. We do not require underlying sets for the category $\mathcal{S}$, because the new notion of "space" is *intended* to be an unknown. However, algebras need carriers. The fundamental idea of *Abstract* Stone Duality is to use the (as yet, unknown) spaces as carriers for the algebras. In particular, in topology, we regard the lattice of open subspaces of a space as *another space*, following another methodological principle due to Marshall Stone: "always topologize" (1938).

For Stone equivalence, $\mathcal{A} \simeq \mathcal{S}^{\mathrm{op}}$. So we are saying that we want $\mathcal{S}^{\mathrm{op}}$ to be a category of algebras over $\mathcal{S}$. For this, we need a way of formulating (potentially infinitary) algebraic theories that works over any category $\mathcal{S}$ and not just over the category of sets. Such an account is provided by the categorical notion of *monad* (Linton, 1969). The textbook on universal algebra using monads has, unfortunately, never been written, but (Manes, 1976) perhaps comes closest to it.

In §6 we shall give some of the formal definitions for monads, as adapted to ASD, and the full treatment is of course given in the papers cited. However, our purpose here is to tell the story behind these ideas.

5.2. Recall that concrete Stone dualities often arise from a "schizophrenic" object Σ, *i.e.* one for which $TX \equiv \mathcal{S}(X, \Sigma)$. On the other hand, by the new hypothesis, TX is also to be an object of $\mathcal{S}$. So it is natural to ask that $\mathcal{S}$ have all *exponentials* of the form Σ^X, by which we mean that $\mathcal{S}$-maps $\Gamma \times X \to \Sigma$ are to be in natural bijection with those $\Gamma \to \Sigma^X$ (§2.5). A pre-requisite for this definition is that $\mathcal{S}$ have finite products, $\mathsf{1}$ and $\times$. Note that we are asking just for powers of Σ (and, consequently, of its powers too), but not of general objects. In other words, we are not saying that the category should be *cartesian closed*—at least as far as *this* paper is concerned.

Nevertheless, *any* object Σ that has powers gives rise to an adjunction

$$\begin{array}{c} \mathcal{S}^{\mathrm{op}} \\ \Sigma^{(-)} \big\uparrow \dashv \big\downarrow \Sigma^{(-)} \\ \mathcal{S} \end{array}$$

which we write vertically in order to avoid confusing it with the one in §4.2 for Stone duality.

This adjunction induces a monad on $\mathcal{S}$, so let $\mathcal{A}$ be its category of Eilenberg–Moore algebras, for which the standard theory gives a comparison functor $\mathcal{S}^{\mathrm{op}} \to \mathcal{A}$ (Eilenberg & Moore, 1965; Taylor, 1999, 7.5.3(c)). Formulating abstract Stone duality as an *axiom* on $\mathcal{S}$, we ask that this be an equivalence, *cf.* §4.7.

As we want to develop a formalism inspired by the λ- and predicate calculi (§2), and to get away from that of set theory, we shall use Greek letters (ϕ, ψ, θ) for terms of type Σ^X, which will represent open subspaces. Then, instead of writing $x \in U$ for membership, we use λ-application, ϕx. (We shall re-employ the set-theoretic notation for more general kinds of subspaces than open ones in §§6.7.)

5.3. Before making things any more abstract, let's link this idea back to the concrete case of set theory instead of topology (*cf.* §2.10). The Lindenbaum–Tarski theorem (§4.5) is strictly classical, but Robert Paré proved a categorical version of it that is valid any elementary topos $\mathcal{S}$ or in any model of intuitionistic Zermelo set theory. The powerset $\mathcal{P}(X)$ is given by the exponential Ω^X, where Ω is the lattice of truth values, which we shall discuss further in §7. Paré showed that the contravariant functor $\Omega^{(-)}$ is monadic in the above sense (Paré, 1974).

We find limits of algebras by computing them for their carriers and lifting the algebraic structure; in categorical jargon, the functor $\mathsf{U} : \mathcal{A} \to \mathcal{S}$ *creates* limits. Since $\mathcal{A} \simeq \mathcal{S}^{\mathrm{op}}$, this derives *colimits* in $\mathcal{S}$ from its limits. (Christian Mikkelsen (1976) had already shown this by other methods.) These results simplified the definition of an elementary topos by removing the need to assert the existence of colimits as an axiom, as had been done previously.

5.4. This result was also the original inspiration for ASD, in 1993, although the results that we discuss in the rest of this section were discovered much later.

How could we set up the monadic situation in topology, within the category of Bourbaki spaces? Ralph Fox's original investigations (Fox, 1945) showed that the function-space Y^X only behaves reasonably well when X is locally compact. The meaning of "reasonably well" crystalised into the notion of a cartesian closure that emerged in the 1960s and was shown to be equivalent to the λ-calculus (§§2.2 & 2.5). It was meanwhile becoming clear that, for a compact subspace, the filter of open neighbourhoods (§5.7) plays a more important role than its set of points (Wilker, 1970).

Then Dana Scott (1972) saw that the crucial case is the function-space Σ^X, where Σ is the Sierpiński space (§7). Σ^X is the lattice of open subsets of X, but it obeys the universal property in §2.5 iff X is locally compact and Σ^X has the non-Hausdorff topology that now bears Scott's name. The study of continuous lattices (Gierz et al., 1980) and domain theory grew out of this. For a historical study of function spaces in topology, see Isbell (1986).

The category of locally compact spaces has products and an object Σ that has powers Σ^X, as we require. Whilst general functions-spaces Y^X exist as Bourbaki spaces when X is locally compact, the result need no longer be locally compact, so we do not have a cartesian closed category. In particular, $\mathbb{N}^{\mathbb{N}}$, known as *Baire space*, is not locally compact, essentially because its open subspaces are rather large, whilst its compact ones are rather small.

Nevertheless, cutting down to sober spaces, we have a model of the monadic situation above, as we show in §§5.8ff.

5.5. Category theory is a very potent drug, which should only be prescribed in small quantities. A theorem of Jon Beck (Taylor, 1999, §7.5) (he never published it himself) characterises when an adjunction $\mathsf{F} \dashv \mathsf{U}$ is monadic in terms of two properties of the right adjoint $\mathsf{U} : \mathcal{A} \to \mathcal{S}$:

First, U must *reflect invertibility*: for $f : A \to B$ in $\mathcal{A}$, if $\mathsf{U}f$ has an inverse in $\mathcal{S}$ then f already had one in $\mathcal{A}$. We shall see in §6.2 that this is equivalent to our abstract definition of sobriety.

Second, U must *create* U-*split coequalisers* (*cf.* §5.3). New students balk at this clause, but it is the active ingredient in our categorical medicine. We need to explain Σ-split *equalisers*, which the coequalisers become under the contravariance. These are subspaces $i : S \subset X$ that have the subspace *topology*, *i.e.* any open subspace of S is the restriction of one of X, but *in a canonical way*. That is, there is a map $I : \Sigma^S \to \Sigma^X$ for which $\Sigma^i \cdot I = \mathsf{id}_{\Sigma^S}$.

Beck's theorem says that, given an idempotent E on Σ^X that *ought* to define a Σ-split subspace of X, then it does: there are i and I for which $E = I \cdot \Sigma^i$. We shall give the formal definitions in §6.5.

5.6. The simplest (and first, (Taylor, 2000a)) examples of Σ-split subspaces are open and closed ones. Any open subsubspace $V \subset U \subset X$ of an *open* subspace is already an open subspace of X. Conversely, the inverse image of $W \subset X$ in U is given by the intersection, $U \cap W$. Then the operation $U \cap (-)$ provides $\Sigma^i : \Sigma^X \to \Sigma^U$ for the inclusion $i : U \subset X$, and its splitting $I : \Sigma^U \to \Sigma^X$ is the inclusion $(V \subset U) \mapsto (V \subset X)$. The composite $E \equiv I \cdot \Sigma^i$ is the idempotent $U \cap (-)$ on Σ^X. In the ASD notation, U and V are represented by their classifiers θ and ϕ, and then E is $\lambda\phi.\, \phi \wedge \theta$.

In the case of *closed* subspaces, any open subsubspace $V \subset C \subset X$ is also the intersection with C of some open subspace of X. But in this case the most convenient choice of representative is the *maximal* one, $V \cup U$, where U is the open complement of C. This provides I, whilst Σ^i is represented by $U \cup (-)$. Now $E \equiv I \cdot \Sigma^i$ is $U \cup (-)$ or $\lambda\phi.\, \phi \vee \theta$ on Σ^X.

5.7. Before giving further examples of Σ-split subspaces, we need to say something about compact ones. Recall that a subspace $K \subset X$ of a Bourbaki space is *compact* if, whenever $K \subset \bigcup_i U_i$ for some family of open subspaces, this already holds for some *finite* subfamily. This says exactly that the predicate $K \subset U$ is Scott continuous as U ranges over open subspaces of X. The result lies in the object of truth values, considered as a topological space. Hence the predicate is a continuous function $A : \Sigma^X \to \Sigma$.

The operator A obeys the Gentzen-style rule (*cf.* §2.1) for universal quantification of any predicate ϕ over K, so long as ϕ denotes an open subspace:

$$\frac{\Gamma,\ x : X,\ x \in K \quad \vdash \quad \phi x \Leftrightarrow \top}{\Gamma \quad \vdash \quad A\phi \Leftrightarrow \top}$$

In particular, if $K \equiv X$ then A is the *universal quantifier* $\forall$, whilst in the case of a subspace we sometimes write A as a *modal operator* $\Box$, which we read as "must" or "necessarily".

5.8. We can show how the definitions of *local* compactness for Bourbaki spaces and locales express the object in question as a Σ-split subspace of Σ^N, where N is any set, which we may think of as $\mathbb{N}$ or a cardinal. Then Σ^N is the powerset of N, or the topology on N considered as a discrete space, where Σ^N is itself equipped with the Scott topology.

According to the definition for spaces that need not be Hausdorff (Hofmann & Mislove, 1981), if a point lies in an open subspace ($x \in U$) then there is a compact subspace K and an open one V such that $x \in V \subset K \subset U$. Moreover, the space has a *family* of such pairs (V^n, K^n), indexed by a set N.

Now let $A_n : \Sigma^X \to \Sigma$ or $A_n : \Sigma^{\Sigma^X}$ be the Scott-continuous operator corresponding to the subspace K^n, for each index $n \in \mathbb{N}$. The open subspaces U and V^n are represented in our notation by $\phi, \beta^n : \Sigma^X$. (We use super- and subscripts to indicate whether these subspaces or terms increase or decrease as the index $n : N$ increases.)

Hence local compactness says that

$$U \;=\; \bigcup\{V^n \mid K^n \subset U\} \qquad \text{or} \qquad \phi \;=\; \exists n.\, \beta^n \wedge A_n\phi,$$

which we call the *basis decomposition*.

Then the maps $i : X \to \Sigma^N$ and $I : \Sigma^X \to \Sigma^{\Sigma^N}$, defined by

$$ix \equiv \lambda n.\, \beta^n x \qquad \text{and} \qquad I\phi \equiv \lambda\xi.\, \exists n.\, \xi n \wedge A_n\phi,$$

satisfy $\Sigma^i \cdot I = \mathsf{id}_{\Sigma^X}$.

5.9. A locally compact *locale* is one whose corresponding frame is a continuous lattice (Johnstone, 1982, §VII 4). This relativises the notion of compactness to a Scott-continuous relation $\beta \ll (-)$ (called *way below*) on open subsets, and then requires that

$$\phi = \bigvee\nolimits^{\uparrow} \{\beta^n \mid \beta^n \ll \phi\}, \quad \text{or} \quad \phi = \exists n.\, \beta^n \wedge A_n\phi, \quad \text{where} \quad A_n\phi \equiv (\beta^n \ll \phi),$$

which is another basis decomposition, giving i and I as before.

These two definitions are actually different, in that

(a) $A_n\top = \top$ and $A_n(\phi \wedge \psi) = A_n\phi \wedge A_n\psi$ in the case of Bourbaki spaces, whilst
(b) the indices $n \in N$ form a lattice in the localic situation.

However, a space only carries a basis with *both* properties if it is itself compact and the intersection of any two compact subspaces is again compact; it is then called *stably locally compact*. See Taylor (2006a), Jung et al. 2001 for more on this approach to local compactness.

5.10. We have shown that every locally compact Bourbaki space or locale can be expressed as a Σ-split subspace of Σ^N. So if we can prove that all such subobjects are locally compact locales then we have a model of the monadic situation. Locally compact locales and sober locally compact Bourbaki spaces are equivalent, if we assume the axiom of choice (Johnstone, 1982, VII 4.5).

Recall from §4.4 that the Scott topology on Σ^N is the free frame $\mathsf{F}N$ on the set N. For a Σ-split subspace $X \subset \Sigma^N$, the topology L on X is a retract of $\mathsf{F}N$, where $H : \mathsf{F}N \twoheadrightarrow L$ and $I : L \rightarrowtail \mathsf{F}N$ are Scott-continuous and satisfy $H \cdot I = \mathsf{id}$. Hence L is a continuous lattice (Johnstone, 1982, VII 2.3).

The maps I and H need not be adjoint, but since H is a frame homomorphism, it has a right adjoint R that is monotone but not necessarily Scott continuous (§4.4). Then $\mathsf{id} \leqslant j \equiv R \cdot H = j \cdot j$ and j preserves meets, making it a nucleus (§4.8), and $H \equiv i^*$ and $R \equiv i_*$, where $i : X \subset \Sigma^N$ is the sublocale that j defines.

5.11. Although we have now related Beck's theorem in category theory to ideas of general topology, these are still rather abstract. We need some more compelling example from the applications of topology to demonstrate that Σ-split subspaces are important to the customers of this subject (*cf.* §3.5). Recall from §§1.8 and 4.7 that the classical Heine–Borel theorem fails in Russian Recursive Analysis but holds in locale theory.

Now, it is possible to construct a "space of Dedekind cuts" in any category that has $\mathbb{N}$, powers of Σ, equalisers and some other properties (Taylor, 2009a, Remark 6.9). However, the "closed interval" $[0, 1]$ within this space need not be compact. The category of dcpos and Scott-continuous functions has the relevant structure, but the order on a dcpo determines its topology, and the "real line" is (uncountable but) discrete in both senses, so its only compact subspaces are finite sets.

5.12. Beck's theorem comes to the rescue.

The classical Heine–Borel theorem provides a Scott-continuous Σ-splitting I for the inclusion of the equaliser $i : \mathbb{R} \rightarrowtail \Sigma^{\mathbb{Q}} \times \Sigma^{\mathbb{Q}}$, where $ix \equiv (\lambda d.\, d < x, \lambda u.\, x < u)$. In traditional notation, this I is given by

$$(V \subset \mathbb{R}) \text{ open } \mapsto \{(D, U) \mid \exists d \in D.\, \exists u \in U.\, (d < u) \wedge [d, u] \subset V\},$$

or

$$\phi : \Sigma^{\mathbb{R}} \mapsto \lambda\delta\upsilon.\, \exists du.\, \delta d \wedge \upsilon u \wedge (d < u) \wedge \forall x{:}[d, u].\, \phi x$$

in the ASD λ-calculus. The idempotent $I \cdot \Sigma^i$ can be shown to be equal to a formula that involves the rationals alone:

$$(I \cdot \Sigma^i)\Phi(\delta, \upsilon) \;\equiv\; \exists n \geq 1.\, \exists q_0 < \cdots < q_n.\; \delta q_0 \wedge \upsilon q_n \;\wedge\; \bigwedge_{k=0}^{n-1} \Phi(\delta_{q_k}, \upsilon_{q_{k+1}}).$$

This satisfies the hypothesis of Beck's theorem, so defines an object R of ASD. In R, the interval [0, 1] is compact, and indeed this object satisfies all of the properties that one could reasonably require of a computable real number object (Taylor, 2009a). Looking more closely at its topology, Σ^R, it can be shown, as in the classical situation, that any open subspace is uniquely expressible as a countable disjoint union of open intervals, although the words "countable" and "interval" require some constructive qualification (Taylor, 2010a, §15).

We now have enough experimental evidence to start writing down some axioms (§3.2), at least for locally compact spaces.

6 The Monadic Framework

In the previous section we described a number of properties in topology that are related to the monadic adjunction $\Sigma^{(-)} \dashv \Sigma^{(-)}$. We also showed that this is satisfied when Σ is the Sierpiński space in the category of locally compact spaces, or the subobject classifier in any topos, and in these cases Σ^X is the topology or powerset of X. In this section we describe the purely categorical structure that we abtract from this situation. Then we shall show how it can can be formulated as a new symbolic language. See Taylor, 2002a, 2002b for the details in ASD and (Barr & Wells, 1985, §3.3) (Mac Lane, 1971, §VI 7) (Taylor, 1999, §7.5) for textbook accounts of Beck's theorem.

6.1. The category $\mathcal{S}$ must have

(a) finite products, $\mathsf{1}$ and $\times$;
(b) an object Σ with a map $\star : \mathsf{1} \to \Sigma$ ($\star$ is only used in §6.10); and
(c) powers Σ^X for all $X \in \mathsf{ob}\mathcal{S}$; then
(d) the adjunction $\Sigma^{(-)} \dashv \Sigma^{(-)}$ must be monadic.

In investigating this structure, we shall need to apply the (contravariant) functor $\Sigma^{(-)}$ repeatedly, giving some unwieldy towers of exponentials. For this reason, we often write

$$\Sigma^2 X \equiv \Sigma^{\Sigma^X} \qquad \Sigma^3 X \equiv \Sigma^{\Sigma^{\Sigma^X}} \qquad \Sigma^4 X \equiv \Sigma^{\Sigma^{\Sigma^{\Sigma^X}}} \ldots$$

The unit of the adjunction, $\eta_X : X \to \Sigma^2 X$, is the transpose of the co-unit ε_X, which is $\mathsf{ev}_X : \Sigma^X \times X \to \Sigma$ and is in turn the other transpose of $\mathsf{id} : \Sigma^X \to \Sigma^X$. Then η is part of the structure of the monad, for which $\Sigma^2 X$ is the free algebra on an object X, with structure map $\mu_X \equiv \Sigma^{\eta_X}$, where μ is called the *multiplication*.

The corresponding symbolic language is the familiar one that has pairing and λ-abstraction, except that the exponential type S^X and corresponding λ-terms $\lambda x.\, s$ may only be formed in the case where S (the type of the body s), is a power of Σ. Then

$$\eta_X x \equiv \lambda\phi.\, \phi x \qquad \text{and} \qquad \mu_X \Phi \equiv \lambda x.\, \Phi(\lambda\phi.\, \phi x)$$

are the unit and multiplication of the monad. Contexts (§2.6) in this first prototype (§3.2) of the language just consist of typed variables, but the study of recursion will oblige us to add equational hypotheses (Taylor, 2005).

6.2. The first clause of Beck's theorem (§5.5(a)) says that if $\Sigma^f : \Sigma^Y \to \Sigma^X$ is invertible then so already was $f : X \to Y$. This is equivalent to requiring, for every object X, that η_X be the equaliser of the parallel pair in the diagram,

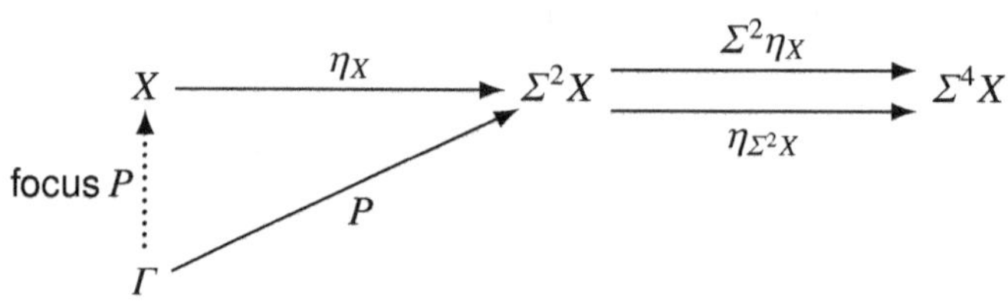

In other words, if P also has equal composites then there is a unique map that makes the triangle commute. P has equal composites iff its transpose, $\tilde{P} : \Sigma^X \to \Sigma^\Gamma$, is a homomorphism of Eilenberg–Moore algebras for the monad. Sobriety then says that $\tilde{P} = \Sigma^{\mathsf{focus}\, P}$ (Taylor, 2002a, §4).

6.3. According to the technique of §2.8, a morphism $P : \Gamma \to \Sigma^2 X$ is the same thing as a term $\Gamma \vdash P : \Sigma^2 X$. This has equal composites with the parallel pair above iff

$$\Gamma,\, \Phi : \Sigma^3 X \quad \vdash \quad \Phi P \;=\; P(\lambda x.\, \Phi(\lambda\phi.\, \phi x)) : \Sigma.$$

We call P *prime* if this holds, extending the usage of this word beyond that in §4.1.

Here we have our first example of the difference between the definitions that we give for the foundations and those that are appropriate for the applications (§3.8): when we add the other topological structure to the theory, P is prime simply iff it preserves $\top$, $\bot$, $\wedge$ and $\vee$ (Taylor, 2006a, §10).

The object X is *sober* iff every prime P has a unique fill-in

$$\Gamma \vdash \mathsf{focus}\, P : X \qquad \text{such that} \qquad \Gamma,\, \phi : \Sigma^X \vdash \phi(\mathsf{focus}\, P) \;=\; P\phi : \Sigma.$$

Following §2.11, we turn this universal property into a new system of symbolic rules. Indeed, we have just given the introduction and β-rules for focus (Taylor, 2002a, §8).

The operation in the elimination rule takes $a : X$ to $\lambda\phi.\ \phi a : \Sigma^2 X$. This is prime: categorically, η_X has equal composites with the parallel pair, by naturality of $\eta_{(-)}$ with respect to η_X, or one could check this by a λ-calculation. Finally, the η-rule is $\mathsf{focus}(\lambda\phi.\ \phi a) = a : X$.

The normalisation theorem (*cf.* §2.12) says that focus may be eliminated from any term $\phi : \Sigma^X$, whilst any term of ground type is provably equivalent to $\mathsf{focus}\, P$, where P does not itself involve focus.

The symbolic calculus extended with focus corresponds (*via* §2.8 again) to a certain category. By the normalisation theorem, this has the same objects (contexts) as the original one, but its morphisms $\Gamma \to \Delta$ are in bijection with the Eilenberg–Moore homomorphisms $\Sigma^\Delta \to \Sigma^\Gamma$ (Taylor, 2002a, §§6-7).

6.4. It is pertinent to ask of this β-rule how much of the expression *surrounding* $\mathsf{focus}\, P$ is to be taken as ϕ, and moved *inside* $P\phi$. That is, for any $F : \Sigma^\Sigma$, does

$$F(\phi(\mathsf{focus}\, P)) \qquad \text{become} \qquad F(P\phi) \qquad \text{or} \qquad P(\lambda x.\ F(\phi x))?$$

So long as P is prime, this doesn't matter, because the two results are equal (consider $\Phi \equiv \lambda Q.\ F(Q\phi)$ in the definition of primality).

Hayo Thielecke (1997) considered an operation called force with the same β-rule, but without the primality side condition. Now it does matter where we draw the boundary of the super-term ϕ: the *computational effect* is to pass ϕ as an argument to P, and then *jump* to the *continuation* F when (if) P returns. In order to study computational effects in general, Thielecke developed categorical machinery that is very similar to our account of abstract sobriety, but independently.

6.5. The second clause of Beck's characterisation (§5.5) brings in *all* algebras that are definable in the sense of Eilenberg and Moore. It says that certain equalisers exist and are taken by $\Sigma^{(-)}$ to coequalisers (Taylor, 2002b, §§2–4).

Our approach to this is to spell out what was meant in §5.5 by "data on Y that ought to define a Σ-split subspace". For this, it is enough to consider equalisers of the form

$$X \overset{i}{\rightarrowtail} Y \underset{\tilde{E}}{\overset{\eta_Y}{\rightrightarrows}} \Sigma^2 Y \qquad \text{(a)}$$

that become *split* by η_{Σ^Y} when we apply $\Sigma^{(-)}$. This means that, in the diagram

$$\Sigma^X \underset{I}{\overset{\Sigma^i}{\leftrightarrows}} \Sigma^Y \quad \overset{\Sigma^{\eta_Y}}{\longleftarrow}\ \ \xrightarrow{\ \eta_{\Sigma^Y}\ }\ \ \underset{\Sigma^{\tilde{E}}}{\longleftarrow} \quad \Sigma^3 Y, \qquad \text{(b)}$$

the equation $\Sigma^{\tilde{E}} \cdot \eta_{\Sigma^Y} \cdot \Sigma^{\tilde{E}} = \Sigma^{\tilde{E}} \cdot \eta_{\Sigma^Y} \cdot \Sigma^{\eta_Y}$ holds.

Besides asking for the *equaliser* $i : X \rightarrowtail Y$ in (a), we further require that the solid lines in diagram (b) form a *coequaliser*. For this we simply need a map I such that

$$I \cdot \Sigma^i \;=\; E \;\equiv\; \Sigma^{\tilde{E}} \cdot \eta_{\Sigma^Y}.$$

Whilst these conditions may look very strange, they are actually adapted from the definition of an Eilenberg–Moore algebra (A, α): with $Y \equiv \Sigma^A$, $\tilde{E} \equiv \Sigma^\alpha$ and $I \equiv \eta_A$, they yield $PA \equiv X$. Conversely, one can show that the category of algebras admits coequalisers *of this form*. However, the category of spaces that we have axiomatised here need not have *all* equalisers, for example **LKLoc** does not.

6.6. Beck's theorem requires the map $\tilde{E}$ to satisfy a certain equation. Writing this in the λ-calculus (§2.8), we call $E : \Sigma^Y \to \Sigma^Y$ a *nucleus* if

$$\Phi : \Sigma^3 Y,\; y : Y \quad \vdash \quad E\,(\lambda y'.\, \Phi\,(\lambda\psi.\, E\psi y'))\, y \;=\; E\,(\lambda y'.\, \Phi\,(\lambda\psi.\, \psi y'))\, y,$$

observing the extra E on the left hand side. Since we don't want to get involved with dependent types, we do not allow E to have parameters.

We have appropriated the name "nucleus" from locale theory (§4.8), since both kinds of nuclei are used to define subspaces. The definitions are not the same, but the use of different letters (E and j) should avoid ambiguity. Like primes (§6.3), nuclei in ASD have another characterisation in the presence of the full topological structure (Taylor, 2006a, §10), namely

$$E(\phi \wedge \psi) \;=\; E(E\phi \wedge E\psi) \qquad \text{and} \qquad E(\phi \vee \psi) \;=\; E(E\phi \vee E\psi).$$

In this setting, ASD nuclei are Scott continuous, whilst those in locale theory need only be monotone; on the other hand, localic nuclei satisfy $\mathsf{id} \leqslant j$, which is not required in ASD. Beware in particular that the ASD nucleus for an open subspace is given by $E\phi \equiv (\theta \wedge \phi)$ (*cf.* §5.6), whilst the localic nucleus is $j\phi \equiv (\theta \Rightarrow \phi)$ (§4.8). Nevertheless, there are some important examples that satisfy all of the conditions of both definitions (Taylor, 2009a, §8) (Taylor, 2004b).

6.7. The symbolic calculus (§2.11) that corresponds to Beck's theorem must take account of both the equaliser of the main types and also the coequaliser of their topologies. It does this using *two* sets of rules (Taylor, 2002b, §8).

A term $\Gamma \vdash a : Y$ has equal composites with the parallel pair in §6.5(a) iff

$$\Gamma,\, \psi : \Sigma^Y \quad \vdash \quad \psi a \;=\; E\psi a : \Sigma,$$

and then we say that it is *admissible*. The introduction rule for the equaliser, *i.e.* its universal property, then allows us to form

$$\Gamma \quad \vdash \quad \mathsf{admit}\, a : \{Y \mid E\},$$

whilst the inclusion i is the elimination rule. The β- and η-rules are

$$\Gamma \vdash i(\mathsf{admit}\, a) \;=\; a : Y \qquad \text{and} \qquad x : \{Y \mid E\} \vdash \mathsf{admit}(ix) = x : \{Y \mid E\}.$$

We also have to ensure that the second diagram (§6.5(b)) is a coequaliser, using another set of rules. The introduction rule comes for free, as the map

$$\Sigma^i : \Sigma^Y \longrightarrow \Sigma^{\{Y|E\}}.$$

The elimination rule says that Σ^i is split by another map I,

$$y : Y,\ \psi : \Sigma^Y \quad \vdash \quad I_{Y,E}\,(\lambda x{:}\{Y \mid E\}.\ \psi(i_{Y,E}x))\,y \;=\; E\psi y,$$

so $I \cdot \Sigma^i = E$ and $\Sigma^i \cdot I = \mathsf{id}_{\{Y|E\}}$. These equations are the β- and η-rules

$$y : Y,\ \psi : \Sigma^Y \quad \vdash \quad I\,(\lambda x.\ \psi(ix))\,y \;=\; E\psi y$$

and $$x : \{Y \mid E\},\ \phi : \Sigma^{\{Y|E\}} \quad \vdash \quad I\phi(ix) \;=\; \phi x.$$

The normalisation theorem for this calculus (*cf.* §2.12) says that every type can be embedded as a subspace of a type formed without comprehension, and terms also normalise in a simple way. This makes the corresponding category equivalent to the opposite of the Eilenberg–Moore category (Taylor, 2002b, §§9–10).

6.8. Any map $f : X \to Y$ that is Σ-split (with $F : \Sigma^X \to \Sigma^Y$ such that $F \cdot \Sigma^f = \mathsf{id}_{\Sigma^X}$ or $F\phi(fx) = \phi x$) agrees up to isomorphism with the subspace $\{Y \mid E\}$ that is defined from $E \equiv \Sigma^f \cdot F$. In particular, f is regular mono. The isomorphism is

$$\begin{array}{rcl} x : X & \vdash & x' \equiv \mathsf{admit}\,(fx) : \{Y \mid E\} \\ x' : \{Y \mid E\} & \vdash & x \equiv \mathsf{focus}_X\,(\lambda\phi{:}\Sigma^X.\ F\phi(ix')) : X. \end{array}$$

To show that these maps are mutually inverse, we need the rules above, together with the lemma

$$i\,(\mathsf{focus}_X\,P) \;=\; \mathsf{focus}_Y\,(\Sigma^2 iP),$$

first proving that if P is prime then so is $\Sigma^2 fP$. This uses sobriety of X and Y.

6.9. Notice that the $\Sigma^{\{\}}\beta$-rule is the only one that introduces E into terms. This is something that we want to avoid if at all possible, because the expressions for nuclei are usually *very* complicated. When E does find its way into a program, it will give rise to a substantial computation.

Actually, this is what we would expect from practical considerations. For example, consider the nucleus that defines $\mathbb{R}$ in ASD (§5.12). The universal quantifier that says that the interval is compact is given by

$$(\forall x{:}[0,1].\ \phi x) \quad \equiv \quad I\phi\,(\lambda d.\ d < 0,\ \lambda u.\ u > 0),$$

which applies the extension $I\phi$ of ϕ to the pseudo-Dedekind cut that represents the interval. The normalisation theorem introduces the expression E, in this case the one given in §5.12. This divides the interval up into sufficiently small parts, on each of which the predicate ϕ must be satisfied, but in the fashion of Ramon Moore's interval analysis (Moore, 1966), *i.e.* by evaluating the arithmetical operations on the (endpoints of) the subinterval, instead of using single values (Taylor, 2009a, 2006b; Bauer, 2008).

6.10. How common is the monadic situation? We may obtain it from *any* category $\mathcal{S}_0$ with an object Σ_0 that has powers Σ_0^X. Let $\mathcal{A}$ be the Eilenberg–Moore category for the monad on $\mathcal{S}_0$; then $\mathcal{S} \equiv \mathcal{A}^{\mathrm{op}}$ has the monadic property. The proof of this is very easy—apart from the most basic requirement, namely that $\mathcal{S}$ have products (coproducts of algebras) (Taylor, 2002b, §7).

Can other categorical structure be defined in $\mathcal{S}$, before we add any specifically topological features? We have already remarked in §5.3 that colimits come for free, as long as we have the corresponding limits, so that $\mathcal{S}$ has coproducts and Σ*-split* coequalisers. In fact, the coproducts are *stable and disjoint, i.e.* the category $\mathcal{S}$ is *extensive* (Carboni et al., 1993; Cockett, 1993). Note that $\mathbf{0} \rightarrowtail \mathbf{1}$ is Σ-split iff Σ has a point (§6.1(b)). Whilst extensivity is a purely categorical *statement*, its *proof* relies heavily on the new λ-calculus (Taylor, 2002b, §11).

6.11. Thielecke argues that Σ (the "answer type" R, in his notation) needs no extra structure. In his view, it may perhaps be seen as a free type variable, or one that is quantified at the outermost level. I once said to him that his work must therefore be either very superficial or very deep. The fact that one can develop quite a lot of theory from just this monadic adjunction $\mathcal{S}^{\mathrm{op}} \leftrightarrows \mathcal{S}$, without any other hypothesis on Σ, leads me to believe more and more in the second possibility.

This structure and its generalisation (Taylor, 2009b) will provide the skeleton on which the more obviously topological structure is the flesh. Dressed in a different way, I also believe that it could be applicable to game semantics of sequential computation, algebraic geometry, differential geometry and quantum logic.

6.12. When we used the traditional notation $\{Y \mid E\}$ for *subset formation* in §6.7 we claimed more than we can deliver on the basis of the monadic hypothesis alone.

Although the symbolic formulation is a *little* easier to handle than the categorical one, it has to be said that devising nuclei in ASD requires a lot of inspired guesswork, because of the need to find splittings. However, this problem is by no means a new one: it is a feature of Jon Beck's theory of monads that he inherited from his own inspiration in homological algebra, where we cannot in general split short exact sequences. This had in turn come from the (mathematically interesting) fact that there are non-split extensions of Abelian groups, starting with the cyclic group of order 4 as an extension of that of order 2.

Like the category of locally compact spaces, $\mathcal{S}$ need not be cartesian closed. However, when we develop induction and recursion in the extended version of this paper, we find that the lack of general *equalisers* is the real handicap to developing the theory. The conjectural extension (Taylor, 2009b) would provide a much more expressive language in the notation $\{y : Y \mid \cdots\}$, without the need for splittings. It would also provide general equalisers and function spaces. Nevertheless, splittings, where they exist, will remain important, for example to prove that $[0,1] \subset \mathbb{R}$ is compact (§5.12).

7 The Sierpiński Space

In order to put some topological flesh on this categorical skeleton, we need some more specific ideas about the object Σ. Beware that, in other subjects, Σ need not be a two-element set or a subobject classifier (*cf.* §4.3). In fact, we shall see that one should instead view it as the space that corresponds under Stone duality to the free algebra on one generator.

7.1. We have already made use of the analogy between topology and set theory, in the form of the Lindenbaum–Tarski–Paré theorem in §§4.5 & 5.3, as part of the motivation of the monadic property as an abstract formulation of Stone duality. Similarly, we begin by looking at the subobject classifier (lattice of truth-values) Ω in a topos, in order to identify the relevant properties of its analogue in topology.

The defining property is that any subobject $U \hookrightarrow X$ is the inverse image of $\top : 1 \hookrightarrow \Omega$ along some *unique* map $\theta : X \to \Omega$. That is, there is a pullback

$$\begin{array}{ccc} U & \longrightarrow & 1 \\ \big\downarrow & \lrcorner & \big\downarrow{\scriptstyle \top} \\ X & \xrightarrow{\ \theta\ } & \Omega \end{array}$$

The map θ is called the *classifier*, and it is crucial that isomorphic subspaces have equal classifiers.

7.2. Relying on this, if we can show that two constructions yield equivalent subobjects then the corresponding classifying maps are equal. Stripping out all of the extensional ideas, this means that we regard two propositions ϕ and θ as *equal* if they are *inter-provable*. Any proof is as good as any other, in contrast to the view of many type theorists, who regard different proofs of the same proposition as terms of an analogous type (*cf.* §2.2).

Although *individual* propositions are not regarded as types in topos theory, the type Ω is that of *all* propositions. Treating it as an object like any other means that we are doing *higher order logic*. There are some strange consequences of the translation of inter-provability into equality, since functions of a logical argument must respect equality.

In particular, to *assert* any ω (where $\omega \in \Omega$) is the same as to say that ω is *equal* to $\top$, from which it follows that $F\omega = F\top$ for any function $F : \Omega \to \Omega$. Now, this equality means that $F\omega \Rightarrow F\top$ and $F\omega \Leftarrow F\top$, still under the hypothesis ω. But ω *quā* hypothesis may be eliminated using Gentzen's introduction rule for $\Rightarrow$ (*cf.* §2.1), in the form

$$\frac{\Gamma,\ \omega \vdash \alpha \Rightarrow \beta}{\overline{\Gamma \vdash \omega \wedge \alpha \Rightarrow \beta,}}$$

giving $\omega \wedge F\omega \Rightarrow F\top$ and $F\omega \Leftarrow \omega \wedge F\top$.

From these we deduce the *Euclidean principle*,

$$\text{for any } F : \Omega \to \Omega \text{ and } \omega : \Omega, \qquad \omega \wedge F\omega \iff \omega \wedge F\top.$$

We shall see that this is more than just a curiosity of higher order logic.

7.3. In a topos, Ω classifies *arbitrary* subobjects, but Giuseppe Rosolini refined the definition to handle restricted classes of monos in other kinds of categories, such as open subspaces in topology and recursively enumerable subsets in the theory of computability. The class of monos, which he called a *dominion*, must include all isomorphisms and be closed under composition and under pullback along arbitrary maps. Then a *dominance* $\top : 1 \to \Sigma$ has the same definition as Ω, but restricted to the given class of monos (Rosolini, 1986). The Euclidean principle holds for any dominance Σ of which all powers Σ^X exist, not just for Ω in a topos, as is shown diagrammatically in (Taylor, 2000a). We shall prove the converse in §8.4 and recover the topos situation in §8.11.

7.4. Now consider the *Sierpiński space* in classical topology. This has two points, $\top$ and $\bot$, of which the former is open and the latter closed, like this:

$$\begin{array}{c} \top : \\ \bot : \end{array} \quad \begin{pmatrix} \odot \\ \bullet \end{pmatrix}.$$

According to our stated methodology, we employ this space because of its universal property. This says that there is a three-way bijective correspondence amongst

(a) open subspaces $U \subset X$,
(b) continuous maps $\theta : X \to \Sigma$ and
(c) closed subspaces $C \subset X$,

where we shall say that θ *classifies* $U \equiv \theta^{-1}(\top)$ and *co-classifies* $C \equiv \theta^{-1}(\bot)$.

Diagrammatically, these inverse images are given by the two pullbacks

$$\begin{array}{ccccc} U & \hookrightarrow & X & \hookleftarrow & C \\ \downarrow & & \downarrow \theta & & \downarrow \\ 1 & \xrightarrow{\ \top\ } & \Sigma & \xleftarrow{\ \bot\ } & 1 \end{array}$$

Either of the subobjects U or C determines θ (and so the other subobject) uniquely, so open inclusions form a dominion for which $\top : 1 \to \Sigma$ is a dominance, whilst closed inclusions form another dominion for which $\bot : 1 \to \Sigma$ is a dominance.

7.5. It is very difficult to assess the significance of a classical statement about the two-element set (*cf.* §3.4), so once again we turn to locale theory for a constructive view. An "open subspace" of a locale is an *element* of the corresponding frame, $a : 1 \to A$, and such elements correspond bijectively to frame homomorphisms $\mathsf{F}1 \to A$. As our candidate for the Sierpiński *locale*, we therefore have the free frame $\mathsf{F}1$ on one generator (Joyal & Tierney, 1984, §IV 3); by §4.4, this is the lattice of Scott-open subsets of $\mathcal{P}(1) \cong \Omega$.

Recall from §4.8 that the open and closed sublocales of X named by $a \in A$ are given by the nuclei $a \Rightarrow (-)$ and $a \vee (-)$ respectively. If $b \in A$ gives rise to an isomorphic sublocale of *either* kind then the corresponding nuclei are equal as endofunctions, but by applying them both to $\bot$, a and b, we deduce that $a = b$. So the Sierpiński locale enjoys the same universal property as its classical analogue, *cf.* (Johnstone, 1982, Lemma II 2.6).

7.6. Hence, in the category of locally compact locales, the Sierpiński locale is a dominance in two ways. It therefore satisfies both the Euclidean principle and its lattice dual. Besides this, any endofunction $F : \Sigma \to \Sigma$ preserves the order on Σ.

Putting all three of these properties together, we deduce that

$$F\sigma \iff F\bot \vee \sigma \wedge F\top,$$

which is called the *Phoa principle*, and had already been observed in computability theory. Indeed, Martin Hyland, who named this principle after his student Wesley Phoa (pronounced "Pwah"), emphasised that Σ should classify both RE and co-RE subsets (Hyland, 1991).

7.7. Once again, it is difficult to see the real meaning of such simple formulae as the Euclidean and Phoa principles. The latter says say that any *function* $F : \Sigma \to \Sigma$ is equal to a *polynomial* in one variable σ and the algebraic operations, $\top$, $\bot$, $\wedge$ and $\vee$. This means that Σ^{Σ} (the topology on the Sierpiński space) is the free algebra $\mathsf{F}1$ on one generator, as we found in locale theory.

This suggests a generalisation to algebraic or differential geometry, that any endo*morphism* $F : R \to R$ of the base ring, *considered as a space* in the new category, is defined by a *polynomial* in one variable (Kock, 1981, I 12.3). Another way of understanding this principle is that it is what is required to make the monad for the "symbolic" algebraic structure (such as $\top$, $\bot$, $\wedge$, $\vee$) agree with that of the λ-calculus ($\Sigma^{(-)} \dashv \Sigma^{(-)}$).

It was in order to provoke investigation of this analogy that I called the earlier equation the *Euclidean principle*. This name was suggested by its resemblance to one step of the Euclidean algorithm for highest common factors. If the Σ^2-monad is really the fundamental idea, then the Phoa principle would say that the symbolic operations agree with *it*, whilst subobjects would just be a manifestation of the way in which polynomials behave that is peculiar to set theory and topology.

7.8. In the next section we shall add the Phoa principle to the monadic calculus from §6, and prove some results from it that look a lot like the familiar properties of open, closed and compact subspaces. Of course, this indicates that the intuitions that we have considered so far are sound and substantial.

What is surprising is how much can be done in the *absence* of an axiom that actually states general Scott continuity. One might expect this to be essential for topology, especially considering its fundamental role in the applications in computer science since Scott's original work in the 1970s.

Even though the theory at this second stage is manifestly incomplete, it is worthy of study as it stands, because it is still fully *lattice dual*. We can rewrite any theorem by interchanging (all of) the following symbols or concepts with their partners, to obtain another valid theorem:

$\leqslant$	$\Rightarrow$	$\top$	$\wedge$	$\exists$	$=$	open	discrete	?	open
$\geqslant$	$\Leftarrow$	$\bot$	$\vee$	$\forall$	$\neq$	closed	Hausdorff	compact	proper

In particular, in §8.5 we shall have a proof rule that *looks* like Gentzen's rule for classical $\neg$, alongside the one in §§2.1 & 7.2 for $\Rightarrow$. The "classical" rule is valid constructively (in particular, in intuitionistic locale theory, §7.5) because it says the same thing for inclusions of closed subspaces that the "intuitionistic" rule says about open ones. Open and closed subspaces are related *via* their common classifiers (§7.4), and not by set-theoretic complementation. Statements in ASD are therefore free of the double negations that plague mathematics that is based on intuitionistic *set* theory, whether they are written in a categorical or type-theoretic style.

7.9. Symmetries like this in a theory are powerful because they are *predictive*. (Consider Mendeleev's periodic table, or Dirac's relativistic quantum mechanics.) When we pair up concepts with their duals, we often find that some things were *missing*. (New elements or particles, such as the positron.)

What is the counterpart of compactness in the table above? Since the symbolic form of one topological concept is the universal quantifier (§5.7), the missing idea must be given by the existential quantifier.

Turning to locale theory for more inspiration, a compact locale is one for which the terminal projection $K \to 1$ is proper, so we are looking for those for which this map is *open*. These were studied in (Johnstone, 1984; Joyal & Tierney, 1984) and called *open locales*. Bourbaki spaces do not throw any light on this property, because they *all* have it. Casting our philosophical net a little wider, however, we find analogous ideas in computability theory (recursive enumerability) and constructive analysis (locatedness (Spitters, 2010)). The common theme seems to be an *explicit presentation* of something.

However, when we look at *sub*spaces that are "open" in this sense, it turns out that they are often (but by no means always) *closed* subspaces, so we need another word for the idea. In English, the word *overt* means "open" in its sense of being explicit, so it captures these ideas very well, although it is a little difficult to find appropriate translations into other languages.

Bourbaki spaces and locales treat infinitary unions and finitary intersections asymmetrically. This is explained in ASD by saying that we may form joins indexed by overt objects and meets by compact ones.

8 Topology Using the Phoa Principle

We can now give a flavour of the new theory of general topology that results from our methodology. This begins the formal development of theorems from axioms in the orthodox way (§3.2) that is developed more fully in the extended version of this paper and in the others in the ASD programme.

The monadic framework that we introduced in §6 plays (part of) the role of the set theory that is taken for granted when we study other subjects such as group theory, but which we have eschewed for computable topology. We now add the lattice structure on Σ and either the Euclidean or the Phoa principle.

See Taylor (2000a) for the details of the results in this section, but, despite the lettering, this was written before the monadic λ-calculus of §6 was devised, so it was expressed in a less clear categorical language. Since the empirical arguments were also not as well developed there as they were in the previous section, we shall give some of the basic results here in detail. On the other hand, (Taylor, 2009a, 2010a) give better treatments of some topics that are closer to the applications, especially the theory of compact subspaces.

8.1. In addition to the axioms in §6.1, we now require the type Σ to have

(a) constants $\top, \bot : 1 \rightrightarrows \Sigma$ (in place of $\star$, §6.1(b)) and operations $\wedge, \vee : \Sigma \times \Sigma \rightrightarrows \Sigma$ that satisfy
(b) the equations for a distributive lattice and
(c) the Euclidean principle $\sigma \wedge F\sigma \Leftrightarrow \sigma \wedge F\top$ if we want to consider set theory or
(d) the Phoa principle $F\sigma \Leftrightarrow F\bot \vee \sigma \wedge F\top$ for general topology,

where $\sigma : \Sigma$ and $F : \Sigma^\Sigma$.

8.2. The lattice structure defines an order relation on Σ and its powers Σ^X. We have, following logical custom, already started to write $\Rightarrow$ and $\Leftrightarrow$ for the order and equality on Σ. We retain $\leqslant$ and $=$ for these relations on its powers (the symbols $\subset$ and $\subseteq$ are inappropriate, for many reasons), but application to an argument changes them into $\Rightarrow$ and $\Leftrightarrow$, so if $\phi \leqslant \psi$ then $\phi a \Rightarrow \psi a$, whilst if $\phi = \psi$ then $\phi a \Leftrightarrow \psi a$.

In the topological case (d), all $\mathcal{S}$-morphisms $\Sigma^Y \to \Sigma^X$ preserve $\leqslant$, so $\mathcal{S}$ is an *ordered category*, which is a special case of a *2-category*. In particular, we may talk about *adjoint* maps $\Sigma^Y \leftrightarrows \Sigma^X$ with respect to this order: $L \dashv R$ means that $\mathsf{id} \leqslant R \cdot L$ and $L \cdot R \leqslant \mathsf{id}$. This is also possible in the set-theoretic case (c), so long as we're careful, as logical negation ($\neg$) reverses the order.

We may extend $\leqslant$ to an order on any object X, where

$$x \sqsubseteq_X y \quad \text{means} \quad (\lambda\phi.\, \phi x) \leqslant (\lambda\phi.\, \phi y).$$

By sobriety (§6.3) and monotonicity, $\sqsubseteq$ is antisymmetric and (when $X \equiv \Sigma^Y$) agrees with $\leqslant$. Wesley Phoa introduced his principle (Phoa, 1990) to make this order equivalent to the existence of a *link* map $\ell : \Sigma \to X$ with $\ell\bot = x$ and $\ell\top = y$, which he used to construct limits of the objects that he was considering.

8.3. The Euclidean principle is a purely *algebraic* property, at least if you are willing to regard λ-calculus as algebra. However, in the context of the monadic framework, it automatically yields the higher order structure of a dominance (§7.3), classifying certain *subobjects*. We shall build up to a characterisation of elementary toposes in §8.11.

First we have to show that the pullback in §7.1 exists. It is equivalent to the equaliser

$$U \xhookrightarrow{\ i\ } X \underset{x \mapsto \lambda\phi.\, \theta x \wedge \phi x}{\overset{x \mapsto \lambda\phi.\, \phi x}{\rightrightarrows}} \Sigma^{\Sigma^X}$$

which is an example of the one in §6.5, with $E \equiv \lambda\phi.\, \theta \wedge \phi$. Notice that this is the same formula that we saw in §5.6 when we showed that open subspaces are Σ-split.

The question is therefore whether E is a nucleus. The λ-calculus definition in §6.7 requires

$$\Phi(\lambda\psi.\, \psi y \wedge \theta y) \wedge \theta y \iff \Phi(\lambda\psi.\, \psi y) \wedge \theta y.$$

This follows from the Euclidean principle, putting $\sigma \equiv \theta y$ and $F \equiv \lambda\tau.\, \Phi(\lambda\psi.\, \psi y \wedge \tau)$. Conversely, by considering $Y \equiv 1$, $\theta \equiv \lambda y.\, \sigma$ and $\Phi \equiv \lambda G.\, F(G\top)$, we see that this principle is also necessary.

Hence the Euclidean principle is exactly what we need to satisfy the *abstract* definition of nucleus. On the other hand, E satisfies the more user-friendly lattice-theoretic characterisation of nuclei that we also mentioned in §6.7, just using the distributive law. Of course, that is because the proof of this characterisation makes use of the Phoa principle (Taylor, 2006a, §10).

By a similar argument, $\lambda\phi.\, \phi \vee \theta$ is a nucleus (for the closed subspace coclassified by θ) iff Σ satisfies the *dual* Euclidean principle.

8.4. We may therefore use the monadic framework in §6 to define the pullbacks in §7.1 and §7.4. Then, bearing in mind that we still have to verify the stronger meanings that these words have outside this paragraph, we shall say that an inclusion $i : U \hookrightarrow X$ *open* if it (is isomorphic to one that) is the pullback of $\top : 1 \to \Sigma$ along some (not *ā priori* unique) $\theta : X \to \Sigma$, which *classifies* U. In particular, $\theta \equiv \top$ classifies $\mathsf{id} : X \hookrightarrow X$.

In order to prove that open inclusions form a dominion with dominance Σ in the sense of Rosolini (§7.3), one of the things that we have to show is that the composite of two open inclusions is open. For this, we use the Σ-splitting I that is provided by the monadic framework.

Suppose that $j : V \hookrightarrow U$ and $i : U \hookrightarrow X$ $V \overset{j}{\hookrightarrow} U \overset{i}{\hookrightarrow} X$ are classified by $\phi : \Sigma^U$ and $\theta : \Sigma^X$ respectively. Since I splits Σ^i, we have $\phi u \Leftrightarrow \Sigma^i(I\phi)u \Leftrightarrow (I\phi)(iu)$, so the bottom right square commutes:

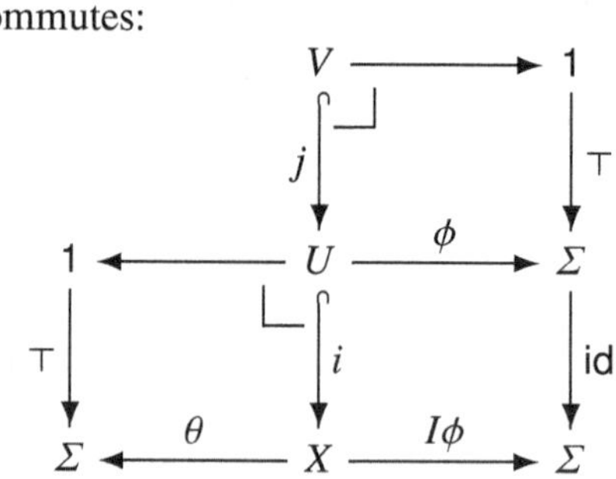

We claim that the *rectangle* on the right is a pullback, so $I\phi$ classifies $V \subset X$.

First, note that $\quad I\phi \;=\; (I\cdot\Sigma^i\cdot I)\phi \;=\; E(I\phi) \;=\; \theta\wedge I\phi \leqslant \theta.$
If $x:\Gamma\to X$ and $\Gamma\to \mathsf{1}$ make a commuting quadrilateral with $I\phi$ and $\top$ then $\top\Leftrightarrow I\phi x\Rightarrow \theta x$. So $x = iu$ for some $u:\Gamma\to U$ since θ classifies U by hypothesis. Then $\phi u\Leftrightarrow (I\phi)(iu)\Leftrightarrow I\phi x\Leftrightarrow\top$, so $u = jv$ for some $v:\Gamma\to V$ since ϕ classifies V. Hence $x = (i\cdot j)v$, and v is unique since $i\cdot j$ is mono. The composition property follows and the inclusions $V\hookrightarrow U\hookrightarrow X$ have classifiers $I\phi\leqslant\theta : X\to\Sigma$.

If $U, V\hookrightarrow X$ are open subobjects of X classified by $\theta,\psi:\Sigma^X$, and $V\subset U$, then by the pullback property,

$$\begin{array}{ccccc}
U\cap V & \xrightarrow{\;\cong\;} & V & \longrightarrow & \mathsf{1}\\
{\scriptstyle j}\downarrow & \lrcorner & \downarrow & \lrcorner & \downarrow{\scriptstyle\top}\\
U & \xrightarrow{\;i\;} & X & \xrightarrow{\;\psi\;} & \Sigma,
\end{array}$$

$V = U\cap V\hookrightarrow U$ is also an open subspace, classified by $\phi\equiv\psi\cdot i\equiv\Sigma^i\psi:\Sigma^U$, and we have the same situation again, with $\psi = I\phi\leqslant\theta$.

Conversely, if $\psi\leqslant\theta : X\to\Sigma$ then $\psi = \psi\wedge\theta = E\psi = I\phi$ where $\phi\equiv\Sigma^i\psi$. So the open inclusion $V\hookrightarrow X$ classified by ψ factors as $V\hookrightarrow U\hookrightarrow X$, where $V\hookrightarrow U$ is classified by ϕ. Hence we have an order-isomorphism between open subobjects $V\subset U$ of X and maps $\psi\leqslant\theta : X\to\Sigma$, from which it follows that classifiers are unique.

We are justified in calling i an "open" inclusion because $I\dashv\Sigma^i$ in the sense of §8.2, and the Frobenius and Beck–Chevalley laws (§2.9) hold (Taylor, 2000a, Prop. 3.11). If we want to use this result in the study of set theory (§8.1(c)) then we must be careful not to *assume* that I preserves $\leqslant$, but the same result also proves this. We write $\exists_i$ for I, for reasons that we shall explain in §8.7.

Reversing the order $\leqslant$ throughout this argument and assuming the dual Euclidean principle, we see that $\bot:\mathsf{1}\to\Sigma$ is a dominance too. We call the inclusions that are expressible as pullbacks of $\bot$ along $\theta:\mathsf{1}\to\Sigma$ *closed*, say that θ *co-classifies* them and write $\forall_i$ for I. Beware, however, that if $\theta\leqslant\phi$ classify C and D respectively then $D\subset C$.

See §8.10(g) for an important converse result.

8.5. In practice, it is often easier to reason with the Gentzen-style rules

$$\frac{\Gamma,\ \sigma\Leftrightarrow\top\ \ \vdash\ \ \alpha\Rightarrow\beta}{\Gamma\ \ \vdash\ \ \sigma\wedge\alpha\Rightarrow\beta}\qquad\qquad \frac{\Gamma,\ \sigma\Leftrightarrow\bot\ \ \vdash\ \ \alpha\Rightarrow\beta}{\Gamma\ \ \vdash\ \ \alpha\Rightarrow\beta\vee\sigma}$$

that are equivalent to the Euclidean principle and its dual, *cf.* §§2.1, 7.2 & 7.8.

To do this, we first have to modify the definition of *contexts* in §6.1, which now consist of both typed variables and *equational hypotheses* of the forms $\sigma\Leftrightarrow\top$ and $\sigma\Leftrightarrow\bot$ for $\sigma:\Sigma$. We proved the Euclidean principle from the Gentzen rule on the left using such contexts in §7.2, and the dual is similar.

This modification is *syntactic sugar* (conservative). A context that consists of several variables and hypotheses of both kinds may be interpreted as a *locally closed* subspace (*i.e.* the intersection of a closed subspace with an open one) of the product of the types of the variables.

Using the Euclidean principle, the open subspace classified by σ is Σ-split with nucleus $E \equiv \lambda\phi.\, \sigma \wedge \phi$. So the inequality $\alpha \Rightarrow \beta$ between subsubspaces is defined as $\alpha \wedge \sigma \Rightarrow \beta \wedge \sigma$ in the ambient space, which is what the Gentzen rule says.

These rules do not force $F : \Sigma^\Sigma$ to preserve the order—this has to be stated separately. Beware that we use this property so often in topology that we usually take it as read.

8.6. Now we can ask when particular subspaces are open or closed, starting with the diagonal $X \subset X \times X$. It is natural to call X

(a) *discrete* if the diagonal is open, its classifier being called *equality*, $=_X$; and
(b) *Hausdorff* if the diagonal is closed, its co-classifier being called *inequality* or *apartness*, $\neq_X$ or $\#_X$.

For example, we expect $\mathbb{N}$ and $\mathbb{Q}$ to have both properties, and $\mathbb{R}$ to be Hausdorff but not discrete. This is in line with the computational fact that we may decide equality of integers or rationals, but only detect when real numbers are *different*.

Beware that, whilst points and the diagonal are open in a discrete space, *arbitrary* subspaces need not be. This is because the *computable* topology on the discrete space $\mathbb{N}$ consists of the *recursively enumerable* subsets, not all of them. Also, our definition of Hausdorffness is slightly different from the one that is used in locale theory (Johnstone, 1982, §III 1.3, Simmons, 1978).

Classically, any discrete space is Hausdorff, but the proof employs a union that is illegitimate for us, and any free algebra with insoluble word problem provides a counterexample. Terms are equal iff they are provably so from the equations, so we observe equality by enumerating proofs, but no such proof need ever turn up. For example, combinatory algebra (with a non-associative binary operation and constants k and s such that $(kx)y = x$ and $((sx)y)z = (xz)(yz)$) encodes the untyped λ-calculus, and therefore arbitrary computation, so equality of its terms is undecidable.

We use a subscript or brackets, $n =_X m$ or $(n = m)$, to indicate the term of type Σ that this definition provides in a discrete space, with *computable* equality. This convention avoids ambiguity of notation with the underlying term calculus, in which any two terms of the same type may be *provably* equal or not.

The proof rules that relate these two notions of (in)equality are

$$\frac{\Gamma \;\vdash\; n = m : X}{\Gamma \;\vdash\; (n =_X m) \Leftrightarrow \top} \qquad\qquad \frac{\Gamma \;\vdash\; h = k : X}{\Gamma \;\vdash\; (h \neq_X k) \Leftrightarrow \bot}$$

The rule for a Hausdorff space is not doubly negated: it is simply the way in which we express membership of a closed subspace, whilst $\neq$ is the natural name for its co-classifier.

From the Gentzen rules in §8.5, we obtain respectively

$$\phi x \wedge (x =_X y) \Longrightarrow \phi y \qquad \text{and} \qquad \phi x \vee (x \neq_X y) \Longleftarrow \phi y.$$

The usual properties of equality (reflexivity, symmetry, transitivity and substitution) follow from the first of these, whilst the dual argument gives those of inequality. See Taylor (2010a, §§4–5) for more detail in the setting of applications.

8.7. Turning to compactness, recall from §5.7 that it can be expressed using an operator $A : \Sigma^{\Sigma^X}$. This is Scott continuous in the *model* (locally compact Bourbaki spaces), but is simply a term or morphism in the axiomatisation.

It is enough to define a *compact object* K as one for which the map $\Sigma \to \Sigma^K$ by $\sigma \mapsto \lambda k.\, \sigma$ has a right adjoint, $\forall_K$, in the sense of §8.2. The two directions of the adjoint correspondence are exactly the introduction and elimination rules of the symbolic formulation, in which it is more convenient to write $\forall k{:}K.\ \phi k$ than $\forall_K(\lambda k{:}K.\ \phi k)$.

However, in topology, we also have the *dual Frobenius law* (*cf.* §2.9),

$$\forall k.\, (\sigma \vee \phi k) \iff \sigma \vee \forall k.\, \phi k,$$

for free. This is deduced from the Phoa (or dual Euclidean) principle (§8.1(d)) by putting $F \equiv \lambda \tau.\, \forall k.\, \tau \vee \phi k$. It does not hold in intuitionistic *set* theory, but it was identified in intuitionistic *locale* theory by Japie Vermeulen (1994).

The first of the familiar properties of compact spaces is that any closed subspace $i : C \subset K$ is also compact. Its quantifier is given by $\forall_C \equiv \forall_K \cdot \forall_i$, where $\forall_i$ is the Σ-splitting of i in §§5.6 & 8.4 [C, Prop. 8.3].

Products and coproducts of compact spaces are again compact, as are equalisers and pullbacks targeted at Hausdorff spaces (Taylor, 2000a, §§8–9).

The Beck–Chevalley condition (§2.9) also comes for free, being just substitution under λ [C, Proposition 8.2]. Its lattice-theoretic interpretation is that any topology Σ^X admits K-indexed meets, and not just finite ones (§7.9).

8.8. When we try to apply these ideas to compact *sub*spaces, we find that the monadic framework in §6 does not provide a sufficiently general theory (so it doesn't extend to proper maps either). This is related to the fact that, in the leading model (§5.4), a compact subspace of a non-Hausdorff locally compact Bourbaki space need not itself be be locally compact. This is discussed in (Taylor, 2006a, §5), which is based on (Hofmann & Mislove, 1981).

The clearest treatment for ASD is that in (Taylor, 2010a, §8), which exploits the fact that the main object of study in that paper is a Hausdorff space ($\mathbb{R}$), by restricting attention to subspaces that are both closed and compact. Since a Hausdorff space H has $\neq$, we expect the operator $A : \Sigma^{\Sigma^H}$ to represent (as a compact space) the closed subspace that is co-classified by

$$\omega \equiv \lambda x.\, A(\lambda y.\, x \neq y) \qquad \text{if} \qquad \frac{\phi \vee \omega \;=\; \top}{\overline{\overline{A\phi \;\Leftrightarrow\; \top}}} \qquad \text{for any } \phi : \Sigma^H.$$

If H is also compact, we obtain conversely

$$A\phi \iff \forall x{:}H.\ \omega x \vee \phi x.$$

Hence closed and compact subspaces coincide in any compact Hausdorff space, because there

$$\phi x \iff \forall y{:}H.\ (x \neq y) \vee \phi y,$$

which is obtained from the formula in §8.6. Notice that A, like θ (§8.4), *decreases* in the order (§8.2) as the compact or closed subspaces get bigger.

8.9. To make sense of this equivalence between closed and compact subspaces, however, we need some way of saying that K is "the same subspace" as C. We do this by defining

$$a \in C \quad \text{as} \quad \theta a \Leftrightarrow \bot \qquad \text{and} \qquad a \in K \quad \text{as} \quad A \leqslant \lambda\phi.\ \phi a$$

for any term $a : X$, possibly including parameters.

The definition of $a \in K$ is motivated by the idea that a compact subspace is the intersection of its open neighbourhoods (Hofmann & Mislove, 1981; Wilker, 1970). In a non-Hausdorff space, however, this intersection may be larger than the original compact subspace, and is called its *saturation*. Nevertheless, C and K do have the same "elements" according to the definitions that we have just given. The account in (Taylor, 2010a, §8) also treats direct images and compact *open* subspaces.

8.10. Because of the lattice duality of the axioms that we have introduced so far (§§7.8f), there is a dual notion to compactness. We call it *overtness*, and it is related to the existential quantifier.

By the same arguments as for compact Hausdorff spaces (Taylor, 2000a, §§6–9),

(a) $\mathbf{0}$ and $\mathbf{1}$ are overt, as are $X \times Y$ and $X + Y$ if X and Y are;
(b) any open subspace of an overt space is again overt;
(c) any equaliser or pullback of overt spaces targeted at a discrete space is overt;
(d) any overt subspace of a discrete space is an open subspace;
(e) any direct image of an overt space is overt;
(f) any map from an overt object to a discrete one is an open map; and
(g) any mono $f : X \rightarrowtail Y$ from an overt object to a discrete one is an open inclusion, classified by $\theta \equiv \lambda y.\ \exists x.\ (fx =_Y y)$.

See (Taylor, 2000a, §§8 & 10) for the proofs, which were inspired by similar results in Joyal & Tierney, (1984).

To justify the last part, define

$$F : \Sigma^X \to \Sigma^Y \qquad \text{by} \qquad F\phi \equiv \lambda y{:}Y.\ \exists x{:}X.\ \phi x \wedge (fx =_Y y).$$

Then $F\phi(fx) \Leftrightarrow \exists x'{:}X.\ \phi x' \wedge (fx' =_Y fx) \Leftrightarrow \exists x'{:}X.\ \phi x' \wedge (x' =_X x) \Leftrightarrow \phi x$ by §8.6. Hence $f : X \rightarrowtail Y$ is Σ-split, and $X \cong \{Y \mid E\}$ by §6.8, where $E \equiv \Sigma^f \cdot F$. Expanding E, we find that $X \subset Y$ is the open subspace classified by θ (§8.3).

8.11. To conclude this brief account of some of the ideas of topology that can be developed from the finitary structure of Σ in the monadic framework, we look at the extreme case in which *every* object of the category $\mathcal{S}$ is overt discrete. This brings us back to the Lindenbaum–Tarski–Paré theorem (§§4.5 & 5.3), as we now have enough structure to complete the characterisation of set theory (§2.10).

Any category $\mathcal{S}$ is an elementary topos iff it satisfies

(a) the axioms for the monadic framework in §6.1;
(b) the Euclidean principle §8.1(c); and
(c) *every* object is overt and discrete.

Since every map goes from an overt object to a discrete one, every mono is an open inclusion, *i.e.* classified by an element of Σ. Also, discreteness of Σ makes it into a Heyting algebra, where $\sigma \Rightarrow \tau$ is $\sigma =_\Omega (\sigma \wedge \tau)$ (Taylor, 2000a, §11).

9 Conclusion

9.1. We argued in §§1–3 that the foundations for a mathematical discipline should be drawn from the headline theorems of that discipline itself, and not from an all-purpose theory of dust. In doing so, we are following a long tradition of so-called "synthetic" reasoning what has always laboured in the face of formidable opposition from the majority view in the mathematical world.

Nevertheless, we have proposed a way of doing this, based on the formulation of the main theorems as universal properties that we have transformed into a specially designed symbolic language. We then applied this method to a computable and constructive form of general topology. In the previous section we saw that this method recovers some of the principal notions of the subject such as compactness, and suggests new ones such as overtness, from a very small number of basic principles.

The extended version of this paper and the others in the Abstract Stone Duality research programme show how the new theory of topology can be developed from these beginnings. Some further discussion of peri-theory (how to design the axioms, §3.3) is still needed (in particular we need to add an explicit axiom for Scott continuity), but to a large extent we can now proceed in the Euclid–Bourbaki style, *i.e.* from axioms to theorems.

9.2. Although we began by rejecting set theory as *foundations*, we still need some account of "sets" as discrete spaces for combinatorial purposes. We do not expect this to be anywhere near as powerful as the traditional one, but category theory provides suitable weaker alternatives, such as the notion of pretopos and André Joyal's (unpublished) theory of arithmetic universes.

Since we only have spaces, the sets have to be chosen from amongst them, but the sense in which we use the word "discrete" in ASD (having equality, §8.6) is too weak for this purpose. In fact, we take *overt discrete* objects as our "sets", *i.e.* those that also have $\exists$ (§8.10). This full subcategory forms a pretopos (Taylor, 2000a,

§10), and indeed an arithmetic universe when we add recursion and Scott continuity (Taylor, 2005).

9.3. However, this study of recursion also shows that the monadic framework in §6 is inadequate, because we need to use equational hypotheses. Categorically, these correspond to equalisers in the category, which our leading model of *locally compact* spaces or locales does not have.

One way in which the theory might be developed is therefore to axiomatise *two* categories:

(a) a larger one that has all equalisers, but not all powers of Σ (let alone of other objects), and
(b) a full subcategory of "locally compact" objects, that contains Σ^X for all locally compact X, but not all equalisers.

Both Bourbaki spaces and locales provide models for this larger category, but only the category of sober Bourbaki spaces (or *spatial* locales) has the richer structure mentioned in 9.6 below.

9.4. Whilst recursion exposes one flaw in the monadic framework, in another way it shows its conceptual strength, specifically with regard to our *abstract* notion of sobriety. When we ask what this means for $\mathbb{N}$, we find that it is equivalent to the rule of definition by description (Taylor, 2002a, §§9–10), which can be developed to general recursion (Taylor, 2000b). For $\mathbb{R}$, sobriety is Dedekind completeness (Taylor, 2009a, §14).

9.5. What would it take to give the subcategory of "sets" in §9.2 the full strength of a set theory? Recall that Bourbaki spaces have *underlying sets*, and that the functor $\mathsf{U} : \mathbf{BbkSp} \to \mathbf{Set}$ has adjoints on both sides that assign either the discrete or indiscriminate topology to a set. We have the analogue of the leftmost adjoint, the inclusion of (overt) discrete spaces in our category of all spaces, so *what if* this inclusion had a right adjoint U? It turns out that this makes the full subcategory of "sets" into an elementary topos (Taylor, 2004a).

This provides a setting in which we are able to compare the category that we have axiomatised abstractly with those of internal locales and Bourbaki spaces in the topos. As peri-theory, we can then distinguish between those features of the traditional axiomatisations of topology that are computable and those that are not, depending on whether the "underlying set" axiom is needed to define them.

9.6. The biggest challenge for ASD at the moment, however, is how to generalise it beyond local compactness, *cf.* §3.7. Hitherto, the theory has effectively relied on the "two categories solution" (§9.3), in which we have exponentials for some purposes and equalisers for others, but not in the same category. Notice that general function-spaces are not the most pressing issue, despite the attention that has been given to them in the literature.

Currently ongoing work (Taylor, 2009b) explores the *equideductive logic* that arises from any equaliser of exponentials,

$$\{x : X \mid \forall y{:}Y.\ \mathfrak{p}(y) \Rightarrow \alpha xy = \beta xy\} \rightarrowtail X \underset{\beta}{\overset{\alpha}{\rightrightarrows}} \Sigma^{\{Y|\mathfrak{p}\}},$$

where the quantifier $\forall$ may now range over any space, not just a compact one. Sober Bourbaki spaces provide a model, in which $\forall$ and $\Rightarrow$ have the obvious meanings.

One can translate the rules of inference of simpler calculi into this logic. For example, a space K is compact if the quantified *predicate* $\forall x{:}K.\ (\phi x = \top)$ above is given by a *term* $\forall x{:}K.\ \phi x : \Sigma$.

As an example of our programme of developing a new symbolic language for mathematical properties that are expressed in categorical form, there is an *existential* quantifier that is related to epis in the category. However, since the latter are not stable under pullback, the quantifier does not obey all of the usual symbolic rules, and in particular does not admit substitution (*cf.* §§2.5&2.9).

The logic can also be used to construct a cartesian closed category similar to Dana Scott's *equilogical spaces* (Bauer et al., 2004).

With the help of this logic we should be in a position to finalise the axioms of ASD. Then we can ask whether, as it did in the case of compactness of $[0, 1] \subset \mathbb{R}$ (§5.12), it provides "the right" topology (whatever that may mean) on $\mathbb{N}^{\mathbb{N}^{\mathbb{N}}}$ or $\mathbb{R}^{\mathbb{R}^{\mathbb{R}}}$ in recursion theory, or on Banach spaces in functional analysis.

The basic form of equideductive logic very probably does not give the answers that we might like for these higher spaces. However, since it is so simple, we might use proof theory to look for interesting alternative models and additional axioms.

Finally, equideductive logic, like the monadic framework in §6, embodies the ideas of Stone duality but depends very little on topology. So we might imagine the development of similar synthetic systems of reasoning for other mathematical disciplines. Equipped with this, we might hope to escape the tyranny of logicians in Cantor's dystopia and return to a new Euclidean paradise in which mathematical arguments actually manipulate the objects that they purport to describe.

Acknowledgements: I would like to thank Andrej Bauer, David Corfield, Anders Kock, Fred Linton, Gabor Lukasz, Andy Pitts, Sridhar Ramesh, Guiseppe Rosolini, Giovanni Sommaruga, Hayo Thielecke and Graham White for their very helpful comments on earlier drafts of this paper.

References

Appel, A. (1992) *Compiling with Continuations*, Cambridge: Cambridge University Press.

Barr, M. and Wells, C. (1985) *Toposes, Triples and Theories*, New York: Springer.

Bauer, A. (2008) Efficient Computation with Dedekind Reals, in Brattka, V., Dillhage, R., Grubba, T. and Klutch, A., eds. *Fifth International Conference on Computability and Complexity in Analysis*, in volume 221 of Electronic Notes in Theoretical Computer Science, Amsterdam: Elsevier.

Bauer, A., Birkedal, L. and Scott, D. (2004) Equilogical Spaces, *Theoretical Computer Science* **315**, 35–59.

Bishop, E. and Bridges, D. (1985) *Constructive Analysis*, Number 279 in Grundlehren der mathematischen Wissenschaften, Heidelberg, Berlin, New York: Springer.

Bourbaki, N. (1966) *Topologie Générale*, Paris: Hermann, English translation, "General Topology" Berlin: Springer (1989).

Bridges, D. and Richman, F. (1987) *Varieties of Constructive Mathematics*, Number 97 in London Mathematical Society Lecture Notes, Cambridge: Cambridge University Press.

Brown, R. (1964) Function Spaces and Product Topologies, *Quarterly Journal of Mathematics* **15**(1), 238–250.

Carboni, A., Lack, S. and Walters, R. (1993) Introduction to Extensive and Distributive Categories, *Journal of Pure and Applied Algebra* **84**, 145–158.

Cardone, F. and Hindley, R. (2006) History of Lambda-Calculus and Combinatory Logic, *Handbook of the History of Logic* **5**.

Church, A. and Rosser, B. (1936) Some Properties of Conversion, *Transactions of the American Mathematical Society* **39**(3), 472–482.

Cleary, J. (1987) Logical Arithmetic, *Future Computing Systems* **2**, 125–149.

Cockett, R. (1993) Introduction to Distributive Categories, *Mathematical Structures in Computer Science* **3**, 277–307.

Corfield, D. (2003) *Towards a Philosophy of Real Mathematics*, Cambridge: Cambridge University Press.

Eilenberg, S. and Moore, J. (1965) Adjoint Functors and Triples, *Illinois Journal of Mathematics* **9**, 381–98.

Fauvel, J. and Gray, J. (1987) *The History of Mathematics, a Reader*, London: Macmillan, Open University.

Fourman, M. and Hyland, M. (1979) Sheaf Models for Analysis, in Fourman, M., Mulvey, C. and Scott, D., eds. *Applications of Sheaves*, number 753 in Lecture Notes in Mathematics, Berlin, New York: Springer, pp. 280–301

Fox, R. (1945) On Topologies for Function-Spaces, *Bulletin of the American Mathematical Society* **51**.

Gentzen, G. (1935) Untersuchungen über das logische Schliessen, *Mathematische Zeitschrift* **39**, 176–210, 405–431, English translation, pages 68–131 in *The Collected Papers of Gerhard Gentzen*, edited by Manfred E. Szabo, Studies in Logic and the Foundations of Mathematics, Amsterdam: North-Holland.

Gierz, G., Hofmann, K., Keimel, K., Lawson, J., Mislove, M. and Scott, D. (1980) *A Compendium of Continuous Lattices*, Berlin, Heidelberg, New York: Springer, second edition, *Continuous Lattices and Domains*, Cambridge: Cambridge University Press (2003).

Girard, J.-Y. (1987) Linear Logic, *Theoretical Computer Science* **50**, 1–102.

Hartshorne, R. (1977) *Algebraic Geometry*, Number 52 in Graduate Texts in Mathematics, Berlin, New York: Springer.

Hofmann, K. and Mislove, M. (1981) Local Compactness and Continuous Lattices, in Banaschewski, B. and Hoffmann, R.-E., eds. *Continuous Lattices*, number 871 in Lecture Notes in Mathematics, Berlin: Springer, pp. 209–248.

Howard, W. (1980) The Formulae-as-Types Notion of Construction, in Curry, H., Seldin, J. and Hindley, R., eds. *To H.B. Curry: Essays on Combinatory Logic, Lambda Calculus and Formalism*, New York: Academic Press, pp. 479–490.

Hudak, P. (1989) Conception, Evolution and Application of Functional Programming Languages, *ACM Computing Surveys* **21**, 355–411.

Hudak, P., Hughes, J., Peyton-Jones, S. and Wadler, P. (2007) A History of Haskell: Being Lazy with Class, in *History of Programming Languages*, New York: ACM Press, pp. 12–55.

Hyland, M. (1991) First Steps in Synthetic Domain Theory, in Carboni, A., Pedicchio, M.-C. and Rosolini, G., eds. *Proceedings of the 1990 Como Category Conference*, number 1488 in Lecture Notes in Mathematics, Berlin, Heidelberg, New York: Springer, pp. 131–156.

Isbell, J. (1986) General Function Spaces, Products and Continuous Lattices, *Mathematical Proceedings of the Cambridge Philosophical Society* **100**, 193–205.

Johnstone, P.(1982) *Stone Spaces*, Number 3 in Cambridge Studies in Advanced Mathematics, Cambridge: Cambridge University Press.

Johnstone, P. (1984) Open Locales and Exponentiation, *Contemporary Mathematics* **30**, 84–116.

Joyal, A. and Tierney, M. (1984) An Extension of the Galois Theory of Grothendieck, *Memoirs of the American Mathematical Society* **51**(309).

Jung, A., Kegelmann, M. and Moshier, A. (2001) Stably Compact Spaces and Closed Relations, *Electronic Notes in Theoretical Computer Science* **45**.

Kock, A. (1981) *Synthetic Differential Geometry*, Number 51 in London Mathematical Society Lecture Notes, Cambridge: Cambridge University Press, second edition, number 333 (2006).

Kuhn, T. (1962) *The Structure of Scientific Revolutions*, Chicago: University of Chicago Press.

Lakatos, I. (1963) Proofs and Refutations: The Logic of Mathematical Discovery, *British Journal for the Philosophy of Science* **14**, 1–25, re-published by Cambridge University Press (1976), edited by John Worrall and Elie Zahar.

Lawvere, B. (1969) Adjointness in Foundations, *Dialectica* **23**, 281–296, Reprinted with commentary in *Theory and Applications of Categories Reprints* **16** (2006), 1–16.

Linton, F. (1969) An Outline of Functorial Semantics, in Eckmann, B., ed. *Seminar on Triples and Categorical Homology Theory*, number 80 in Lecture Notes in Mathematics, Berlin, Heidelberg, New York: Springer, pp. 7–52.

Mac Lane, S. (1963) Natural Associativity and Commutativity, *Rice University Studies* **49**, 28–46.

Mac Lane, S. (1971) *Categories for the Working Mathematician*, Berlin, Heidelberg and New York: Springer.

Mac Lane, S. (1988) Categories and Concepts in Perspective, in Duren, P., Askey, R. and Merzbach, U., eds. *A Century of Mathematics in America*, vol. 1, pp. 323–365, Providence, RI: American Mathematical Society. Addendum in vol. 3, pp. 439–441.

Manes, E. (1976) *Algebraic Theories*, Number 26 in Graduate Texts in Mathematics, Berlin, Heidelberg, New York: Springer.

McLarty, C. (1990) The Uses and Abuses of the History of Topos Theory, *British Journal for the Philosophy of Science* **41**, 351–375.

Mikkelsen, C. (1976) *Lattice-Theoretic and Logical Aspects of Elementary Topoi*, PhD thesis, Århus: Århus Universitet, Various publications, number 25.

Moore, R. (1966) *Interval Analysis*, Automatic Computation, Englewood Cliff, NJ: Prentice Hall, second edition, *Introduction to Interval Analysis*, with Baker Kearfott and Michael Cloud, Society for Industrial and Applied Mathematics (2009).

Paré, R. (1974) Colimits in Topoi, *Bulletin of the American Mathematical Society* **80**(3), 556–561.

Phoa, W. (1990) *Domain Theory in Realizability Toposes*, PhD thesis, Cambridge: University of Cambridge. University of Edinburgh Dept. of Computer Science report CST-82-91 and ECS-LFCS-91-171.

Pitts, A. (2001) Categorical Logic, in *Handbook of Logic in Computer Science*, vol. 5, Oxford: Oxford University Press.

Rosolini, G. (1986) *Continuity and Effectiveness in Topoi*, D. Phil. thesis, Oxford: University of Oxford.

Russell, B. and Whitehead, A.N. (1910–1913) *Principia Mathematica*, Cambridge: Cambridge University Press.

Scott, D. (1972) Continuous Lattices, in Lawvere, F.W., ed. *Toposes, Algebraic Geometry and Logic*, number 274 in Lecture Notes in Mathematics, Berlin, Heidelberg, New York: Springer, pp. 97–136.

Scott, D. (1993) A Type-Theoretical Alternative to ISWIM, CUCH, OWHY, *Theoretical Computer Science* **121**, 422–440. Written in 1969.

Seldin, J. (2002) Curry's Anticipation of the Types Used in Programming Languages, *Proceedings of the Canadian Society for History and Philosophy of Mathematics* **15**, 148–163.

Simmons, H. (1978) The Lattice-Theoretic Part of Topological Separation Properties, *Proceedings of the Edinburgh Mathematical Society (2)* **21**, 41–48.

Smolin, L. (2006) *The Trouble with Physics*, New York: Houghton Mifflin, republished by Penguin (2008).

Spitters, B. (2010) Located and overt sublocales, to appear in *Annals of Pure and Applied Logic*, **162**, 36–54.

Steenrod, N. (1967) A Convenient Category of Topological Spaces, *Michigan Mathematics Journal* **14**, 133–152.

Stone, M. (1937) Applications of the Theory of Boolean Rings to General Topology, *Transactions of the American Mathematical Society* **41**, 375–481.

Stone, M. (1938) The Representation of Boolean Algebras, *Bulletin of the American Mathematical Society* **44**, 807–816.

Tarski, A. (1935) Zur Grundlegung der Boole'schen Algebra, *Fundamenta Mathematicae* **24**, 177–198.

Taylor, P. (1999) *Practical Foundations of Mathematics*, Number 59 in Cambridge Studies in Advanced Mathematics, Cambridge: Cambridge University Press.

Thielecke, H. (1997) *Categorical Structure of Continuation Passing Style*, PhD thesis, Edinburgh: University of Edinburgh.

Vermeulen, J. (1994) Proper Maps of Locales, *Journal of Pure and Applied Algebra* **92**, 79–107.

Vickers, S. (1988) *Topology Via Logic*, volume 5 of *Cambridge Tracts in Theoretical Computer Science*, Cambridge: Cambridge University Press.

Wilker, P. (1970) Adjoint Product and Hom Functors in General Topology, *Pacific Journal of Mathematics* **34**, 269–283.

Paul Taylor's papers on the Abstract Stone Duality programme, including the extended version of this one, are available from www.paultaylor.eu/asd/

Taylor, P. (2000a) Geometric and Higher-Order Logic in Terms of Abstract Stone Duality, *Theory and Applications of Categories* **7**, 284–338.

Taylor, P. (2000b) Non-Artin Gluing in Recursion Theory and Lifting in Abstract Stone Duality.

Taylor, P. (2002a) Sober Spaces and Continuations, *Theory and Applications of Categories* **10**, 248–299.

Taylor, P. (2002b) Subspaces in Abstract Stone Duality, *Theory and Applications of Categories* **10**, 300–366.

Taylor, P. (2002c) Scott Domains in Abstract Stone Duality, `http://www.paultaylor.eu/asd/pcfasd.pdf`

Taylor, P. (2004a) An Elementary Theory of Various Categories of Spaces and Locales.

Taylor, P. (2004b) Tychonov's Theorem in Abstract Stone Duality.

Taylor, P. (2005) Inside Every Model of Abstract Stone Duality Lies an Arithmetic Universe, *Electronic Notes in Theoretical Computer Science* **122**, 247–296.

Taylor, P. (2006a) Computably Based Locally Compact Spaces, *Logical Methods in Computer Science* **2**, 1–70.

Taylor, P. (2006b) Interval Analysis Without Intervals, *Real Numbers and Computers*.

Taylor, P. (2009a) With Andrej Bauer, The Dedekind Reals in Abstract Stone Duality, *Mathematical Structures in Computer Science* **19**, 757–838.

Taylor, P. (2009b) Equideductive Categories and Their Logic.

Taylor, P. (2010a) A lambda Calculus for Real Analysis, *Journal of Logic and Analysis* **2**(5), 1–115.

Taylor, P. (2010b) Computability for Locally Compact Spaces.

Conclusion: A Perspective on Future Research in FOM

Giovanni Sommaruga and John Bell

This book has presented a partial account of the history of studies in FOM, as well as a (by no means complete) survey of contemporary studies in FOM. It is natural now to ask what future studies in FOM ought or might look like. Here is a sketch of a research agenda in foundational studies.

(i) Determining what foundational studies are all about, analyzing the sense of the term 'foundations of mathematics', and at the same time trying to determine something like the proper sense of FOM etc: these have always been part of foundational studies, but they play a more acute or prominent role, it would seem, at the present time. One can distinguish at least two approaches to these issues:

- A bottom-up approach: this means investigating what has been done under the heading of FOM, describing it, and then subjecting it to analysis, classification, etc. This bottom-up approach can be pursued either purely descriptively, or in an evaluative way. The latter involves taking a certain standpoint and on this basis judging what is meaningful, relevant, or interesting, or not, as the case may be. The evaluative approach is fairly common and it is exemplified by the articles of Maddy, Shapiro and to a certain extent also Hale.
- A top-down approach: this involves the attempt to determine the conditions to be satisfied by or the criteria of what is to count as FOM. This approach is less common and is exemplified by Hellman's article and to a certain extent also by Feferman's.

The analysis associated with the bottom-up approach is already quite thorough and complete, and as a result, so it seems to us, may not admit much further

G. Sommaruga
Dept. of Humanities, Social and Political Sciences, Chair for philosophy,
ETH, Zurich, Switzerland
e-mail: sommarug@ethz.ch

G. Sommaruga (ed.), *Foundational Theories of Classical and Constructive Mathematics*, The Western Ontario Series in Philosophy of Science 76,
DOI 10.1007/978-94-007-0431-2_15,

development and investigation. The top-down approach, on the other hand, is more recent and requires further elaboration. What is sought here is some sort of formalization in Hao Wang's sense of the concept of FOM: this may also imply a certain amount of axiomatization. Another possibility would be to reflect on what it means to try to combine the bottom-up approach with the top-down approach. Hints in this direction can, we believe, be found in Awodey's article and its distinction between "logical foundations" and "categorical foundations" as the hard core of the logical foundations.

(ii) The debate between category-theoretical FOM and set-theoretical FOM continues, as is witnessed by the articles of Awodey, Feferman (with references to Hellman) and McLarty. The relationship between these two kinds of FOM has been considerably clarified, as we see from the articles of Awodey, McLarty, Lambek and Scott and others. Still, it is probably premature to claim that all has been said on this issue. One major question lurking in the background of this debate is that of priority: should, for example, set theory or category theory be accorded primary foundational status[1]? If this question is given an ontological twist, one may ask, as Taylor does in his article, whether FOM are to be foundations of discrete mathematics—in which case set theory may well be an excellent candidate—or rather of continuous mathematics—in which case category theory would fare far better.

This gives rise to a further, and perhaps even more fundamental question, namely: does it really make sense to pursuing the question of priority? This question amounts to asking: should FOM be a monism or a pluralism? Should one strive to find the ultimate foundation (whatever that may be), the foundation of foundations so to speak, or does it make more sense, is it more fruitful (or perhaps even unavoidable) to consider a plurality of foundations, each of which possesses its own insights, its own advantages and disadvantages, each of which, as Hao Wang put it, contributing in its own way to the better understanding of the grand picture of mathematics[2] ? These latter questions of course are not limited to set theory and category theory as FOMs, but extend equally to type theory, to Modal-Structural mathematics, to the Scottish neo-logicist program and others.

Apart from this more philosophical "macro-debate" there also are more technical "micro-debates" in category theory, set theory and in other foundational theories. One of these micro-debates is, for example, the debate whether ETCS or CCAF or a type-theoretical formulation is the appropriate foundation for category theory. Another such a micro-debate concerns the issue of how to formulate Replacement in category theory. There are analogous micro-debates in set theory.

---

[1] In this connection it is worth mentioning Jean-Pierre Marquis's (1995) useful classification of foundations into "logical", "cognitive", "epistemic", "semantic", "ontological" and "methodological".

[2] See Hellman and Bell (2006) for a discussion of pluralism in the foundations of mathematics.

(iii) As our retrospective look at foundational studies shows very clearly, the clash between constructive and classical mathematics has been a major theme on the research agenda of FOM (according to Bernays just about the most important one). Not much of a conflict remains (cf. Parsons's comment in this regard): nevertheless the relationship between constructive and classical mathematics is still an active subject of investigation, as, e.g., the articles of Bell and Lambek and Scott demonstrate. The clarification of the relationship between classical and constructive mathematics and their respective foundations is making progress, but has by no means come to an end. The question of priority of the one over the other has fallen away and has been replaced by a pluralist view which studies and investigates constructive mathematics as an instructive and interesting alternative to classical mathematics. This pluralist approach to the foundations of classical mathematics and those of constructive mathematics still offers a wide range of open problems and issues, of a more technical than philosophical nature. But as Tieszen's article highlights, philosophical questions surrounding constructive and classical mathematics are still very much alive.

(iv) Even though the "macro-debate" between classical and constructive mathematics is over, there still are plenty of "micro-discussions" and issues within the foundations of constructive mathematics: There are the more technical questions raised in Aczel's article: Which are the most appropriate or useful foundations of constructive mathematics? Is there some sort of unification to be found within the plurality of foundational theories of constructive mathematics? And there is McCarty's more philosophical concern in his article that the foundations of constructive mathematics might be "polluted" by a number of erroneous (philosophical) views and myths which ought to be discarded in order to identify the proper core or foundations of constructive (intuitionistic) mathematics. Mayberry in his article proceeds in a totally different direction: he explores a radically new approach to constructive mathematics, an approach which is constructive only in an unorthodox sense, somewhere on the border between constructive and classical mathematics. His approach also shows that there is room for (radically) new and yet to be discovered foundational approaches to constructive mathematics. Various questions concerning the foundations of constructive mathematics will continue to play a vital role within future studies in FOM.

(v) As we said at the beginning of this conclusion: The survey of contemporary studies in FOM presented in this book is partial and by no means complete. But even if we were to attempt to enumerate the further foundational topics and subjects and open problems to be examined, solved or dealt with in the future, this would only serve to confirm "the essential incompleteness" of the research agenda in FOM whose outlines have just been sketched.

References

Hellman, G. and Bell, J.L. (2006) Pluralism and the Foundations of Mathematics, in Waters, C.K., Longino, H. and Kellert, S., eds. *Scientific Pluralism*, Minnesota Studies in the Philosophy of Science, vol. 19, Minneapolis and London: University of Minnesota Press, pp. 64–79.
Marquis, J.-P. (1995) Category Theory and the Foundations of Mathematics: Philosophical Excavations, *Synthese* **103**, 421–447.

FSC
www.fsc.org
MIX
Papier | Fördert
gute Waldnutzung
FSC® C083411

Kolibri 360 GmbH
Ferdinand-Jühlke-Straße 7
99095 Erfurt, Deutschland
produktsicherheit@kolibri360.de